Springer Proceedings in Mathematics & Statistics

Volume 227

Springer Proceedings in Mathematics & Statistics

This book series features volumes composed of selected contributions from workshops and conferences in all areas of current research in mathematics and statistics, including operation research and optimization. In addition to an overall evaluation of the interest, scientific quality, and timeliness of each proposal at the hands of the publisher, individual contributions are all refereed to the high quality standards of leading journals in the field. Thus, this series provides the research community with well-edited, authoritative reports on developments in the most exciting areas of mathematical and statistical research today.

More information about this series at http://www.springer.com/series/10533

Cira Perna · Monica Pratesi
Anne Ruiz-Gazen
Editors

Studies in Theoretical and Applied Statistics

SIS 2016, Salerno, Italy, June 8–10

Società
Italiana di
Statistica

Springer

Editors
Cira Perna
Dipartimento di Scienze
Economiche e Statistiche
Università degli Studi di Salerno
Fisciano, Salerno
Italy

Anne Ruiz-Gazen
Toulouse School of Economics
University of Toulouse
Toulouse Cedex 6
France

Monica Pratesi
Dipartimento di Economia e
Management
Università degli Studi di Pisa
Pisa
Italy

ISSN 2194-1009 ISSN 2194-1017 (electronic)
Springer Proceedings in Mathematics & Statistics
ISBN 978-3-030-08883-5 ISBN 978-3-319-73906-9 (eBook)
https://doi.org/10.1007/978-3-319-73906-9

Mathematics Subject Classification (2010): S11001, S17010, S12008, S17040

Printed on acid-free paper

This Springer imprint is published by the registered company Springer International Publishing AG part of Springer Nature
The registered company address is: Gewerbestrasse 11, 6330 Cham, Switzerland

Preface

This book contains a selection of the papers presented during the 48th Scientific Meeting of the Italian Statistical Society (SIS2016), held in Salerno on June 8–10, 2016.

This biennial conference is a traditional national and international meeting for connecting researchers in statistics, demography, and applied statistics in Italy. The conference aims at bringing together national and foreign researchers and practitioners to discuss recent developments in statistical methods for economics, social sciences, and all fields of application of statistics.

The Scientific Programme Committee provided a balanced and stimulating program that appealed to the diverse interests of the participants.

This book of selected papers is organized in chapters each related to a theme discussed in the meeting. In the editing process, we reordered the themes and collapsed some of them. The result still resembles the large variety of topics addressed in Salerno.

From the modern data sources and survey design issues to the study of the measures of sustainable development, the reader can find a large collection of research topics in Statistical Methods and in Applied Statistics and Demography.

In this respect, the papers included in this volume provide a comprehensive overview of the current Italian scientific researches in theoretical and applied statistics.

1. Open data and big data in public administration and official statistics
2. Survey sampling: theory and application
3. A recent debate in Statistics
4. Statistical algorithms
5. Ordinal and symbolic data
6. Statistical models and methods for network data
7. Forecasting time series
8. Spatial analysis
9. Issues on ecological and environmental statistics

10. Statistics and the education system
11. Economic and financial data analysis
12. Sustainable development: theory, measures, and applications

The Scientific Programme Committee, the Session Organizers, the local hosting University, and many volunteers have contributed substantially to the organization of the conference and to the referee process to obtain this book. We acknowledge their work and the support of our Society. Particularly, we wish to thank Marcella Niglio for her continuous support and assistance in the editing of this book.

Wishing you a productive and stimulating reading,

Salerno, Italy — Cira Perna
Pisa, Italy — Monica Pratesi
Toulouse, France — Anne Ruiz-Gazen
October 2017

Contents

About the Editors

Cira Perna is currently Professor of Statistics and Head of the Department of Economics and Statistics, University of Salerno (Italy). Her research work mainly focuses on nonlinear time series, artificial neural network models, and resampling techniques. She has published a number of papers in national and international journals on these topics, and she has been a member of the scientific committees of several national and international conferences.

Monica Pratesi is Professor of Statistics, University of Pisa, and holds the Jean Monnet Chair "Small Area Methods for Monitoring of Poverty and Living Conditions in the EU" 2015–2017. She is the Director of the Tuscan Interuniversity Centre—Advanced Statistics for Equitable and Sustainable Development, entitled to Camilo Dagum. Her research interests include methods for survey sampling and analysis of survey data, small area estimation, and design-based population inference. She has published a number of papers in national and international journals on these topics and has been a member of the scientific committees of several national and international conferences.

Anne Ruiz-Gazen is Professor of Applied Mathematics, specializing in statistics, and a member of the Toulouse School of Economics—Research at University Toulouse 1 Capitole. Her areas of research include multivariate data analysis, survey sampling theory and, to a less extent, spatial econometrics and statistics. She has published more than fifty articles in refereed journals and books and has been a member of the scientific committees of several conferences.

Part I
Advances in Survey Methods and New Sources in Public Statistics

Robustness in Survey Sampling Using the Conditional Bias Approach with R Implementation

Cyril Favre-Martinoz, Anne Ruiz-Gazen, Jean Francois Beaumont and David Haziza

Abstract The classical tools of robust statistics have to be adapted to the finite population context. Recently, a unified approach for robust estimation in surveys has been introduced. It is based on an influence measure called the conditional bias that allows to take into account the particular finite population framework and the sampling design. In the present paper, we focus on the design-based approach and we recall the main properties of the conditional bias and how it can be used to define a general class of robust estimators of a total. The link between this class and the well-known winsorized estimators is detailed. We also recall how the approach can be adapted for estimating domain totals in a robust and consistent way. The implementation in R of the proposed methodology is presented with some functions that estimate the conditional bias, calculate the proposed robust estimators and compute the weights associated to the winsorized estimator for particular designs. One function for computing consistently domain totals is also proposed.

Keywords Conditional bias · Design-based · Huber-function · Mean squared error

C. Favre-Martinoz (✉)
Direction de la Méthodologie et de la Coordination Statistique et Internationale, INSEE, 18 Boulevard Adolphe Pinard, 75014 Paris, France
e-mail: cyril.favre-martinoz@insee.fr

A. Ruiz-Gazen
Toulouse School of Economics, University Toulouse Capitole, Toulouse, France
e-mail: anne.ruiz-gazen@tse-fr.eu

J. F. Beaumont
Statistical Research and Innovation Division, Statistics Canada, Ottawa, ON K1A 0T6, Canada
e-mail: jean-francois.beaumont@canada.ca

D. Haziza
Département de mathématiques et statistique, Université de Montréal, Montréal, QC H3C 3J7, Canada
e-mail: haziza@dms.umontreal.ca

C. Perna et al. (eds.), *Studies in Theoretical and Applied Statistics*, Springer Proceedings in Mathematics & Statistics 227,
https://doi.org/10.1007/978-3-319-73906-9_1

1 Introduction

Robustness in survey sampling differs from robustness in classical statistics for several reasons. The main one is conceptual. In the classical robust statistics framework, the data generating process for the majority of the data differs from the process generating the remaining part of the data that are considered as outliers. In this context, the distribution of the main bulk of the data is the object of interest. In survey sampling, errors are corrected at the editing stage. Thus, the use of classical robust estimators that may drastically downweight some valid data and lead to biased estimators is not recommended. However, in business surveys for example, the distribution of some variables may be highly skewed and some influential units may lead to unstable estimators. Hence, robust estimation in survey sampling means protection against influential observations in order to avoid highly variable estimators. The definition of influential units has to be adapted to the survey sampling context. The well-known influence function, for example, does not take into account the sampling weights but only the variable of interest. Beaumont et al. [2] propose to use the conditional bias as a unified measure of influence. In the present paper we propose to recall this approach together with some recent extensions and applications. We also present some R functions to implement the methodology. The parameters of interest are finite population totals or domain totals. In what follows, we focus exclusively on the design-based approach, which means that the randomness comes from the sample selection process only, but the conditional bias approach is unified and also involves the model-based framework.

In the second section, we review the notion of conditional bias introduced by Muñoz-Pichardo et al. [10] and Moreno-Rebollo et al. [9] and revisited by Beaumont et al. [2] in the finite population framework. The conditional bias has proved useful in measuring the influence of observations on estimators such as the Horvitz–Thompson estimator. It also gives a general way to define robust estimators that will be detailed in Sect. 3. These robust estimators can be related to the well-known winsorized estimators as detailed in Favre-Martinoz et al. [7] and recalled in Sect. 4 of the present paper. Estimating domain totals is also a problem of importance in surveys and Favre-Martinoz et al. [7] proposed to adapt the conditional bias approach in order to obtain robust and consistent estimators. The method is recalled in Sect. 5. In Sect. 6, we present some `R` functions to implement the proposed methodology for three different well-known sampling designs and their stratified counterparts.

2 The Conditional Bias as a Measure of Influence in Survey Sampling

Consider a finite population U of size N. We are interested in estimating the total of a variable y. We assume that we cannot have access to the whole population U but only to a sample S drawn at random from U given a sampling design P. The results in Beaumont et al. [2] are for general one-stage designs. Among the well-known one-stage sampling designs, we consider simple random sampling without replacement, Poisson sampling, rejective sampling, also known as the maximum entropy design, or the conditional Poisson design (see [12]) and their stratified counterparts. Except for the Poisson design, the other designs belong to the class of fixed-size sampling designs. Stratification, Poisson or conditional Poisson designs allow for unequal probabilities.

Let π_i be the first-order inclusion probability for $i \in U$ and π_{ij} the second-order inclusion probability for $i,j \in U$. We assume that these inclusion probabilities are strictly positive and we consider the well-known Horvitz–Thompson (HT) estimator of the total $t_y = \sum_{i \in U} y_i$, defined by:

$$\hat{t}_y^{\mathrm{HT}} = \sum_{i \in s} d_i y_i, \tag{1}$$

where $d_i = 1/\pi_i$ denotes the sampling weight of unit $i \in U$. The variance of $\hat{t}_y^{\mathrm{HT}}$ is:

$$\mathrm{Var}(\hat{t}_y^{\mathrm{HT}}) = \sum_{i \in U} \sum_{j \in U} \left(\frac{\pi_{ij}}{\pi_i \pi_j} - 1 \right) y_i y_j. \tag{2}$$

The HT estimator is an unbiased estimator of t_y and its variability is measured by its variance which depends on the variable of interest y and on the sampling design P through the inclusion probabilities. The conditional bias of a unit is a measure of the influence of a unit on the variance of an estimator.

Let I_i the inclusion variable which equals 1 if unit i belongs to the sample and 0 otherwise. The conditional bias of an estimator $\hat{\theta}$ of a parameter θ, for a unit i in U, has been introduced in the context of survey sampling by Moreno-Rebollo et al. [9] and is defined as:

$$B_i^{\hat{\theta}}(I_i = 1) = E_P(\hat{\theta} - \theta \mid I_i = 1), \tag{3}$$

$$B_i^{\hat{\theta}}(I_i = 0) = E_P(\hat{\theta} - \theta \mid I_i = 0). \tag{4}$$

While the bias of an estimator $\hat{\theta}$, $E_P(\hat{\theta} - \theta)$, is the mean of the estimation errors over all the possible samples, the conditional bias for a sampled unit defined as the mean of the errors over the samples that contain this unit. The conditional bias for a non-sampled unit is defined as the mean of the errors over the samples that do not contain this unit.

For the HT estimator, the conditional bias can be written as:

$$B_i^{\mathrm{HT}}(I_i = 1) = \sum_{j \in U} \left(\frac{\pi_{ij}}{\pi_i \pi_j} - 1 \right) y_j, \tag{5}$$

$$B_i^{\mathrm{HT}}(I_i = 0) = -B_i^{\mathrm{HT}}(I_i = 1)/(d_i - 1).$$

The following expression establishes the link between the conditional bias (for the sampled units) and the variance of the HT estimator:

$$\mathrm{Var}(\hat{t}_y^{\mathrm{HT}}) = E_P(\hat{t}_y^{\mathrm{HT}} - t_y)^2 = \sum_{i \in U} B_i^{\mathrm{HT}}(I_i = 1)\, y_i. \tag{6}$$

From (6), it is possible to interpret the conditional bias as a contribution of unit i to the variance which is an average of the y_i values weighted by the conditional bias of each observation. Note that, if $\pi_i = 1$, the conditional bias of the unit is zero regardless of is the value y_i. Exact or approximate expressions of the conditional bias are given in Beaumont et al. [2] for simple random sampling without replacement, the Poisson and the conditional Poisson sampling.

3 Robust Estimators

The following property is true for Poisson sampling and approximately true for the other three designs provided that the population size N is large enough:

$$\hat{t}_y^{\mathrm{HT}} - t_y = \sum_{i \in S} B_i^{\mathrm{HT}}(I_i = 1) + \sum_{i \in U-S} B_i^{\mathrm{HT}}(I_i = 0). \tag{7}$$

It states that when summing the conditional biases of the sampled and non-sampled units, we obtain the sampling error. It gives a motivation for the following robust estimator:

$$\hat{t}_y^{\mathrm{RHT}} = \hat{t}_y^{\mathrm{HT}} - \sum_{i \in S} \hat{B}_i^{\mathrm{HT}}(I_i = 1) + \sum_{i \in S} \psi(\hat{B}_i^{\mathrm{HT}}(I_i = 1)), \tag{8}$$

where $\hat{B}_i^{\mathrm{HT}}(I_i = 1)$ is an estimator of the conditional bias of the robust estimator and the function ψ is the Huber function defined by:

$$\psi(x) = \mathrm{sign}(x) \times \min(|x|, c)$$

where $c > 0$ and $\mathrm{sign}(x) = 1$, for $x \geq 0$, and $\mathrm{sign}(x) = -1$, otherwise.

Note that the estimator proposed by Kokic and Bell [8] for a stratified sampling design belongs to this class of robust estimators with a ψ function that differs from the Huber function. The robust estimator (RHT) is biased but we expect to reduce the mean squared error by reducing the influence of some units.

The Huber function depends on a constant c that needs to be determined. Beaumont et al. [2] propose to choose the value of c that minimizes the maximum absolute estimated conditional bias with respect to $\hat{t}_y^{\mathrm{RHT}}$:

$$\max\{|\hat{B}_i^{\mathrm{RHT}}(I_i = 1)|; i \in s\}. \tag{9}$$

One can prove that the resulting estimator is given by:

$$\hat{t}_y^{\text{RHT}} = \hat{t}_y^{\text{HT}} - \frac{1}{2}\left(\hat{B}_{min}^{\text{HT}}(I_i = 1) + \hat{B}_{max}^{\text{HT}}(I_i = 1)\right), \tag{10}$$

where $\hat{B}_{min}^{\text{HT}}(I_i = 1)$ (resp. $\hat{B}_{max}^{\text{HT}}(I_i = 1)$) is the minimum (resp. maximum) estimated conditional bias among all the units of the sample. In the sequel, we will denote $\hat{B}_{min}^{\text{HT}}(I_i = 1)$ (resp. $\hat{B}_{max}^{\text{HT}}(I_i = 1)$) by $\hat{B}_{min}^{\text{HT}}$ (resp. $\hat{B}_{max}^{\text{HT}}$).

Beaumont et al. [2] prove that this estimator converges to the true total under some assumptions. It is also possible to estimate its mean squared error using some bootstrap procedure. Moreover, the estimator (10) belongs to a more general class of estimators of the form

$$\hat{t}_y^{\text{RHT}} = \hat{t}_y^{\text{HT}} + \Delta, \tag{11}$$

where Δ is a particular random variable. As we will see in Sect. 4, the winsorized estimators can be written in the form (11). Therefore, the conditional bias can be used to determine the winsorization threshold.

4 Application to Winsorized Estimators

Estimator (10) can be written with alternative expressions, which can make it easier to implement in some cases. In this section, we consider the well-known winsorized estimators. Such estimators have been studied in the literature and we can distinguish two winsorized forms: the standard one and the Dalén-Tambay winsorization described by Dalén [3] and Tambay [11].

The first form of winsorization involves trimming the value of units that are above a given threshold, taking their weight into account. If $\tilde{y}_i$ denotes the value of variable y for unit i after winsorization, we have

$$\tilde{y}_i = \begin{cases} y_i & \text{if } d_i y_i \leq c \\ c/d_i & \text{if } d_i y_i > c \end{cases} \tag{12}$$

where $c > 0$ is the winsorization threshold. The standard winsorized estimator (13) of the total t_y can be written in the form (11):

$$\begin{aligned} \hat{t}_y^{s} &= \sum_{i \in S} d_i \tilde{y}_i \\ &= \hat{t}_y^{\text{HT}} + \Delta(c), \end{aligned} \tag{13}$$

where

$$\Delta(c) = -\sum_{i \in S} \max\left(0, d_i y_i - c\right).$$

An alternative is to express $\hat{t}_y^s$ as a weighted sum of the initial values using modified weights:

$$\hat{t}_y^s = \sum_{i \in S} \tilde{d}_i y_i,$$

where

$$\tilde{d}_i = d_i \frac{\min\left(y_i, c/d_i\right)}{y_i}. \tag{14}$$

If unit i is not influential ($\min\left(y_i, c/d_i\right) = y_i$), then its weight is not modified ($\tilde{d}_i = d_i$). For influential units, the modified weight $\tilde{d}_i$ is less than d_i and may be less than 1. Note that if $y_i = 0$, the contribution of unit i to the estimated total, $\hat{t}_y^s$, is zero and an arbitrary value can be assigned to the modified weight $\tilde{d}_i$.

In the case of the second form of winsorization, the values of the variable of interest after winsorization are defined by

$$\tilde{y}_i = \begin{cases} y_i & \text{if } d_i y_i \leq c \\ \dfrac{c}{d_i} + \dfrac{1}{d_i}\left(y_i - \dfrac{c}{d_i}\right) & \text{if } d_i y_i > c \end{cases} \tag{15}$$

The winsorized estimator of the total t_y is then given by:

$$\hat{t}_y^{DT} = \sum_{i \in S} d_i \tilde{y}_i. \tag{16}$$
$$= \hat{t}_y^{\mathrm{HT}} + \Delta(c),$$

where

$$\Delta(c) = -\sum_{i \in S} \frac{(d_i - 1)}{d_i} \max\left(0, d_i y_i - c\right).$$

Estimator (16) can also be written in the form (11). And as in the case of $\hat{t}_y^s$, an alternative expression for $\hat{t}_y^{DT}$ is a weighted sum of the initial values using modified weights:

$$\hat{t}_y^{DT} = \sum_{i \in S} \tilde{d}_i y_i,$$

where

$$\tilde{d}_i = 1 + (d_i - 1)\frac{\min\left(y_i, \frac{c}{d_i}\right)}{y_i}. \tag{17}$$

As in the case of the standard winsorized estimator, the weight of a non-influential unit is not modified. But unlike standard winsorization, the Dalén-Tambay winsorization guarantees that the modified weights will not be less than 1.

Since the standard and Dalén-Tambay winsorized estimators are of the form (11), the optimal constant c_{opt} that minimizes (9) is obtained by solving

$$\Delta(c) = -\frac{1}{2}(\hat{B}^{\text{HT}}_{min} + \hat{B}^{\text{HT}}_{max})$$

or

$$\sum_{j\in S} a_j \max\left(0, d_j y_j - c\right) = \frac{\hat{B}^{\text{HT}}_{min} + \hat{B}^{\text{HT}}_{max}}{2}, \tag{18}$$

where $a_j = 1$ for the standard case and $a_j = (d_j - 1)/d_j$ for the non standard winsorization. Favre-Martinoz et al. [7] have shown that a solution to equation (18) exists under the two following conditions:

(C1) $\pi_{ij} - \pi_i\pi_j \le 0$, for all $i,j \in U$,
(C2) $\hat{B}^{\text{HT}}_{min} + \hat{B}^{\text{HT}}_{max} \ge 0$.

(C1) is satisfied for most one-stage designs used in practice, such as stratified simple random sampling and Poisson sampling. (C2) implies that $\hat{t}^{\text{RHT}}_y$ must be less than or equal to $\hat{t}^{\text{HT}}_y$ which is quite intuitive, since by construction, a winsorized estimator should be smaller than the Horvitz–Thompson estimator. It is generally expected that (C2) will be satisfied in most skewed populations encountered in business surveys and social surveys. It can be shown that the solution to Eq. (18) is unique if the above conditions are met and if $y_i \ge 0$ for $i \in S$.

The value of the constant c_{opt} differs for each type of winsorization but the resulting robust estimators are identical. More precisely, we have

$$\hat{t}^{s}_y(c_{opt}) = \hat{t}^{DT}_y(c_{opt}) = \hat{t}^{\text{RHT}}_y = \hat{t}^{\text{ HT}}_y - \frac{\hat{B}^{\text{HT}}_{min} + \hat{B}^{\text{HT}}_{max}}{2}. \tag{19}$$

We can cite two real applications of winsorization using the conditional bias at the French National Institute for Statistics and Economic Studies (INSEE). The first one is an experimentation in the Information and Communication Technologies Survey in 2011 while the second is a comparison with the method by Kokic and Bell [8] in the Wage Structure and Labor Cost Survey [5].

5 Robust Estimation of Domain Totals

In practice, we usually want to produce estimates for population totals at the global level but also at some subpopulation levels called domains. Let $t_{y,g} = \sum_{i\in U_g} y_i$ be the y-variable total in domain g. We assume that the domains form a partition of the population such that $t_y = \sum_{i\in U} y_i = \sum_{g=1}^{G} t_{y,g}$, where G is the number of domains.

Let S_g be the set of sampled units in domain g. The expansion estimator of $t_{y,g}$ is given by $\hat{t}^{\mathrm{HT}}_{y,g} = \sum_{i \in S_g} d_i y_i$. We have the consistency relation

$$\sum_{g=1}^{G} \hat{t}^{\mathrm{HT}}_{y,g} = \hat{t}^{\mathrm{HT}}_{y}.$$

When influential values are present, we can apply a robust procedure separately for each domain using the method described in Sect. 3, and obtain robust estimators, $\hat{t}^{\mathrm{RHT}}_{y,g}$, $g = 1, \ldots, G$. A robust estimator of the total at the population level, $\hat{t}^{\mathrm{R}(agg)}_{y}$, is easily obtained by aggregating the robust estimators $\hat{t}^{\mathrm{RHT}}_{y,g}$. Thus we have

$$\hat{t}^{\mathrm{R}(agg)}_{y} = \sum_{g=1}^{G} \hat{t}^{\mathrm{RHT}}_{y,g}$$

and the consistency relation between the domain-level estimators and the population-level estimator is satisfied. However, aggregating G robust estimators, each suffering from a potential bias, may produce a highly biased aggregate robust estimator. In most cases, the bias of $\hat{t}^{\mathrm{R}(agg)}_{y}$ will be negative, since each of the $\hat{t}^{\mathrm{RHT}}_{y,g}$ estimators has a negative bias.

To avoid having an estimator with an unacceptable bias, we first compute the robust estimator, $\hat{t}^{\mathrm{RHT}}_{y,g}$, for each domain using (19). Then we independently compute a robust estimator $\hat{t}^{\mathrm{R0}}_{y}$ of the total t_y in the population, using (19). In this case, the consistency relation is no longer necessarily satisfied. In other words, we may have $\hat{t}^{\mathrm{R0}}_{y} \neq \sum_{g=1}^{G} \hat{t}^{\mathrm{RHT}}_{y,g}$.

In order to force consistency between the robust domain estimators and the aggregate robust estimator, Favre-Martinoz et al. [7] propose a method similar to calibration [6]. The idea is to compute final robust estimators $\hat{t}^{\mathrm{RHT}*}_{y,g}$, $g = 0, 1, \ldots, G$, that are as close as possible to the initial robust estimators $\hat{t}^{\mathrm{RHT}}_{y,g}$, using a particular distance function, and that satisfy the calibration equation

$$\sum_{g=1}^{G} \hat{t}^{\mathrm{RHT}*}_{y,g} = \hat{t}^{\mathrm{R0}*}_{y}. \tag{20}$$

In the case of the generalized chi-square distance function, the final robust estimators, $\hat{t}^{\mathrm{RHT}*}_{y,g}$, are such that

$$\sum_{g=0}^{G} \frac{\left\{ \hat{t}^{\mathrm{RHT}*}_{y,g} - \hat{t}^{\mathrm{RHT}}_{y,g} \right\}^2}{2 q_g \hat{t}^{\mathrm{RHT}}_{y,g}} \tag{21}$$

is minimized subject to (20). The coefficient q_g in the above expression is a weight assigned to the initial estimator in domain g, $\hat{t}^{\text{RHT}}_{y,g}$, and can be interpreted as its importance in the minimization problem. Using the Lagrange multipliers method, a solution to this minimization problem can be easily derived and is given by

$$\hat{t}^{\text{RHT}*}_{y,g} = \hat{t}^{\text{RHT}}_{y,g} - \frac{\sum_{h=0}^{G} \delta_h \hat{t}^{\text{RHT}}_{y,h}}{\sum_{h=0}^{G} q_h \hat{t}^{\text{RHT}}_{y,h}} \delta_g \, q_g \, \hat{t}^{\text{RHT}}_{y,g}, \tag{22}$$

where $\delta_0 = -1$ and $\delta_g = 1$, for $g = 1, \ldots, G$.

As in the previous section, we can express $\hat{t}^{\text{RHT}*}_{y,g}$ as a weighted sum with the initial weights using modified y values or as a weighted sum of the initial y values using modified weights (see [7], for details).

The choice of the values q_g, $g = 0, \ldots, G$, is crucial. A small value of q_g ensures that the initial estimator in domain g, is not modified excessively. If $q_g = 0$, then the final robust estimator $\hat{t}^{\text{RHT}*}_{y,g}$ is identical to the initial robust estimator $\hat{t}^{\text{RHT}}_{y,g}$. Note in particular that the initial robust estimate at the population level, $\hat{t}^{\text{R0}}_{y}$, can also be modified as soon as $q_0 > 0$. One natural choice for the q_g is

$$q_g = \widehat{CV}(\hat{t}^{\text{HT}}_{y,g}) / \sum_{g=1}^{G} \widehat{CV}(\hat{t}^{\text{HT}}_{y,g}),$$

where $\widehat{CV}(\hat{t}^{\text{HT}}_{y,g})$ is the estimated coefficient of variation (CV) associated with domain g. More details concerning this choice and other alternatives can be found in Favre-Martinoz et al. [7].

It is also possible to find winsorization thresholds $c_g, g = 1, \ldots, G$, such that the standard winsorized estimator or the Dalén-Tambay winsorized estimator is equal to $\hat{t}^{\text{RHT}*}_{y,g}$ by following a procedure similar to the one in Sect. 4. A necessary condition for the existence of a solution is that $\hat{t}^{\text{HT}}_{y,g} - \hat{t}^{\text{RHT}*}_{y,g} \geq 0$ for all g.

6 Implementation with the R Language

The robust estimators proposed above together with the optimal constant c and the different associated modified weights are implemented in R functions which are available upon request from the first two authors of the present paper. An R package project is under progress. Some aspects are similar to the functions written by Tillé and Matei [13] for the package `sampling`. In particular, the functions for estimating the conditional bias and calculating the robust estimator for a given sample depend on whether the sampling design is stratified or not. The functions can deal with the three particular sampling designs we introduced above (simple random sampling without replacement, Poisson and rejective sampling) for both the non stratified and the stratified cases.

The function for estimating the conditional biases for the HT estimator of a total and the non-stratified designs is called `HTcondbiasest.r` while the one for the stratified sampling designs is called `strata_HTcondbiasest.r`. For several given variables of interest as input, these functions return a vector with the estimates of the conditional bias of the HT estimator where each row corresponds to a sample unit. There are also two functions called `robustest.r` and `strata_robustest.r` that compute the robust estimate we propose in Sect. 3. These functions return the robust estimates of the total for the given variables of interest. We also propose two functions, called `tuningconst.r` for the non-stratified designs and `strata_tuningconst.r` for the stratified sampling designs, that compute the optimal tuning constant associated to the proposed robust estimator.

Two other functions, called `weightswin.r` and `strata_weightswin.r`, compute the tuning constant and the modified weights associated to the winsorized estimator as recalled in Sect. 4 for the non-stratified designs and the stratified designs respectively.

For domain estimation, a function called `domain_weights.r` computes the modified weights that satisfy the consistency relation from the initial robust weights following the methodology of Sect. 5. In this function, the user has to specify the initial robust weights computed for instance using the function `weightswin.r` or `strata_weightswin.r`, the set of coefficients q, the calibration method and the bounds, if needed for the calibration method. These modified weights are computed by using the function `calib` available in the package `sampling`.

References

1. Beaumont, J.-F., Rivest, L.-P.: Dealing with outliers with survey data. In: Rao, C. R., Pfeffermann, D. (eds.) Handbook of Statistics, Sample Surveys: Theory Methods and Inference, vol. 29, pp. 247–279 (2009)
2. Beaumont, J.-F., Haziza, D., Ruiz-Gazen, A.: A unified approach to robust estimation in finite population sampling. Biometrika **100**, 555–569 (2013)
3. Dalén, J.: Practical estimators of a population total which reduce the impact of large observations. R and D Report, Statistics Sweden (1987)
4. Demoly, E.: Les valeurs influentes dans lénqute TIC 2011: Traitements et expérimentations en cours. Acte du sminaire de méthodologie statistique de l'INSEE, juillet (2013)
5. Deroyon, T., Favre-Martinoz, C.: A comparison of Kokic and Bell and conditional bias methods for outlier treatment. In: Work Session on Statistical Data Editing. The Hague, Netherlands, Unece (2017)
6. Deville, J.-C., Särndal, C.E.: Calibration estimators in survey sampling. J. Am. Stat. Assoc. **87**(418), 376–382 (1992)
7. Favre-Martinoz, C., Haziza, D., Beaumont, J.-F.: A method of determining the winsorization threshold, with an application to domain estimation. Surv. Methodol. **41**(1), 57–77 (2015)
8. Kokic, P.N., Bell, P.A.: Optimal winzorizing cutoffs for a stratified finite population estimator. J. Official Stat. **10**, 419–435 (1994)
9. Moreno-Rebollo, J.L., Muoz-Reyez, A.M., Muoz-Pichardo, J.M.: Influence diagnostics in survey sampling: conditional bias. Biometrika **86**, 923–968 (1999)

10. Muñoz-Pichardo, J., Muñoz-Garcia, J., Moreno-Rebollo, J.L., Piño-Mejias, R.: A new approach to influence analysis in linear models. Sankhya Series A **57**, 393–409 (1995)
11. Tambay, J.-L.: An integrated approach for the treatment of outliers in sub-annual surveys. In: Proceedings of the Survey Research Methods Section, American Statistical Association, Alexandria, Virginia, pp. 229–234 (1988)
12. Tillé, Y.: Sampling Algorithms. Springer (2011)
13. Tillé, Y., Matei, A.: The sampling package. Software manual, CRAN (2015). https://cran.r-project.org/src/contrib/Descriptions/sampling.html

Methodological Perspectives for Surveying Rare and Clustered Population: Towards a Sequentially Adaptive Approach

Federico Andreis, Emanuela Furfaro and Fulvia Mecatti

Abstract Sampling a rare and clustered trait in a finite population is challenging: traditional sampling designs usually require a large sample size in order to obtain reasonably accurate estimates, resulting in a considerable investment of resources in front of the detection of a small number of cases. A notable example is the case of WHO's tuberculosis (TB) prevalence surveys, crucial for countries that bear a high TB burden, the prevalence of cases being still less than 1%. In the latest WHO guidelines, spatial patterns are not explicitly accounted for, with the risk of missing a large number of cases; moreover, cost and logistic constraints can pose further problems. After reviewing the methodology in use by WHO, the use of adaptive and sequential approaches is discussed as natural alternatives to improve over the limits of the current practice. A simulation study is presented to highlight possible advantages and limitations of these alternatives, and an integrated approach, combining both adaptive and sequential features in a single sampling strategy is advocated as a promising methodological perspective.

Keywords Spatial pattern · Prevalence surveys · logistic constraints
Poisson sampling · Horvitz-Thompson estimation

1 Introduction and Motivational Example

When knowledge pertaining a trait that is particularly rare in the population is of interest, for example the estimation of its prevalence, the collection of survey data

F. Andreis (✉)
Carlo F. Dondena Centre for Research on Social Dynamics and Public Policy,
Università Commerciale Luigi Bocconi, Milan, Italy
e-mail: federico.andreis@unibocconi.it

E. Furfaro · F. Mecatti
Università degli Studi di Milano-Bicocca, Piazza dell'Ateneo Nuovo 1, 20126 Milan, Italy
e-mail: emanuela.furfaro@unimib.it

F. Mecatti
e-mail: fulvia.mecatti@unimib.it

C. Perna et al. (eds.), *Studies in Theoretical and Applied Statistics*,
Springer Proceedings in Mathematics & Statistics 227,
https://doi.org/10.1007/978-3-319-73906-9_2

presents various challenging aspects. For instance, in order to obtain a reasonably accurate estimate, very large sample sizes are needed, thus inflating survey costs; moreover, when cases are not only rare but also unevenly distributed throughout space, i.e. they present specific spatial patterns such as, for example, clustering, traditional sampling designs tend to perform poorly [11]. The inspirational example for this paper is an epidemiological undertaking of international relevance: the estimation of tuberculosis (TB) prevalence in countries considered to bear a high burden, and where notification data obtained through routine surveillance are incomplete or of unproven accuracy [3]. In these countries, typically developing countries in Sub-Saharan Africa and South-Eastern Asia, the prevalence of TB is measured by means of nationwide, population-based surveys that are carried out by the World Health Organisation (WHO) with the support of local agencies. In this setting, an accurate estimation of the true TB prevalence is of paramount importance to be able to inform public health policies aimed at reducing the burden; moreover, due to the presence of medical doctors on field during these surveys, every TB case that can be found can and will be cured. This paper is concerned with the most critical aspects of the current practice and possible research lines to overcome its limitations. The main purpose is to discuss some recently proposed strategies, not yet applied to the inspirational example, as possible alternatives possessing desirable features, such as (*i*) the ability to increase the number of detected TB-positives, sample size being equal, (*ii*) the possibility to control the final costs at the survey planning stage, also accounting for possible country-specific logistic constraints, and (*iii*) allowing reliable estimation of the population prevalence.

The paper is structured as follows: in Sect. 2 we give an overview of the method currently implemented by WHO in the actual practice of TB prevalence surveys and present as alternative strategies the adaptive and sequential approaches. In Sect. 3 some empirical evidence is given from a simulation study comparing the methods discussed in Sect. 2 in terms of key aspects for an efficient sampling strategy for surveying a rare and clustered trait, namely (*i*) number of cases detected, (*ii*) total survey costs, and (*iii*) estimation. We discuss the potential of integrating features both from the adaptive and the sequential approaches and conclude the paper with some final remarks in Sect. 4.

2 Sampling Strategies

In this section, we discuss currently available sampling strategies to deal with estimation of the prevalence of a rare and clustered trait in a finite population. After giving an overview of the sampling strategy currently suggested in WHO's guidelines for TB prevalence surveys, the use of adaptive cluster sampling and of a sequential approach for surveying a rare and clustered trait such as TB are introduced and comparatively discussed.

2.1 Current Practice in TB Prevalence Surveys

The sampling strategy currently suggested by WHO's guidelines is a multistage procedure where at the first stage a probability-proportional-to-size (π-ps) design is implemented. The population is divided into a certain number of areas, defined as geographical regions of as homogenous population size as possible; this working hypothesis allows keeping the final sample size in control thus helping, to some extent, the planning of the survey. All eligible individuals within the sampled areas are invited to show up at a moving lab, where a medical examination takes place, and spotted TB positives can be treated immediately. The number of areas to be sampled is determined by using a standard function of (*i*) a prior guess of the true prevalence, (*ii*) the desired estimation precision (usually around 25%), and (*iii*) an estimate of the variability existing between the areas' prevalences [12]. A classic Horvitz-Thompson (HT) approach is employed to estimate the population prevalence. This suggested sampling strategy, Unequal Probability Cluster Sampling (UPCS, from now on), is easy to implement and understand, as expected of an approach suggested in official guidelines, however it has limitations. The rarity of TB positives and their uneven distribution over the inspected areas, in particular, lead to the need for a very large sample size to obtain an accurate estimate of the true prevalence. Available information on between areas variability is only accounted for in the sample size determination, i.e., it is only marginally exploited. It is well understood that traditional designs tend to miss cases when they are clustered: it is speculated that information concerning between areas variability should be exploited to inform unit selection itself, in order to be able to concentrate surveying efforts in areas where spotting a case is more likely. Finally, in addition to fixed costs such as those for the equipment and the staff, other sources of expenditure may arise due to logistic issues typical of the surveyed country; for instance, the type of terrain and the need for additional staff as well as possibly additional means of transportation may be area-specific features, hence leading the costs to be strongly dependent on which areas have been selected for sampling. The required equipment itself may vary according to the country in which the survey is being carried out: different technologies are available for screening survey participants and their use is usually determined by national regulations on radiation exposure. Staff costs may vary because in some countries local staff may be able to conduct and manage the survey, while in others help from WHO staff is needed; moreover, country-specific additional expenditures may present for raising awareness and increase locals' participation.

WHO's practice for TB prevalence surveys may then draw benefit from a more refined sampling strategy able, for instance, to lead to an oversampling of cases, the sample size being equal, and explicitely allow to control at the design level of the survey for variable costs and possible logistic constraints.

2.2 *Adaptive Cluster Sampling as a Tool for Enhancing Case-Detection Ability*

Adaptive strategies have been developed to deal with populations presenting spatial patterns of the trait of interest, as well as to deal with hidden and hard to sample populations [10]. Different approaches are possible under the general idea of adaptive sampling: among these, we deem most suitable for our epidemiological example the so-called adaptive cluster sampling (ACS, [9]). Under this approach, the country at study is divided into a regular grid of M non-overlapping areas called quadrats; ACS then requires (*i*) a proximity measure between quadrats so that a neighbourhood can be defined for each one, (*ii*) a condition, typically of the form $y_j > c, c \in \mathbb{N}$ where y_j is the number of cases in area $j = 1, \dots, M$, and (*iii*) an initial sampling step with given inclusion probabilities π_j to draw a first sample of areas. The initial sample is drawn using a simple π-ps design and for those areas that satisfy the condition, i.e. for which a certain prescribed number of cases has been found, the neighbourhood is chosen to be included in the final sample as well. The sampling procedure stops when no more neighbouring areas satisfy the condition. This method has proven to be more efficient than traditional non-adaptive sampling strategies when the population is rare and clustered and when the within quadrats variability in terms of prevalence is lower than the between quadrats variability [10]. As compared to traditional designs, ACS would provide unbiased estimation of the population prevalence while most likely returning a larger amount of cases. However, in its basic form the final sample size is random, thus making it difficult to plan survey costs. Moreover, there is no specific accounting for logistic constraints and variable costs.

2.3 *List Sequential Sampling as a Tool for Dealing with Logistic Constraints*

Renewed interest has recently arisen on sequential methods (a notable example can be found in [2]), that apply when population units can be ordered in some way, leading to interesting developments in the field of spatial sampling. Such methods allow the user to specify a flexible weighting system used to sequentially adjust inclusion probabilities on the basis of auxiliary information (not necessarily available beforehand). As opposed to adaptive designs, these methods offer some way to control the final sample size but do not possess the specific feature of being able to over-sample study cases. Bondesson and Thorburn [2] developed a general sequential method for obtaining a π-ps sample that well suits our inspirational example of TB prevalence surveys: this very broad approach considers populations whose units can be ordered somehow and selected sequentially in order to make a decision about their inclusion/not inclusion in the sample. Moving from some preliminary choice, the inclusion probabilities are revised after each unit has been visited by means of an updating procedure that allows correlation to be purposively introduced between successive sample membership indicators (we refer the reader to the seminal paper [2] and to [4, 5] for details on the updating algorithm); we focus in particular on Spatially Correlated Poisson Sampling (SCPS, [6]), that provides a spatial extension to the method.

The sequential approach can naturally accomodate for a predefined route: this might be particularly relevant when planning TB prevalence surveys, in that specific knowledge might lead to individuate a path along which transportation costs are minimized and logistic constraints can be taken into account beforehand. An HT approach to unbiased estimation can be undertaken with this sequential strategy as well, where sample cases are weighted according to the conditioning mechanism induced by the updating rule of the inclusion probabilities (cfr. [2]). Differently from ACS, however, the list-sequential setting does not allow, in its current formulation, for over-detection of cases nor to adaptively incorporate sample evidence on the response on the run.

3 Some Empirical Evidence

A simulation study inspired by the TB prevalence surveys example is presented. The main aim is to compare UPCS, ACS and SCPS with respect to the following key aspects:

1. *estimation*: unbiased and accurate estimation of the prevalence is of paramount importance for epidemiological and public health purposes
2. *cases detection rate*: a crucial challenge in TB prevalence surveys (as well as in surveying any rare and clustered trait) is the enhancement of the detection power of the sampling strategy, since every found case, is a case that can potentially be treated
3. *costs*: from an operational point of view, reducing costs could lead to the opportunity of furthering the survey and thus managing to spot (and treat) more cases, as well as to improve accuracy in estimating the true prevalence.

The simulation has been carried out completely in the `R` environment [8], and the packages `spatstat` [1] and `BalancedSampling` [7] have been used to implement spatial patterns generation and SPCS, respectively. For ACS, the code was based on preliminary (unpublished) work by Kristen Sauby and Mary Christman, from University of Florida, Gainesville.

For the sake of illustration and to highlight limitations and advantages of each of the considered strategies, two artificial populations assumed as a possible TB prevalence survey scenarios have been simulated. Both populations are composed by $N = 250{,}000$ units evenly spread over a two-dimensional space. The study variable has value 1 for population units that are TB-positive cases and value 0 otherwise, with the actual population TB prevalence $p \approx 0.01$. In the first simulated scenario, the cases are mostly clustered in 3 groups homogeneous in terms of prevalence, mimicking an outbreak situation in three well separated geographical location. In the second scenario, the cases are clustered in 30 small homogenous groups, ideally mimicking 30 small villages with high prevalence of cases spread throughout the area. Figures 1 and 2 depict both populations and the respective results concerning estimation, detection rate and survey costs.

The sample size is determined as a function of (*i*) a guess of the true prevalence, (*ii*) of the desired estimation precision and (*iii*) of an estimate of the variability existing between the areas' prevalences, that should reflect the spatial pattern of TB-positive cases. The sample size determination considers the classic simple random sampling approach, as well as an additive factor that depends on spatial features. We assume a perfect guess of the true prevalence and rely on WHO's guidelines recommendations for all the involved quantities [12]. Under these working assumptions, the prescribed sample size is $\approx$28 areas to be sampled to reach the desired number of individuals. Note that this translates into a fairly large sample fraction of approximately 28% of the whole population.

In order to compare sampling strategies with respect to costs, we consider a standard linear cost function. The actual total cost for the survey is computed as a simple linear function of the total number of selected population, the total number of sampled areas (implying that the cost for moving the lab from one area to another is a significant component of the total survey cost) and of some fixed costs. We set the fixed costs at 100,000 (essential staff, equipment, advertising, ...), the unit cost per sampled individual to 10, and the unit cost per sampled area at 1000 (cost for transportation and installation of the moving lab in the new location). We ignore, for simplicity, other sources of cost such as unexpected events or area-specific issues that can be expected in an actual implementation. Moreover, we encode the advantage of a careful route planning, easily accomodated by the sequential approach, by reducing by 20% the area sampling cost for SCPS.

All the individuals living in a selected area are invited to participate into the final sample. In our scenarios, based on empirical results from previous simulation studies (not shown here), we decide to select a reduced initial sample of $\approx$19 areas for ACS, in order to contrast to some extent the possibility of sampling too many areas; the adaptive rule is set to include nearby areas in the sample if the number of cases y_j in a selected area j exceeds the threshold of 25 TB-positives. SCPS is set up so to provide a spatially balanced sample by means of the maximal weights strategy [6], and with the list of areas ordered along a regular path from the lower left to the lower right passing through all quadrats. Finally, we will throughout assume for simplicity a second-stage units 100% participation rate to the survey; note that this is a best scenario choice according to what suggested by WHO guidelines, in that the actual participation rate for TB prevalence surveys is usually in the range 85–90% [12].

Figures 1 and 2 summarize the results of 20,000 Monte Carlo runs on the two simulated populations. The actual population prevalence is unbiasedly estimated by all methods (as expected from the theory) with comparable variability in both

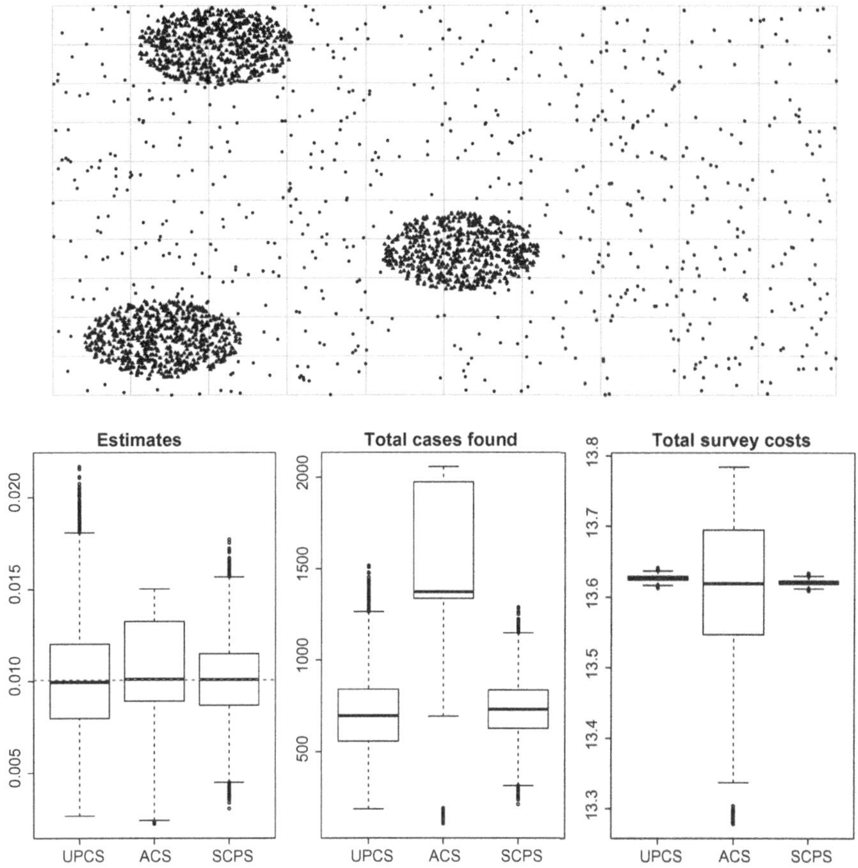

Fig. 1 First scenario. Upper panel: cases (black dots), non-cases not shown. Lower panels, from left to right, Monte Carlo distributions of: the estimators under the three sampling designs (the dashed line indicates the true prevalence), the number of detected cases, and the total costs (plotted on the log scale for better readability)

scenarios. The shape of the Monte Carlo distribution of the ACS estimates appears to be scenario-dependent, as opposed to UPCS and SCPS, the reason being that the sampling outcome for this procedure strongly depends on the specific spatial pattern of the cases. The characteristic oversampling feature of adaptive sampling is immediately evident (center panel in both figures), as it is the out of control impact this has on the total costs (measured on a log scale for better readability). On the other hand, the sequential approach (SCPS) presents a very stable behaviour in terms of detection power and costs, also highlighting the gain in planning a cost-minimizing route beforehand, but is unable, as expected, to spot more cases than the current practice (UPCS).

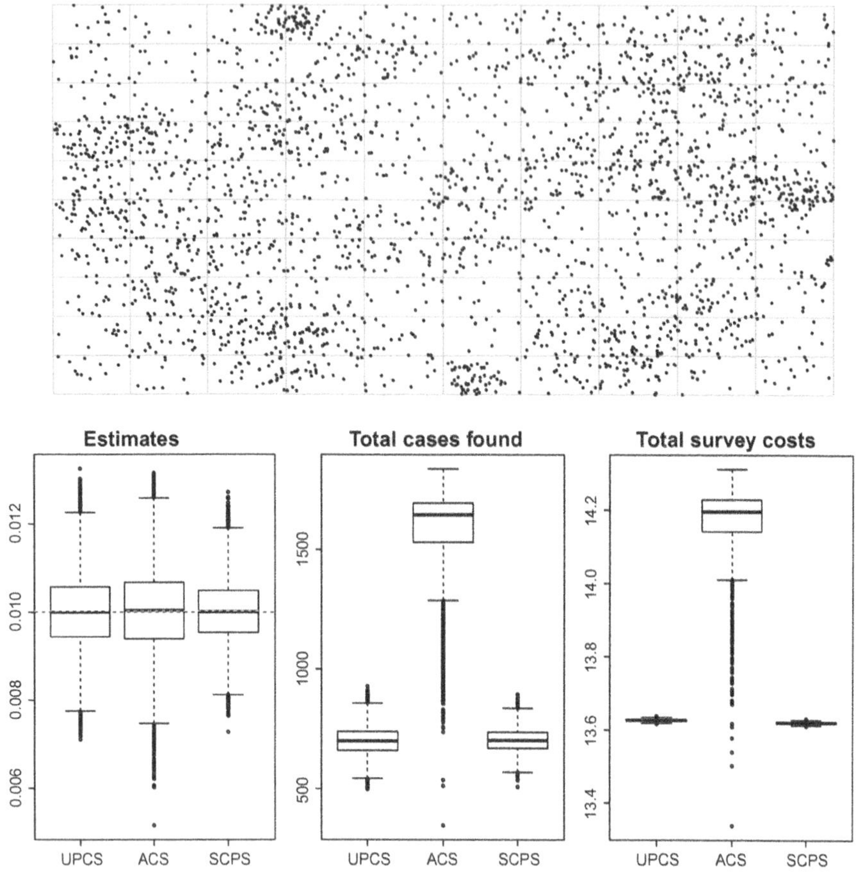

Fig. 2 Second scenario. Upper panel: cases (black dots), non-cases not shown. Lower panels, from left to right, Monte Carlo distributions of: the estimators under the three sampling designs (the dashed line indicates the true prevalence), the number of detected cases, and the total costs (plotted on the log scale for better readability)

4 Discussing Promising Perspectives: A First Step Towards an Integrated Strategy

The use of adaptive cluster sampling has been suggested as a natural alternative to the current methodology used by WHO in TB surveys when a primary aim, along with unbiased estimation of the population TB prevalence, would also be to over-detect TB cases. However, ACS lacks of control over both logistic constraints, which can be a challenge in national TB prevalence surveys typically in poorer countries, and on final sample size, which can lead to unpredictably inflated survey costs. One way to account for such issues may be to sequentially visit units following a prescribed route that minimizes costs, using a sequential sampling approach such as the one outlined in Sect. 2.3. On the other hand, this last approach would not allow for possibly enhancing the case-detection rate, a desirable feature in sampling rare traits such as TB cases. As a consequence, a mixed approach based on a synergic integration of both adaptive and list sequential sampling strategies appears promising, each method presenting interesting strong points.

If the weighting system of the sequential method could be suitably modified to account for sample evidence, then the sampling mechanism would be able to adaptively adjust inclusion probabilities on the run, while still retaining a prescribed visiting order and possibly keeping sample size under some control. In addition to this, an integrated approach would be able to overcome another major issue arising from the use of ACS, that is the definition of a neighbourhood, usually a non-trivial problem in practice. An integrated approach would then have to address (*i*) logistic and cost issues, (*ii*) oversampling of cases, and (*iii*) estimation of the quantity of interest via a suitable system of data-weighting able to adjust for the selection bias due to over-detection of positive cases. With reference to the inspirational example of WHO's TB surveys, once a route that minimizes transportation costs and satisfies logistic constraints had been decided, point (*i*) in the previous list would have been addressed. This would of course require a deep knowledge of the country where the survey is to take place, and possibly a suitable algorithm to choose the best route. The next step towards the integrated approach would then be to extend dependence of the weights in the list-sequential framework to the observed response to tackle point (*ii*). The idea is to employ the observed number of TB cases to update the inclusion probabilities at each step, by explicitely making the weights a function of this information, whenever available. Moreover, area-specific information available only after the unit had been selected could be also taken into account. In order for the updated inclusion probabilities to be proper probabilities (i.e. real numbers in [0, 1] or subsets thereof), the weights must fulfill some conditions (cfr. [2]) but can, in general, be either negative or positive. This is of interest for the problem at hand, since a positive weight induces a negative correlation between successive sample membership indicators, and viceversa. This feature has been successfully exploited, for example in [5, 6], in order to construct spatially balanced designs. By letting the sign of the weights adaptively change throughout the sampling, it would be possible to adjust its behaviour given sample evidence; the strategy would then lead to select with higher inclusion probabilities areas close to where cases have been found, ideally more likely to contain cases, given the spatially clustered structure of TB. Finally, point (*iii*) addresses estimation of the true TB prevalence given sample evidence. An integrated proposal would naturally allow for HT-type estimation as for ACS and SCPS respectively. However, preliminary simulations, not presented here, suggest that although it is indeed possible to construct an unbiased estimator by properly keeping into account the complex conditioning structure induced by both the sequential and the adaptive components, the resulting estimator might be highly unstable. This is fairly expected of an HT-type functional, highlighting how the true estimation challenge would be reducing the estimator variability, possibly relaxing the exact unbiasedness requirement. In order to reduce instability, modifications via post-stratification adjustments (Hájek's rescaling) or a regression approach via exploitation of possibly available auxiliary information might lead to a solution and will be subject to further research. It is worth noting that the development of a sampling strategy that allows to enhance cases-detection while controlling survey costs as well as accounting for logistic constraints, has the potential to effectively apply in a wider range of practical fields besides the epidemiological example that motivated this paper.

References

1. Baddeley, A., Turner, R.: spatstat: an R package for analyzing spatial point patterns. J. Stat. Softw. **12**(6), 1–42 (2005). http://www.jstatsoft.org/v12/i06/
2. Bondesson, L., Thorburn, D.: A list sequential sampling method suitable for real-time sampling. Scand. J. Stat. **35**, 466–483 (2008)
3. Glaziou, P., van der Werf, M.J., Onozaki, I., Dye, C.: Tuberculosis prevalence surveys: rationale and cost. Int. Tuberc. Lung Dis. **12**(9), 1003–1008 (2008)
4. Grafström, A.: On a generalization of Poisson sampling. J. Stat. Plann. Infer. **140**(4), 982–991 (2010)
5. Grafström, A., Lundström, N.L.P., Schellin, L.: Spatially balanced sampling through the pivotal method. Biometrics **68**(2), 514–520 (2011)
6. Grafström, A.: Spatially correlated Poisson sampling. J. Stat. Plann. Infer. **142**(1), 139–147 (2012)
7. Grafström, A., Lisic, J.: BalancedSampling: balanced and spatially balanced sampling. R package version 1.5.1. http://CRAN.R-project.org/package=BalancedSampling (2016)
8. R Core Team: R: A language and environment for statistical computing. R Foundation for Statistical Computing, Vienna, Austria. http://www.R-project.org/ (2015)
9. Thompson, S.K.: Adaptive cluster sampling. J. Am. Stat. Assoc. **85**, 1050–1059 (1990)
10. Thompson, S.K., Seber, G.A.F.: Adaptive Sampling. Wiley (1996)
11. Thompson, W.L.: Sampling Rare or Elusive Species. Island Press, New York (2004)
12. WHO: Tubercoulosis Prevalence Surveys: A Handbook. WHO Press, Geneva (2011)

Age Management in Italian Companies. Findings from Two INAPP Surveys

Maria Laura Aversa, Paolo Emilio Cardone and Luisa D'Agostino

Abstract The aim of this paper is to analyze the behavior of the Italian companies and the solutions adopted for keeping and reintegrating ageing workers in the labour market, as well as the strategies implemented for their professional enhancement, starting from the results of two INAPP surveys. The first one, is a quantitative survey on the attitude of the small and medium-sized enterprises (SMEs) employers towards the ageing workers; the second one is a qualitative research on age management best practices in the large companies. As a result, emergent trends will be underlined in the great enterprises, with a focus on similarities and differences with respect to the SMEs, in order to individualize a feasible development perspective.

Keywords Age management · Workforce ageing · Logistic regression model

National Institute for Public Policy Analysis, former ISFOL - National Research Institute for Vocational Education and Training Employment, that changed its company name in INAPP on December 1st 2016 (www.inapp.org).

The paper is the result of a common work, but the main author of Sect. 2 is Cardone P.E., and the main authors of Sects. 3–4 are Aversa M. L. and D'Agostino L. Usual disclaimer applies.

M. L. Aversa · L. D'Agostino
INAPP-Labour Market Union, Rome, Italy
e-mail: l.aversa@inapp.org

L. D'Agostino
e-mail: l.dagostino@inapp.org

P. E. Cardone (✉)
INAPP-Statistical Office/Sapienza University of Rome, Rome, Italy
e-mail: p.cardone@inapp.org; paoloemilio.cardone@uniroma1.it

C. Perna et al. (eds.), *Studies in Theoretical and Applied Statistics*,
Springer Proceedings in Mathematics & Statistics 227,
https://doi.org/10.1007/978-3-319-73906-9_3

1 Introduction

Since the last century, the most advanced countries all over the world are facing unprecedented demographic challenges, such as a population ageing due to two joint demographic factors: a greater life expectancy and a decreased fertility [1]. Therefore, population ageing is one of the most important challenges facing western countries and it is an irreversible event because the proportion of older people is growing faster than any other age group. Rapid ageing in western countries is accompanied by strong changes in family structures and roles, as well as in labour patterns. Since recent forecasts predict a relevant increase in participation rate of older workers from now on, due also to the pension reform, specific attention must be paid to the older workers personal and organizational resources for facing the (forced) extension of their working life (in terms of health conditions, skills maintenance, resilience to work-related stress factors etc.). The strategy of promoting active ageing in the workplace implies a system of business strategies called "Age Management". This strategy is aimed at leading the competence transfer between generations, monitoring and supporting the time and procedures of retirement, exchanging knowledge between youngest and oldest workers (reverse mentoring) and other good practices such as life-long learning, and monitoring of worker health.

The aim of this paper is to reflect on the implications of demographic ageing on Italian businesses workforce and its management. It consists of two sections. The first one describes the results of the quantitative survey "*Indagine campionaria presso gli attori del sistema produttivo sulla gestione della forza di lavoro matura*" and presents a logit model estimation with the perceived increase in the workforce average age as the dependent variable. The second section illustrates the qualitative research, "*Rilevazione delle Buone Pratiche realizzate da imprese private per fronteggiare il problema dell'invecchiamento della forza lavoro*", the emergent trends about age management strategies in large enterprises, highlighting similarities and differences with respect to the SMEs. Close the paper some remarks about work-force ageing and business strategies against the economic crisis.

2 Ageing Workforce in SMEs

The aim of the quantitative research "*Indagine campionaria presso gli attori del sistema produttivo sulla gestione della forza di lavoro matura*" is to study the relationship among the enterprises development strategies and the solutions adopted for the maintenance, the professional promotion and the possible reintegration of ageing workers, according to the recent pension reforms too. This research is a sample survey on 2.000 private SMEs (except agriculture) with 10–249 employees.

A nationally representative stratified probability sample of enterprises has been drawn out from the Register of active enterprises of Italian National Institute of

Table 1 Incidence of workers per age class and gender. Absolute values and %

	Age < 29	Age 30–49	Age 50+	Total employees
Total employees	988.374	3.261.240	1.216.177	5.465.791
Female	310.895	1.114.853	344.922	1.770.670
Perc. age group/total	18.1	59.7	22.3	100.0
Perc. Female within age group	31.5	34.2	28.4	32.4

Source Inapp (former Isfol), 2015

Table 2 Incidence of workers within age class and size class enterprise. Values %

	Age < 29	Age 30–49	Age 50+	Total employees
10–19 employees	20.5	58.2	21.3	100.0
20–49 employees	16.5	62.9	20.6	100.0
50–249 employees	17.3	58.4	24.3	100.0
Total SMEs	18.1	59.7	22.3	100.0

Source Inapp (former Isfol), 2015

Statistics,[1] considering: geographical area (North-West, North-East, Center, South and Islands), economic sector (Industry, Construction, low added-value Services and high added-value Services) and size class (10–19, 20–49, 50–249 employees) [2].[2]

As shown in Table 1, out of a total of 5.465.791 employees, 1.216.177 are over 50 (22.3%). Their distribution by gender shows a significant prevalence of men over women for all age class, in particular for age 50+ (women are only 28.4%).

Regarding gender analysis within age class, the incidence of female workers over 50 is very low for building sector (only 8.4%), high for low added-value services (33.9%) and very high for high added-value services (41.9%) where women incidence is elevated in every age class.

As shown in Table 2, the incidence of over 50 is greater for medium enterprises where is over 50 about one worker every 4 of them (24.3%). Regarding small companies is 20.6% and smallest is 21.3%. This is confirmed for female analysis: the incidence of women age 50+ is higher in medium (30.4%) than small (26.9 and 26.7%) with exactly 2 points up to total value (28.4%).

The incidence of workers over 50 is balanced within economic sectors because it is between 20.8% for high added-value services and 23.8% for industry (Table 3).

[1]ISTAT-ASIA Archivio Statistico delle Imprese Attive.

[2]In every stratification's level of the basic and backup sample drawn by INAPP, the statistical units have been managed using a procedure software that has allowed the respect of the preset quotas inside every strata.

Table 3 Incidence of workers for age class and economic sector. Values %

	Age < 29	Age 30–49	Age 50+	Total employees
Industry	15.7	60.4	23.8	100.0
Building	18.0	59.5	22.5	100.0
Services (low added-value)	20.7	58.3	21.1	100.0
Services (high added-value)	18.2	61.0	20.8	100.0
Total employees	18.1	59.7	22.3	100.0

Source Inapp (former Isfol), 2015

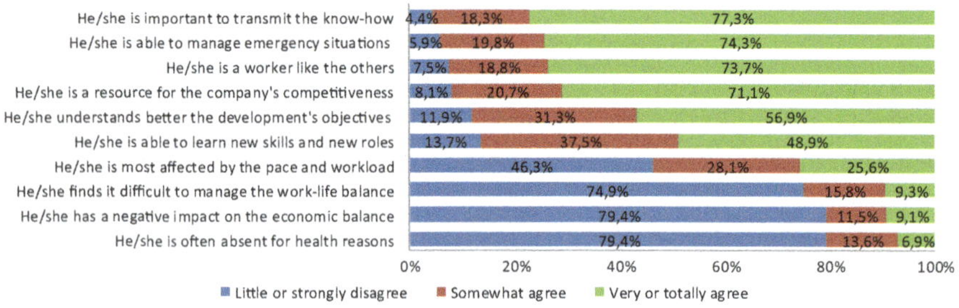

Fig. 1 Image of older workers among SMEs. *Source* Inapp (former Isfol), 2015

Over 70% of SMEs consider older workers as a resource for the company's competitiveness and business, almost 80% think they are important to preserve and send outward its core know-how, as a sort of trade mark (Fig. 1).

56% of SMEs adopt flexible working time (e.g. part-time) as the main age management tool, while other types of measures, such as teleworking, job rotation or adaptation of the workplace are much less provided. In particular, except in few cases (16%), intergenerational work groups should be strengthened in order to involve older workers in the intergenerational transmission of knowledge as tutor or coaches because only 20% does this always or often. Finally, in very few cases age management issue has been debated among social partners and also this propensity clearly decreases in smaller enterprises [3].

The recent crisis has highlighted the importance of education and training at all stages of life, in particular for older adults to avoid unemployment. This requires older people to maintain and update the skills they have, particularly in relation to new technologies. In INAPP survey, participation in lifelong learning for workers over 50 is 54%, but it is only formal classroom training: there is not a significant trace of training on the job, participation in congresses (e.g. exhibitions, trade fairs, conferences, workshop or seminars), participation in learning or quality circle, distance learning course program or study visits.

The survey on SMEs leads quantitative data on the theme of the difficult relationship between the market strategies and those of the human resources. 35% of surveyed enterprises perceive an ageing of their workforce: 10% of them looks this ageing as a disadvantage, over 20% as an advantage and almost 70% have a neutral position. The perception of ageing is reported with higher frequency in the medium-sized companies (44.8% in those between 50 and 249 employees), in the industrial sector (39.9%) and in general in the North East (38.8%). It is a weaker "perceived" matter between the firms in the South (29.7%) and among companies of smaller size, especially in the field of basic services and construction where over 50 workers represent a major component of human resources.

The majority of the companies that had seen an increase (68% of 35%) tend not to consider the working ageing a factor beneficial nor disadvantageous. Those who consider ageing a value and an opportunity are about twice those who consider the increase in the average age as a problem for their business (21.6% vs. 10.4%), and these statements are slightly more frequent in the case of enterprises operating in industry (24.1%), in the South (24.1%) and the North West (23.4%).

Companies active in the industrial sector report a negative relationship between the workers ageing and the possibility to introduce new technologies (14%), the difficult adaptability of older people to new jobs (12.2%), the demand for flexible working time (9.9%) and more generally to organizational change (9.6%). Among the advanced services also, the increase in the average age of the workforce is defined as critical.

Among the Italian SMEs practices of age management are quite rare. For example, as far as the skills and knowledge transfer among generation is less developed than expected. With increasing size, the practice to use older workers for the transmission of knowledge and skills increases. In the enterprises between 50 and 249 workers, those over 50 are found to exert this role *always* or at least *often* in 28.7% of cases. Among the various productive sectors, services with high added values record the lowest percentage of recourse to the "silver workers" as trainers or tutors, while the other three economic sectors appear practically aligned each other [4].

Improving working conditions for older workers is the other key where companies have begun (or should begin) to take measures. Facilitating access to part-time jobs and developing flexible work arrangements are ways to give older workers choice and smooth work-retirement transitions.

More needs to be done to remove age-barriers to employment and to encourage greater participation after the age of 50. Age-neutral human resources management strategies would reduce age discrimination, and should be applied (e.g. in the context of recruiting processes). However, in order to increase employability, productivity and working conditions, age management strategies need to be developed. First of all, it should develop tools for a "demographic check" and for an "age structure analysis" in order to help companies build strategies adapted to their specific situations. General guidelines for age management have their limitations because of the heterogeneity of older workers as well as the more limited strategic human resources management capacities of SMEs. Guidance for SMEs, as well as coaching, mentoring and guidance for older workers and older job seekers, should be provided in a wider scale.

2.1 A Logistic Regression Model

The 2.000 SMEs in the sample analysis generated 207.675 Italian enterprises using a weight calibration variable (see Sect. 2). These were analyzed using a logistic regression model (in Stata release 13). In order to achieve this goal, we have used the perceived increase in the workforce average age as the dependent variable. Concretely, in our study analyzed variables are:

- *Geographical distribution*. Categorical. Four areas: North West, North East, Centre, South and Islands (reference cat.).
- *Economic sector*. Categorical. Four sectors. Industry (reference cat.), Building, Services (low added-value), Services (high added-value).
- *Size enterprise*. Categorical. Three intervals. From 10 to 19 (reference cat.), between 20 and 49, between 50 and 249.
- *Career development* (*for over 50*). Categorical. Three levels. Marginal measure (reference cat.), As others employees, More than others employees.
- *Bonus and financial incentives* (*for over 50*). Categorical. Three levels. Marginal measure (reference cat.), As others employees, More than others employees.
- *Tutoring or coaching* (*by over 50*). Categorical. Five levels. Never (reference cat.), Rarely, Sometimes, Often, Always.
- *Technical skills* (*competences to develop*). Dummy variable: Yes, No (reference cat.).
- *Intergenerational work groups or job rotation* (*competences already developed*). Dummy variable: Yes, No (reference cat.).

Some covariates seem to have a direct relation with the dependent variable, but the descriptive analysis show that SMEs which perceive increase in the workforce average age (35%) adopt career development in marginal measure, or like the other employees, in 97% of the cases (and only 3% more than others employees); 96% of SMEs adopt bonus and financial incentives programs for over 50 in marginal measure or like the other employees. In addition, 79% of them do not promote intergenerational work groups or job rotation. In other words, SMEs propose particular programs not because of perceiving an increase in their workforce average age.

Table 4 shows coefficients (*Beta*) and odds ratios of logistic model. This means that the coefficients in logistic regression are in terms of the log odds, that is, the North East coefficient 0.398 implies that a one unit change in geographical area results in a 0.398 unit change in the log of the odds.

It can be expressed in odds by getting rid of the natural log. This is done by taking the exponential for both sides of the equation, because there is a direct relationship between the coefficients produced by logit and the odds ratios produced by logistic: a logit is defined as the natural log (base *e*) of the odds.

This fitted model says that, holding covariates at a fixed value, the odds of perceiving increase in the workforce average age for North East over the odds of perceiving increases for South and Islands (reference category) is exp

Table 4 Logistic model: coefficients and ODDS ratios

Variables		Coef (B)	Sign.	Exp (B)
• **Geographical distribution**	North West	0.262	0.090	1.300
South and Islands (ref.)	North East	0.398	0.015	1.489
	Center	0.244	0.133	1.277
• **Economic sector**	Building	−0.261	0.107	0.769
Industry (ref.)	Services (low added-value)	−0.298	0.029	0.741
	Services (high added-value)	−0.209	0.180	0.811
• **Size enterprise**	20–49 empl.	0.275	0.026	1.316
10–19 empl. (ref.)	50–249 empl.	0.307	0.030	1.359
• **Career development (for over 50)**	As others empl.	0.353	0.050	1.424
Marginal measure (ref.)	More than others empl.	1.527	0.024	4.604
• **Bonus and financial incentives (for over 50)**	As others empl.	0.417	0.022	1.518
Marginal measure (ref.)	More than others empl.	0.280	0.646	1.323
• **Tutoring or coaching (by over 50)**	Rarely	0.430	0.017	1.537
Never (ref.)	Sometimes	0.431	0.003	1.540
	Often	0.777	0.000	2.175
	Always	0.953	0.000	2.594
• **Technical skills (competences to develop)**	Yes	−0.290	0.017	0.748
No (ref.)				
• **Intergenerational work groups or job rotation**				
(competences already developed)	Yes	0.314	0.039	1.370
No (ref.)				
	Intercept	−1.541	0.000	0.214

Note. Weighted model
Source Inapp (former Isfol), 2015

(0.398) = 1.489. In terms of percent change, we can say that the odds for North East are 49% higher than the odds for South and Islands. In other words, the hazard to perceive an increase in the workforce average age is higher for North East enterprises rather than firms from South Italy (p values of North West and Center are not significant because $p > 0.05$).

Regarding economic sector, the hazard to perceive an increase in the workforce average age for low added-value services is lower of 26% than the odds for industry (OR = 0.741). The size enterprise play an important role: with reference very small enterprise (10–19 empl.), odds of small with 20–49 empl. (OR = 1.316) and medium with 50–249 empl. (OR = 1.359) are higher. It means the odds increase as the size increases: small and medium firms have 32% and 36% hazard higher than very small firms (10–19 empl.) to perceive an increase in the workforce average age respectively.

Furthermore, enterprises who have career development initiatives for over 50 have an hazard almost 5 times higher than enterprises which adopt them in marginal measure (OR = 4.604). Hazard enterprises who adopt often and always programs of tutoring and coaching by workers over 50 are double than enterprises for these who never adopt those programs (OR = 2.175 and OR = 2.594). Also firms who adopt those programs rarely have hazard higher to perceive an increase in the workforce average age. Finally, hazard for companies with intergenerational work groups or job rotation is higher than firms without them (OR = 1.370). Substantially, an active role for older workers increases the hazard to perceive an increase in the workforce average age.

3 Age Management in Large Companies

The aim of the qualitative research is to describe and to analyze the most meaningful age management experiences realized by the great enterprises that operate in some specific segments of the industrial and services sectors, to face the problem of the ageing workers and the possible obsolescence of their skills and competences. This research is a sample survey on 152 large enterprises (15 of them identified as age management "good practices", with 8 deepening case-studies) extracted from ASIA on the base of the geographical area (North-West, North-East, Center, South and Islands), size class (over 250 employees) and economic sector.[3]

The chosen research methodological approach is based on the latest proposals, traceable in the literature, for the collection, classification and analysis of good practices (Best Practices Research—BPR), referring to the criteria of effectiveness, efficiency and sustainability, replication and transferability and mainstreaming (Fig. 2).[4] Reference to the main studies on the subject, is also made for the definition of age management dimensions.[5]

The 152 large enterprises surveyed are mainly manufacturing (65.8%); followed by the companies working in financial service (15.8%), telecommunications, publishing and information technology (10.5%), electricity, gas, water supply and waste management (5.3%) and building (2.6%). They are predominantly localized in

---

[3]Manufacturing, building, electricity, gas, water and waste management supply, telecommunications, publishing information technology, financial services.

[4]Ministero del Lavoro, della Salute e delle Politiche Sociali – D.G. per le Politiche per l'Orientamento e la Formazione, *La catalogazione delle buone pratiche FSE: lo scenario europeo*, [5].

[5]Alan Walker e Gerhard Naegele, Malpede e Villosio [6], progetto "Age Management" della Regione Veneto (Iniziativa comunitaria Equal, 2006).

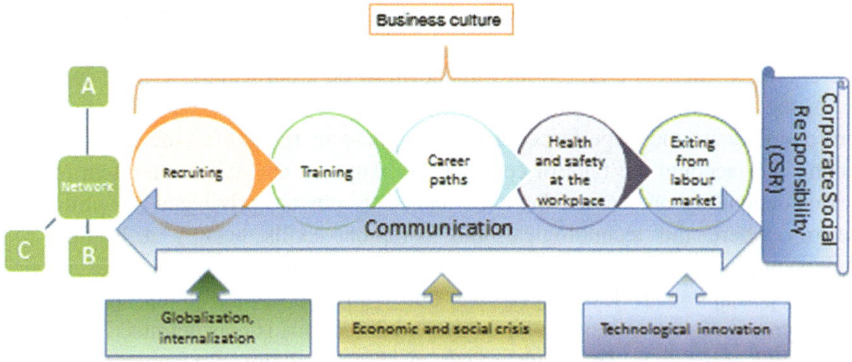

Fig. 2 Logical setting research. *Source* Inapp (former Isfol), 2015

northern Italy (Northwest 41.4%, Northeast 32.2%), followed by the Centre (16.4%), and South and Islands (9.9%).

Companies with over 500 employees represent 57.2% while companies up to 499 employees represent 42.8% and they operate in 59.9% of cases at international level, in 26.3% of cases at national level and in 13.8% of cases at interregional and local levels. Enterprises that are not part of a group represent 23.0%, while in 23.7% of cases the group is present in Italy and in 38.2% abroad.

Out of a total of 196.133 employees, 47.566 are over 50 (24.3%). Their distribution by gender shows a significant prevalence of men (27.1%) over women (18.8%).

The number of the workers over 50, by company size, does not have significant deviations from the statistical average, a fact that occurs both for the economic sector and for geographical area. Mature workers are concentrated in the services sector (electricity, gas, water supply and waste management, and financial services) rather than in the production of goods (manufacturing and building) and in companies located in the south and in the center of Italy.

The results show, beyond more structured and consolidated experiences identified as "good practices", a significant presence of "promising practices", actions just started but not recognised as age management strategies yet. They demonstrate the growing and widespread attention paid by major Italian companies to a more responsible management of the aged workforce and professional cycle on the whole.

From a purely quantitative reading, it was found that 11 companies have declared to carry out interventions on all 5 dimensions of age management, 39 on 4, 65 on 3, 29 on 2 dimensions, only 6 companies on one dimension, while two companies do not perform any type of operation that refers to the dimensions under analysis. The enterprises of the sample, therefore, undertake measures and actions aimed the improvement either in multiple dimensions, or in more aspects of the same dimension.

Generally, large companies appear oriented toward finding solutions not strictly related to ageing workers, but rather to the life cycle management in the company; the age of workers, therefore, does not seem to be a particularly critical factor.

In the process of personnel recruitment, large companies mainly evaluate the professionalism and experience of the candidate in relation to the required task, taking into account the type of contract too. Anyway the staff recruitment method varies depending on sectors: the experienced figures are privileged in building; in sectors where the technological and computer skills occupy a leading role (telecommunications, publishing, information technology) young people are preferred, perhaps with less experience but with appropriate skills. In fact, the employment of young people with higher qualifications is concentrated in the most innovative companies. The results of survey show a relationship between enterprise innovation, internalization, high-tech sector and staff recruitment.

The large enterprises, with a predominant presence of over 50 employees, implement actions to improve this target's working life quality: they carry out different medical examinations on age bases, encourage conciliation practices but, at the same time, they support the admission of young people during selections. Besides, they pay more attention to the exit from working activity, connecting the exit of mature workers with the admission of young people, or providing a reduction of the working hours. Enterprises which are mainly composed by young staff, carry out wide ranging/far-reaching policies, paying particular attention on professional training, on task redesigning and on the research of experienced staff during the recruiting phase.

It is worth remembering that even from the quantitative survey on the SMEs emerged that the demographic composition doesn't represent an obstacle to the development of enterprises. The SMEs don't consider the age as a relevant parameter to determine the efficiency of a worker (especially for clerical and managerial tasks), while during staff's recruitment both profession competences and experiences are privileged.

In this context, the upgrading of skills plays a key role. Formative policies show a double value for the enhancement of mature workers: as an essential instrument in the continuous process of professional growth and as an opportunity for the intergenerational competences transfer, so that the aged workers become active part of the process. The survey recorded many examples, both in traditional ways (coaching, tutoring, mentoring) and in new recent modality (reverse mentoring).

Training, also, plays a key-role in the process of development of age management measures that support the permanence of mature workers within the company by improving the work organization (Table 5).

The topic of work organization is related also to the development of career. By means of age management, organizational strategies take into account the evolution of workers' motivations, expectations and needs throughout his whole professional career, according to the company's mission.

In companies that subsidize the work-family balance as an organizational set-up, the demographic structure shows that over 50 s rate is rather substantial (41.3%), quite far from the average value; in detail, males represent 46.3%, female 27.8%.

Table 5 Companies that have implemented specific actions to develop the skills of over 50 workers. Absolute values and % firms total in a specific class of employees

Actions	Up to 499 employees		500 or more employees	
	a.v.	%	a.v.	%
Internal training measures for over 50 workers	7	36.8	5	23.8
External training schemes for over 50 workers	3	15.8	2	9.5
Training courses for mobile workers	1	5.3	1	4.8
Intergenerational skills transfer	11	57.9	19	90.5
Educational exchanges with other companies	3	15.8	2	9.5
Budget, recognition and/or skills certification	1	5.3	3	14.3
Personal projects that develop new skills	2	10.5	1	4.8
Other	4	21.1	2	9.5
Total companies	19	–	21	–

Source Inapp (former Isfol), 2015

Equally evident, the over 50 presence in large enterprises that subsidize organizational practices like team job or job rotation (33.7%; Table 6).

Regarding these issues, it should be stressed that the results of the SMEs survey, underline an inadequate system for the definition of the employee's career paths, in contrast with large companies policy. When these systems are present, they are focused on firms of upper-middle-size (50–249 employees) and they are rarely addressed to target over 50 s. They are mainly achieved by changing roles and tasks, connected to new skills acquisition or seniority, but sometimes also to new responsibilities as coach or tutor for intergenerational skills transferring. Practices of skills verifying and potential checking, or periodic vocational guidance and assessment, are also not common.

The large enterprises that have implemented best practices in age management present some common features. All of them are very large companies (+500 employees) with more than 30% of workers over 50 and located in the north of Italy. They belong to international corporation groups and most part of them produce in the services sector, where working-life extension creates re-motivation and skills upgrade needs; whereas, in manufacturing sector, the most affected by economic crisis, the need to manage the life cycle of workers, seniors in particular, is linked mainly to the strenuous and arduous work, or to higher risk of on-the-job injury. The corporate culture of these 15 companies appears shaped around the Corporate Social Responsibility (CSR) approach and the human resource development (training is considered strategic), as well as around the propensity for innovation.

As you can deduce analyzing the 15 good practice observed by qualitative survey, especially by case-studies, the most important actions of age management

Table 6 Over 50 rate of personnel in companies that activate innovative organizational set-up (%)

Innovative organizational set-up		Over 50 personnel		
		Total	Male	Female
Subsidizing practices of organizational set-up as team job and job rotation	No	22.4	24.8	17.3
	Yes	33.7	41.7	24.1
Introducing new technologies	No	24.2	27.1	18.6
	Yes	28.2	25.5	32.2
Developing practices for work-family balance	No	23.7	26.3	18.5
	Yes	41.3	46.3	27.8
Promoting practices of fidelization	No	24.3	27.1	18.8
	Yes	19.1	19.5	18.4
Other	No	24.2	27.0	18.8
	Yes	29.1	31.3	23.3
Total		24.3	27.1	18.8

Note To compensate the lack or incomplete answer provided by respondents on the demographic structure of enterprises (number of employees by age and gender) missing data were imputed using the following criteria

- for "Total employees" were used the data in the ISTAT ASIA 2011
- for total employees over 50 were used average values by sector (ATECO2 digit), derived from ad hoc processing ISTAT survey "Labour Force 2013"
- for the distribution of employees by gender were average values by sector (NACE 5 digits), number of employees and geographical distribution of the survey Indigo-CVTS

Source Inapp (former Isfol), 2015

can be grouped into three priority areas: training, work-experience enhancing and intergenerational dialogue sustaining. The general aim is maintaining the overall company profitability by improving worker's productivity, particularly that of older employees, normally related to highest costs.

Furthermore, actions for motivation support, for generational diversity enhancing and work-life balance promoting, are very important. Particularly, are included: interventions aimed to get old employees involved in new activities by enhancing their experience (both work and life); needs analysis realized paying attention to intergenerational dialogue and workers involvement; test of participatory working method, mixing working groups by gender and age.

Generally, large companies dealing with workers ageing, follow the same path characterized by some inevitable steps: (1) awareness of the problem; (2) analysis of the demographic structure of the company; (3) recognition of the mature staff needs; (4) design and launching of pilot projects; (5) results verification and evaluation for proper corrective actions implementation. This often involves the introduction of several organizational innovations, such as a new business function (e.g. Inclusion Division), or a specific responsibility position (e.g. Diversity Manager).

The main elements of strength of age management good practices are related with a positive approach that better interpret the right meaning of "lifecycle management", instead of "older workers management" (positive valuation approach of ageing; intergenerational approach considering the complexity of company's demographic system). Some key-factors moreover are very important in order to facilitate the initiative sustainability and implementation, such as clear communication in all project phases, convenient aims for both staff and company, assimilation in plant-level bargaining, economic sustainability, periodic results verification and final evaluation. At last, a favorable external environment, CSR friendly, free from clichés, stereotypes and prejudices that can compromise the results of corporate policies against discrimination, is essential. For such reason, a business network could be useful, by best practices knowledge too, to spread out a new corporate culture based on management's innovation standard, still not very diffused in Italian work organizations.

4 Conclusions

The Italian enterprise actions are influenced not only by actual economic crisis, but also by the demographic ageing process and its impact on the workforce's system evolution.

The quantitative survey reveals that the SMEs have implemented a defensive strategy against the crisis, with no relevant differences with regard to the size. The smallest one mainly have been hit by negative economic trends and their profitability and productivity decreased more than medium sized enterprises that invested in training and skills improvement in a more effective way.

Conversely, age management's best practices survey shows that large companies implemented differentiated strategies. On the one hand you can find enterprises taking a defensive attitude (cutting costs, staff reduction, resorting to layoffs); on the other hand there are firms that have worked towards the internationalization, the innovation of product and process, the search for new markets, the expansion and the promotion of competitiveness also through human capital enhancement.

Qualitative in-depth analysis results show that demographic ageing is a field still unexplored also by largest companies. The focus on this phenomenon is quite recent and often connected to specific needs imposed by contingent circumstances or by external solicitations.[6] In any case you can observe some solidify

[6]For example: economic crisis increasing, retirement system reforming, 2012 as European Year for Active Ageing and Solidarity between Generations declaring.

experiences too, where enterprises have been committed for long time in developing a new corporate welfare, based on working well-being and diversity management. With increasing size, enterprises implement virtuous actions in a long-term development prospect of working-life cycle and diversity management.[7] Therefore, in large companies it is more likely to detect policies and structured age management's interventions, as well as experiences inspired by CSR and developed in the long term, contributing to build a corporate identity within the territorial system.

In summary, the extension of working life is influenced by a factors combination: regulatory environment, labour market rules and the different business strategies aimed to facing the work-force ageing. Companies can contribute to keep older workers longer in employment by changing their structure and management strategies, while labour policies can make available financial incentives. Preserving older workers employability requires, therefore, multidimensional actions both in labour and welfare policies and in enterprises organizational set-up. Only the establishment of collaborative working relationships between all stakeholders involved (workers, enterprises, trade unions, local networks and the society as a whole) can improve both the older workers opportunities to keep their job and a longer work-life's good quality. This will be possible if national and regional active labour market policies contribute to create a good working environment in order to develop a new corporate welfare inspired to diversity management criteria. In this context, the right answer to workforce ageing is the intergenerational cooperation rather than the generational turnover in labour market. If age represent an opportunity, instead of a critical factor, generational turnover mechanism should be definitively abandoned to increase the employment level.

References

1. Central European Ageing Platform: Green Paper (2013)
2. Checcucci, P., Fefè, R., Scarpetti, G. (eds.): Età e invecchiamento della forza lavoro nelle piccole e medie imprese italiane. INAPP, Rome (2017) (forthcoming)
3. Checcucci, P., Fefè, R.: Old Hopelessy Inn. The role of public policies in supporting the extension of working life. Isfol. http://isfoloa.isfol.it/handle/123456789/1251 (2016)
4. Checcucci, P. (ed): L'anno europeo dell'invecchiamento attivo e della solidarietà fra le generazioni: spunti di riflessione. ISFOL, Roma. http://sbnlo2.cilea.it/bw5ne2/opac.aspx?WEB=ISFL&IDS=18499 (2012)

[7]These themes are in step with European Union guidelines suggesting working policies based on life-cycle, rather isolated target-group, and life-long age management and diversity management approach, aimed to considering relationship evolution between persons, labour-market and family-life.

5. Ministero del Lavoro, della Salute e delle Politiche Sociali – D.G. per le Politiche per l'Orientamento e la Formazione: La catalogazione delle buone pratiche FSE: lo scenario europeo (2008)
6. Malpede, C., Villosio, C.: Dal Lavoro al pensionamento. Angeli (2009)
7. Walker, A.: Managing an Ageing Workforce: A Guide to Good Practice. Eurofound, Dublin (1999)

Generating High Quality Administrative Data: New Technologies in a National Statistical Reuse Perspective

Manlio Calzaroni, Cristina Martelli and Antonio Samaritani

Abstract Statistical reuse of administrative data is limited by serious issues about data quality: these concerns are particularly serious in innovative contexts, in which only administrative data could provide the required granularity and fit to the real processes. Administrative data semantic and meta-information, in particular, are hardly ascribable to general notation standards: PA, as a services purchaser, is often not able to provide to its suppliers efficient clues about the way data must be denoted in their statistical reuse perspective. In this paper we discuss how new technologies and semantic web, in particular, may provide unprecedented methods and instruments to support PA in orienting their suppliers to high quality administrative data: the joint role of the National Statistical Institute and Public bodies, as owners of administrative data, will be discussed.

Keywords Data reuse • Statistical and administrative information systems
Semantic web • Statistical and PA ontologies

1 Administrative and Statistical Data: A Problematic Convergence

A complex social and economic environment requires a network approach both at administrative and statistical level [1].

M. Calzaroni
ISTAT, Rome, Italy
e-mail: manlio.calzaroni@istat.it

C. Martelli (✉)
Dipartimento DiSIA, Università Firenze, Florence, Italy
e-mail: cristina.martelli@unifi.it

A. Samaritani
AGID, Rome, Italy
e-mail: samaritani@agid.gov.it

C. Perna et al. (eds.), *Studies in Theoretical and Applied Statistics*,
Springer Proceedings in Mathematics & Statistics 227,
https://doi.org/10.1007/978-3-319-73906-9_4

The main strategy, in this perspective, is data integration and reuse [2]. At administrative level, this means efficiency in services delivery, bureaucracy burden reduction and a more integrated administrative vision.

As regards computing, this means less redundant data systems and a highest quality level. Statisticians, when able to rely on good administrative data sources, achieve a granularity description level, hardly attainable with traditional survey protocols.

These results may be achieved only with a joint effort of legislation, administration, IT services and statistical agencies protocols.

1.1 The European Vision

The European Commission advocated on data reuse both in statistical and administrative framework.

Following the Commission guideline on production of EU statistics (Brussels, 10.8.2009 COM (2009): "Any development in the area of statistics is determined by two main drivers: on the on hand the need to deal with new and emerging needs for statistics, and on the other the need to reduce the burden on respondents, as well as the costs for producing statistics. In addition, the circumstances in which statistics are produced have changed following developments in the information technology.... Users increasingly need integrated and consistent data, as the phenomena that are being measured become more complex and interrelated. ... Efficiency gains can be obtained by the re-use of these administrative data for statistical purposes.... But efforts are needed to ensure the quality of the data, because very often the administrative are not available in the form needed for statistics."

Over the past thirty years, the European Statistical System (ESS) has developed a conceptual system of classification for economic and social statistics, useful to compare statistical data from member countries [3]. This scheme is an important starting point as it represents a relevant experience widely tested, also in non-statistical environments.

In the administrative and management context, reuse is an extremely sensitive topic to deal with [4, 5]. The European agenda for digital and collaborative economy may be, in fact, a strategical lever for data reuse: in the European vision, digital infrastructures activity aims at empowering researchers with easy and controlled online access to facilities, resources and collaboration tools, bringing to them the power of ICT for computing, connectivity, data storage and access to virtual research environments.

Digital Single Market, for instance, is one of the Commission's top priorities and it requires Europe to overcome barriers related to infrastructure, broadband accessibility, copyright and data protection, by enhancing the use of online services and digital technologies. If properly realized, also by a semantic point of view, it could be a remarkable driver for an administrative and informational/statistic joint language.

As far as it regards public administrations and services, the European Commission is taking concrete actions for the development of Cross-border Digital Public Services. These include, but are not limited to, the creation of European interoperable platforms such as a common framework for citizens' electronic identity management (eID), and the fostering of innovation through the Competitiveness and Innovation Programme (funding Large Scale Pilots and eParticipation projects).

1.2 The National Vision: Opportunities and Concerns

Taking into account the European regulatory environment, many are the initiatives that have been taken by national bodies (jointly or individually) to cope the new and emerging information needs. The exigency to reduce the burden on respondents, as well as the costs for producing statistics, is driving National Statistical Agencies (and Istat) to (re)use integrated administrative data.

An important premise: up to date, it is easier to reuse, for statistical purposes, administrative data on private entities than on public bodies. Some example will highlight these problems.

In 2011, Istat developed, in support of economic censuses, a system of integrated administrative data useful to realize a "virtual census" on economic entities, to keep updated on a yearly base. It was impossible, in that occasion, to obtain the same product for public bodies: available administrative data did not have the required statistical quality. For this in April 2016 Istat succeed in launching a survey to gather the figures useful to update the 2011 census on PA.

A similarly unbalanced situation has been found when reusing electronic invoice data. As well known, the review on public spending is a major policy aim. In this area, electronic invoice could represent a strategical data source. An analysis of these data is ongoing with the ultimate prospect of their integration in the PA Statistical informative system.

Nevertheless, the task is not easy, as expenditures classification and conceptualization is quite uneven.

Public bodies expenditure data may be derived from balance sheets. Different administrative acts create administrative data potentially useful to analyzing specific subjects (ex-ante or ex post, as described). In the following, the classification environment:

(a) Balance sheets are conceptualized along two main concepts: *program* and *mission*; missions are coded with Cofog (Classification of the functions of government), at different levels, depending on institutions. In balance sheet, the concept of *chapter*, to describe specific expenditure, is characterized by more than 5000 items, with inconsistency problem on the same *chapters* in different agencies;

(b) PA expenditure are documented by SIOPE, a computerised system which records the receipts and payments made by public administration treasurers/cashiers in order to ensure that public accounts comply with the conditions provided for in EU legislation. SIOPE uses an own classification (for central administration 352 codes);
(c) Electronic invoice classify expenses using (not always) CPV (Common Procurement Vocabulary) standard that use about 9.500 voices. This coding is required only for expenditure greater than 40.000 euros.
(d) Istat defined a specific classification in support of 2011 census to gather information on services performed by PA to privates and firms.

The minimum goal should be an expenditure analysis from an aggregate level (balance sheet) to the single voice (electronic invoice); with the described classifications, in practice, this objective is quite impossible to achieve.

These examples show that, despite European and national high-level initiatives, the convergence between administrative and statistical environments is far for being optimal. The examples previously discussed show a convergence of different reasons: (i) some are linked to long standing established practices, not suitable anymore, to data reuse; (ii) other are linked to innovative administrative processes. Real world evolves, in fact, at a very high pace and, jointly, administrative data go forward. Statistical agencies could be obliged chasing, often late, the stakeholders' adopted language. In this perspective, the contribute of digital agenda achievement is paramount, as it is the only channel really connected to most innovative processes and stakeholders profiles.

As far as it regards implementing the Italian Digital Agenda, AGID, a legally mandated organization under the supervision of the President of the Council of Ministers, is developing technical requirements and guidelines to enable and empower cooperation among information systems of governmental bodies and the European Union. In this perspective, it is strategical to design the national framework for interoperability of services for the technical and semantic interoperability.

One of the most strategical proposal is The Public Digital Identity System (SPID), the Italian government framework compliant with the EU eIDAS regulatory environment, aimed at implementing electronic identification and trust services in e-government and business applications. eIDAS regulation aims at boosting the user convenience, trust and confidence in the digital world, while keeping pace with technological developments, promoting innovation and stimulating competition.

Another important product is Pago PA, which provides customers and non-customers alike with its electronic payment solutions in favour of the Public Administration and public service providers. It is immediate to see the level of overall knowledge enhancement that these new services may provide at descriptive granularity level, as they address single, specific items. The descriptive semantic has, however, to be governed. In this sense, for instance, it would be desirable to go even further on electronic invoice, for instance linking the service to e-procurement platforms and trying to identify the purchased item and not only the class to which it belongs.

A convergence path among different institutions is not only desirable, but possible, as witnessed by the successful example provided, some years ago by the economic classification (NACE). During the changes connected with the new NACE 2007 classification, Istat Internal Revenue Service, Chambers of Commerce, INPS worked in an specific committee (where were represented the main PA interested in it, more than 40) to develop a unique Italian version of the new classification; four different economic classification (one for each bodies before mentioned) were acting in Italy before this coding transfer.

In the following some issues to consider to ensure convergence among PA (services holders and providers) and IT (technology suppliers).

1.3 Public Administrations

Reusable (administrative and statistical) data critically depend on public services computerization.

Unfortunately, public administrations are traditionally linked to administrative dialects, particular semantics, specific regulations and habits. Moreover, the network and reuse perspective is not common in directives implementing regulations: the priorities are more focused on specific objectives. Among the standards that have to be meet when computerizing a public service, reuse is not frequent and often semantics and coding protocols are not mandatory. When a service has to be automatized, procurement in Public administration is often completely demanded to IT services, usually not involved in dominion issues: the result is different administrative languages and semantics difficult to harmonize.

1.4 IT Services

Usually IT professionals are called to transform specific requirements in solutions, which, on their turn, generate data repositories: IT specialists are not specifically trained on reuse problem and usually they are not particularly sensitive to the issue, especially if this approach could be perceived as a technological performances downgrading. Sometimes reuse approach is considered uneconomical by IT companies, as emancipates purchasing bodies and increases competitiveness. It is very usual, in fact, that IT companies, who has designed and built an operational system, are involved again to produce also the informational level: it is usually a problem, in fact, to switch to a new supplier when the documentation level, often drawn up only for system maintenance is low or merely technological.

2 Current Strategies Toward Network Administrative and Statistical System

To overcome the problems previously outlined, a joint national bodies' role is paramount. In the framework of this paper, in particular, we will focus on statistical agencies and on those responsible for implementing and developing technical requirements and digital guidelines: for Italy ISTAT and AGID.

Both the bodies, from different perspectives, have the mandate to enable and empower the cooperation among governmental bodies' information systems and to promote and disseminate information initiatives. In charge of both also training for citizens and civil servants. A strong coordination among the two is the necessary precondition for any reuse politics.

The objective may be pursued at different levels.

(i) From a minimum level, in which specific languages (classification and definition) are framed in a more general classification systems (for example, among others ESS). In this way it would be possible to integrate information generate by any specific classification in a general schema;
(ii) To the level in which the same language is adopted and used.

The ultimate objective is to stimulate an evolutive virtuous circle between operational and informational systems to bridge the (i) level to the (ii).

Given the general framework, an operational system, needing a higher descriptive granularity level, may adopt a more detailed and specialized language. Data may be used by reference to the general level for a wider stakeholders' community, or shared with bodies interested at the same granularity level: this latter is only feasible if national bodies provide diffusion of best practices. A similar schema has been applied, for instance, in information systems in support of work accidents risk profile detection; administrative data, referred to very specific work processes has been also linked to the more general National Insurance taxonomy.

3 The Contribute of New Technologies

New technologies represent, together with laws, the feasibility framework to address the issues previously discussed.

Many innovative proposal and solutions have been advanced to diffuse and reuse administrative data repositories: ETL, big data, open data, linked data. All these approaches operate on administrative system *output data* and for all of them important concerns on data quality hamper their statistical reuse. Different bodies, who want to harmonize and integrate their data, are nowadays obliged to perform complex ETL procedure: transparency and quality level of the integrated system is often damaged. This issue is even more severe if ETL results are diffused as open data and reused in statistical analysis whose bias is not clear. It would make a great difference if new technologies became involved also in the process of administrative sources generation, along the whole data biography, from creation, to usage and diffusion.

Ontologies, for instance, if produced and diffused at official level jointly by national authorities, may represent an important leverage: when a new operational system has to be produced, they could be used also to generate shared schemas and coding tables.

An important trade-off between the exigencies of concurrency and standard languages is characterizing modern societies: the increase of system players often implies the proliferation of administrative or technical languages. A modern response cannot consist in centralisation, bureaucracy and hierarchical informative structures: the answer is in semantic technologies, able to provide a widespread, shared communication structure among different and concurrent players.

It must be noted that technologies support is not properly considered when drafting new regulations; for instance when addressing transparency in public procurements and electronic invoicing, exceptions in data specification are granted for lower amounts.

The ratio is to lower the bureaucracy burden, which is often such only for the system bad organization and conceptualization: a good use of new technologies in support of operational systems would result in a better users experience and in a richer and reusable data repository.

4 Conclusions

Administrative and statistical data reuse is considered, both at Community and at national level, a mandatory condition for efficient public information dissemination. Politics evaluation and transparency promotion actions would be hardly feasible without a full exploitation of the data generated by the administrative processes involved in the analyzed processes.

In many fields, standard semantic and coding procedures are available, but too often data reuse protocols are activated only when the data have already been generated and produced: in other terms, in many strategic fields, only ex-post measures are applied, seriously affecting, in this way, the overall information quality.

Different is the situation when the data reuse objective is considered along the whole information biography, as it exerts its effect both ex ante (at the policy formulation stage, providing information on dominion and scope of governance action; generating and coordinating administrative acts) and ex post, when new rules have been put into effect, to evaluate their consequences and outcomes.

Semantic homogeneity, base for effective, efficient, sustainable and economic data reuse, has to be strived and achieved at every governance stage, from law-making and policies organization to services computerization

In this perspective, in the framework of a more general discussion of the role of major public bodies in the generation of a common, general and public administrative language, a specific reflection has been developed on new semantic technologies and on ontologies. In this work perspective, in fact, they may provide a tangible support to national agencies to develop a joint and harmonized language, functional both to governance actions and to an open, competitive and innovative society.

References

1. Mauthner, N.S., Parry, O.: Open access digital data sharing: principles, policies and practices☆. Soc. Epistemol. **27**(1), 47–67 (2013)
2. Wallgren, A., Wallgren, B.: Register-Based Statistics: Administrative Data for Statistical Purposes, vol. 553. Wiley (2007)
3. Zhang, L.C.: Topics of statistical theory for register-based statistics and data integration. Stat. Neerl. **66**(1), 41–63 (2012)
4. Nam, T., Pardo, T.A.: Conceptualizing smart city with dimensions of technology, people, and institutions. In: Proceedings of the 12th Annual International Digital Government Research Conference: Digital Government Innovation in Challenging Times, June 2011, pp. 282–291. ACM
5. Yoo, Y., Henfridsson, O., Lyytinen, K.: Research commentary—the new organizing logic of digital innovation: an agenda for information systems research. Inf. Syst. Res. **21**(4), 724–735 (2010)

Exploring Solutions for Linking Big Data in Official Statistics

Tiziana Tuoto, Daniela Fusco and Loredana Di Consiglio

Abstract Official statistics has acknowledged the value of big data and has started exploring the use of diverse sources in several domains. Sometimes, big data objects can be easily connected to statistical units. If a unit identifier is available, the opportunity to link big data to existing statistical micro data can allow enlarging the content, the coverage, the accuracy and the timeliness of official statistics, for example Internet-scraped data could be used with this aim. In this setting, new challenges arise in data integration with respect to linking administrative data. In this work, we describe a real case of integration of web scraped data and a statistical register of agritourisms specifying the novelties and challenges of the procedure.

Keywords Big data · Internet-scraped data · Data integration
Data linkage · Farm register

1 Introduction

In the recent years, Official Statistics has acknowledged the value of Big data and has started exploring the use of diverse sources in several domains. For some of these external sources, the object can be easily associated with a statistical unit of the target population. In those cases, if a unit identifier is available and shared with the NSI, the opportunity to link Big data to already existing statistical data at micro-level can allow enlarging the content, the coverage, the accuracy and the timeliness of official statistics. In the case of the Internet-scraped data, there is a

T. Tuoto (✉) · D. Fusco · L. Di Consiglio
Istat, Via Cesare Balbo 16, Rome, Italy
e-mail: tuoto@istat.it

D. Fusco
e-mail: dafusco@istat.it

L. Di Consiglio
e-mail: diconsig@istat.it

C. Perna et al. (eds.), *Studies in Theoretical and Applied Statistics*,
Springer Proceedings in Mathematics & Statistics 227,
https://doi.org/10.1007/978-3-319-73906-9_5

great potential to enrich the information of the traditional statistics. In this setting, new challenges arise for data integration experts in official statistics, due to the deep differences with respect to the familiar framework in which administrative data have been integrated for a long time in order to produce statistical outputs, e.g. statistical business and population registers. In this study, exploiting a real case as a starting point, we describe novelties and challenges in integrating Internet-scraped data with traditional statistical datasets, from the entity extraction and recognition phases to the unit matching algorithms. As case study, we propose the linkage of Internet-scraped information with data related to agritourisms, as reported in the statistical Italian farm register, that is obtained by the integration of several administrative sources. In order to overcome some limits of the so far well-established linkage procedures, we propose to explore new techniques not yet introduced in the official statistics production system. Finally, we devote the due attention to the output quality evaluation, to better analyse benefits and risks of the integration and to allow the analysts to take into account potential integration errors in subsequent analyses.

2 Standard Solutions in Official Statistics

The combined usage of data coming from different sources is a current practice in official statistics since some decades. Administrative data is largely used for building up statistical registers, as benchmark at aggregated level, as auxiliary information and for replacement of survey variables [4]. When data are available at micro level, record linkage techniques are widely applied in order to recognize in different sources the same real world entity even in absence of a unique identifier.

The most widespread statistical approach to record linkage is based on the Fellegi and Sunter [5] theory and several statistical institutes have designed and developed tools to be able to face record linkage problems arising in linking survey and administrative data. An up-dated critical review of methods and software for record linkage is in Tuoto et al. [15] where the proliferation of methodologies and tools in recent years is interpreted according to assigned criteria, namely flexibility of the tools with respect to the support to input/output formats, extensibility, maturity, supported language and coverage of functionalities related to identified sub-phases in which it is possible to organize a record linkage process to reduce its recognized complexity.

Since 2006, the Italian National Institute for Statistics (Istat) has designed and developed RELAIS, an open source toolkit providing a set of techniques for facing record linkage projects [14]. As principal feature, RELAIS is designed and developed to allow the combination of different techniques for each of the record linkage phases, so that it can provide solutions for different linkage problems on the basis of application and data specific requirements.

In ten years, the RELAIS toolkit has been applied in several situations, i.e. when dealing with:

- data of different quality coming from administrative sources (as resident permits, hospital admissions, road traffic accidents, etc.);
- huge amount of data (i.e. the 60 million of people involved in the Italian population census at the time of the post-enumeration survey)

Previous experiences with data related to agriculture have already shown that these data have some peculiarities with respect to the record linkage task, essentially related to the inaccuracy of the "name" of the statistical units and the difficulties in identifying detailed address in rural areas. In fact, in the agricultural field the "name" could be the farm name (like the business name) or the farm operator name. This is particularly the case when such information come from administrative sources and refer to units that only indirectly are associated to the target statistical units, for instance because the sources were compiled for different purposes, e.g. fiscal and benefit purposes. Moreover, when extended large families operate farms, different family members could appear as reference for tax, benefits or as land owners. Furthermore farm addresses in rural areas are often the generic name of the land and are often shared for all the units and houses located, increasing the difficulties in linking sources and resulting in imperfect matching.

3 Issues in Linking Big Data

The definition of a linkage strategy requires selecting the set of the most discriminant common variables. Linkage methods typically compare different variables of the entities using a set of distance measures. The resulting similarity scores may be combined using different aggregation functions. If the data sources use different variable value representation formats, values have to be normalized by applying transformations prior to the comparison. Designing a good linkage strategy is a non-trivial problem as the linkage expert needs to have detailed knowledge about the data sources in order to choose appropriate variables, data transformations, distance measures together with good thresholds as well as aggregation functions.

Currently, a very large number of information is available from the web. Official statistics has acknowledged the value of Internet-scraped data and has started exploring their use in several domains (for instance in statistics on ICT use in enterprises [1] and tourism [6, 11]). In most cases, data coming from the web is not directly comparable with data collected and organized by National Statistical Institutes and a lot of work is needed to create integrated data.

In fact, data can be scraped directly from the website where several information at unit level can be harvested (company, agritourism in this case). On the other hand, this model requires first identifying the websites and then it has to face with different queries and different formats obtained from each website. Otherwise, data can be scraped from "hub", website hosting and describing a plurality of units (for instance, hotel, agritourism, etc.), in this case the number of information that can be achieved is smaller than in the previous case and conditional on the available information on the hub.

In addition, it is likely that the Internet-scraped data are not standardized or codified. In some sense, official statistics are already prepared to face this kind of problems but the pre-processing phase is still a very time-consuming process and a lot of work is needed to identify models that can easily support the data reconciliation, the management of the complexity and to allow the data integration step. Other typical problems can be related to the low quality of data and to changes in the data model of the external source.

So, in order to integrate web scraped data in the official statistics production process, a system is required for data ingestion and reconciliation that allows managing a big data volume of data coming from a variety of sources. The statistical production system needs to produce the ontology and the big data architecture, and the mechanisms for the data verification, reconciliation and validation.

The main issue of this task seems to be the definition of the reconciliation algorithms and their comparative assessment and selection. A reconciliation step is needed to deal with the variety and variability of data, e.g. with the presence of several different formats, with the scarce (or non-existing) interoperability among semantics of the fields of the several datasets. Generally, in order to reduce the ingestion and integration cost, by optimizing services and exploiting integrated information at the needed quality level, a better interoperability and integration among systems is required [2, 8, 12].

This is desirable but not always possible, due to lack of communication, agreement, operability between actors. This problem can be partially solved by using specific reconciliation processes to make these data interoperable with other ingested and harvested data. On the other hand, this approach does not solve the problem since instances can be not interoperable and linked together. For example, a street names coming from two different sources physically identifying the same street may be written in different manner creating a sematic miss-link. These problems have to be solved as well with reconciliation processes. On the other hand, the web data sources may report a "commercial" name as "name" for identifying unit, something appealing for web searchers and potential customers, while in the official data the unit is reported with different name unrelated with the previous one.

The whole linkage activity includes processes of data analysis for ontology modeling, data mining, formal verification of inconsistencies and incompleteness to perform data reconciliation and integration. Among the several issues, the most critical aspects are related to the ontology construction that enables deduction and reasoning, and on the verification and validation of the obtained model.

The reconciliation phase may find advantages in relying on a repository or dictionary, for example of existing Street in a Town, but this kind of products is expensive to build up or to acquire. In fact, a relevant process of data improvement for semantic interoperability is related to the application of reconciliations among the entities associated for instance with locations as streets, civic numbers and localities. Unfortunately, it is not always possible to perform reconciliation at street number level, i.e. connecting an instance that uniquely identifies a street number on a road, the finest level of localization; sometimes, the reconciliation is only at street-level, with less precision, or at the worst at municipality level. On this regard, there are different types of inconsistencies for which reconciliation did not return results, both for the lack of references into the scraped data (some streets and civic numbers can be missing or incomplete) and for lack of correspondences into the repository/dictionary. Finally, it could happen to have location entities which result as wrong and not reconcilable due to (i) the presence of wrong values for streets and/or locations, and (ii) the lack of a consistent reference location.

To summarize, the reconciliation phase requires different steps, for instance starting from searching for correspondences at the finest level (i.e. the street number), but allowing for correspondences at higher level of disaggregation (i.e. street, municipality) in the further steps. Finally, applying a manual correction and cleaning or manual search of non-identified matches into a list of probable candidates suggested by the previous disregarded results.

During the reconciliation step, there may be cases where no connection among the data are caused by a different encoding of the instances, for instance the name of the municipality. To support the reconciliation process each time new data are available, they should be automatically completed with the correct Istat municipality code.

Another very common issue in integration is related to the existence of multiple ways to express the toponymy qualifiers, (e.g. Piazza and P.zza) or parts of the proper name of the street (such as Santa, or S. or S or S.ta): this issue can be overcome thanks to support tables, inside which the possible change of notation for each individual case identified are inserted.

At the end of this huge and complex reconciliation process, it is possible to define the most appropriate linkage strategy, i.e.:

- Choose appropriate variables.
- Specify and tune the parameters: the distance measures together with good thresholds.
- Define the proper model for aggregating scores functions.

However, definitively, the effectiveness of the linkage process is dramatically affected by the output quality of the pre-processing phase.

An alternative to standard linkage methods (Fellegi and Sunter, [5, 9]) is represented by supervised learning algorithm which employs genetic programming in order to learn linkage rules from a set of existing reference links. For instances, [10] proposed GenLink, which is capable of matching entities between heterogeneous data sets which adhere to different schema. By employing an expressive linkage rule representation, the algorithm learns rules which:

- Select discriminative properties for comparison.
- Apply chains of data transformations to normalize property values prior to comparison.

- Apply multiple distance measures combined with appropriate distance thresholds.
- Aggregate the result of multiple comparisons using linear as well as non-linear aggregation functions.

Following genetic programming, the GenLink algorithm starts with an initial population of candidate solutions which is iteratively evolved by applying a set of genetic operators. The basic idea of GenLink is to evolve the population by using a set of specialized crossover operators. The applicability and efficacy of this kind of solution in official statistics context has to be studied and tested.

Finally, whatever will be the linkage strategy, for a proper usage in official statistics context, the verification and validation of results is a key aspect, as well as the provision of the quality indicators of process and products, in order to allow other data analysts to perform the correct analyses on the integrated data. This task is often very expensive in terms of time and resources.

4 A Case Study: Linking Internet-Scraped Data of Farmhouse to the Farm Register

As case study we propose an integration of Internet-scraped data regarding agritourism with data reported in the Farm Register built up by Istat. The final aim of this integration is the use of Internet information for statistical purpose, in particular to update and integrate some agricultural data collected in the Farm Register.

In fact, in order to produce harmonized and comparable statistics, one of the Eurostat core recommendations is to define and set-up a Statistical Farm Register (SFR). SFR represents a key element for the Agricultural Statistical System and the basis for sample selection. In Italy, it is built by integrating several administrative sources (Integrated Administration and Control System, Animal register, Tax declaration on agricultural land, land cadaster, Chamber of Commerce, Tax on Value Added on agricultural income) and some statistical sources (Business Register, Agricultural Census, Agritourism survey, Quality product survey).

Big data could be used to update the SFR, permitting the production and the periodical dissemination of statistics related to the activities and to the services offered by the agritourism farms, at a minimum cost. The choice of agritourism topic depends on the presence of portals in the web, an important perspective in this experimental phase.

So, the initial and most important target of web scraping is represented by the different websites acting as "hubs" (hosting and describing a plurality of Agritourism), in general maintained by Regions, or by private organizations, and containing information regarding name, address, geographic coordinates, telephone, e-mail, prices, offered services, etc. A specific scraping application is developed for each hub; this permits to collect all the semi-structured information so to compare it to the official data set. At the end, three hubs are considered, among the most popular websites providing this kind of information.

As specified in the previous section, this unrefined information requires a laborious pre-processing activity because they are unusable for integration in the way they are scraped. Then it is necessary to standardize the variables for each hub dataset. In the internet-scraped data the several information are not delimited by separators or fixed length of the fields, so it is necessary to recognize address, province, town, region, country distinguishing those by the agritourism name. It is necessary to recognize the different variables and then codify municipalities with the Istat code. Addresses are validated using the following software http://www.egon.com/en/solutions/address-validation.html

In some cases it is not possible to identify a correct address and the observations without a normalized address have been deleted (about 35%).

Furthermore, to increase the significance of the company names in each file the most common words (i.e. Agriturismo, Azienda Agricola, Affittacamere, ect.) have been eliminated.

In addition, it is necessary to make a first linkage process between the three datasets from the three different hubs in order to obtain a single deduplicated dataset comparable with the Farm Register.

The linkage activities have been performed using RELAIS. Several linkage strategies have been applied, the most effective in revealing matches without introducing false matches is based on the following step:

1. Data cleaning—preparation of the input files (pre-processing) as above described; as well known in official statistics the preparation of input files is the first phase and requires 75% of the whole effort to implement a record linkage procedure, in this case the pre-processing step is particularly huge and expensive, requiring almost the 95% of the whole time.
2. Choice of the common identifying attributes (matching variables); after the previous phase, it is important to choose matching variables that are as suitable as possible for the considered linking process. We use: company name, address, province, municipality and region.
3. Search space creation/reduction; to reduce the complexity it is necessary to reduce the number of pairs to compare. We choose Blocked Simhash function [13, 3], which combines the Blocking Union method (using region variable) with SimHash (using company name/address variable).
4. Choice of decision model; in absence of a unique identifier we decide for a probabilistic model according to the Fellegi and Sunter [5] theory.
5. Choice of comparison functions; Comparison functions measure the "similarity" between two fields. We use equality for province and municipality variables and, to overcome non identical strings, the Inclusion3Grams for address and company name variables. This function have been chosen because it takes into account the number of 3-length grams in common between the two strings, and the target is the 3-grams amount of the shortest string.

To obtain the best result, the linkage strategy has been designed in two iterations: at the beginning we used the company name and region for the search space creation with Blocked Simhash function. So only the company name has been used as match variable with province and municipality.

In the second process, on the set of unlinked records of the first step, the address has been used instead of the company name, with the same model, functions and thresholds.

The chosen strategy allowed to link 2765 units, 37.8% of the smaller file, represented by the 7301 farms from the web. The farms in the Farm register are 13503.

The result is still adequate because there are some aspects to be taken into account. The frame of SFR Agritourism is 13000 units on a total of 20000 existing for the Agricultural Ministry. The difference may be caused by the failure of the address normalization we have described. The SFR was built in 2013, while the websites are updated frequently and the number of agritourism in the portal is likely to be increasing. On the other hand, agritourism obtained by web scraping the portals might be false farms and for this reason not included in the SFR.

The double linkage process has allowed to overcome two types of difficulties. Addresses referred to the headquarter in a source and to the farm manager residence in the other one could be resolved. Similarly, the issue of different names in the two sources was overcome.

The comparison function Inclusion 3 Grams was very functional to this issue. In fact, often in the company name string was also the name of the farm manager in one file. With other similarity functions it could easily result in a link failure. This function was also very useful for addresses, especially addresses in German (province of Bolzano) written in different ways in the two files.

5 Concluding Remarks and Future Work Direction

The first attempt to link web-scraped and traditional data in official statistics allows underlining potentials and issues connected to this operation.

The first evidence highlights the role played by the pre-processing phase and the data cleaning/reconciliation activity. Traditionally these procedures require a lot of work, 75% of the whole effort to implement a record linkage procedure, according to [7] but in this case the time and work devoted to this task were even more than ¾ of the whole effort. Ignoring this task, however, may compromise the effectiveness of the following steps. In fact, an attempt on raw not-preprocessed data resulted in no matches identified.

The achieved match rate may seem low but the knowledge of data can explain these results. Moreover, the comprehension of the reasons why the match rate is low in this application may provide the main leverage to improve it. The main issues in this specific field are:

- Very often the farm names are different in the sources. Generally in internet we can find the name of the agritourism, instead in administrative or statistical sources the farm is registered with the company name. So, even with an accurate data cleaning, it is impossible to identify the same unit if the address is not accurate and does not provide enough linking power;
- However, very often the farm addresses are inaccurate. They are main roads, shared by large lands, like *contrada*, *regione*, etc., generally without a street number, so it is not possible to identify an exact and unique position.
- In some sources the farm headquarter address is recorded, while in other ones we find the farm manager residence address. When the residence of the farm manager does not correspond to the farm headquarter address, it is not possible to improve the matching results even after address normalization.

It is quite clear that some of the previous issues are not specific of the Internet-scraped data but they could generally arise with data coming from any kind of sources not designed for statistical purposes. Moreover, in this specific field the Internet data itself may provide a strong improvement to the linkage procedure by means of the easy availability of further information on the geographical references of the interest units. The use of geoinformation in linkage procedure seems a promising direction to explore.

Finally, it seems that this kind of linkage activity needs first of all to define an incremental process able to analyze, integrate and validate each added data sources. The main point to facilitate this task is related to the availability of mechanism for automatizing the data interpretation, verification, reconciliation and coding. In the Big data context, whit several sources different with respect to many aspects, this step cannot be managed with traditional solution, requiring a hard human resource involvements. In parallel, techniques for data linkage, different with respect to the traditional ones, may allow the comparison of less structured data. At the end, the statistical validation of the linkage results and the measurement of output quality need to be assessed.

References

1. Barcaroli, G., Scannapieco, M., Nurra, A., Scarnò, M., Salamone, S., Summa, D.: Internet as Data Source in the Istat survey on ICT in Enterprises. Austrian J. Stat. **44**, 31–43 (2015)
2. Bishop, B., Kiryakov, A., Ognyanoff, D., Peikov, I., Tashev, Z., Velkov, R.: OWLIM: a family of scalable semantic repositories. Semant. Web J. **2**(1) (2011)
3. Charikar, M.S.: Similarity estimation techniques from rounding algorithms. In: Proceedings ACM STOC 2, pp. 380–388. Montreal, Quebec, Canada (2002)
4. Citro, C.:(2014) From multiple modes for surveys to multiple data sources for estimates. Surv. Method.
5. Fellegi, I.P., Sunter A.B.: A theory for record linkage. J. Am. Stat. Soc. **64** (1969)
6. Fuchs, M., Höpken, W., Lexhagen, M.: Big data analytics for knowledge generation in tourism destinations—a case from Sweden. J. Destination Mark. Manage. (2014)

7. Gill, L.: Methods for Automatic Record Matching and Linking and their Use in National Statistics. National Statistics Methodological Series No. 25. London: Office for National Statistics. http://www.statistics.gov.uk/StatBase/Product.asp?vlnk=9224 (2001)
8. Gupta, S., Szekely, P., Knoblock, C., Goel, A., Taheriyan, M., Muslea, M.: Karma: a system for mapping structured sources into the semantic web. In: Proceedings of the 9th Extended Semantic Web Conference (ESWC2012)
9. Jaro, M.: Advances in Record Linkage Methodologies as Applied to Matching the 1985 Census of Tampa. Fla J. Am. Stat. Soc. **84**(406), 414–420 (1989)
10. Isele, R., Bizer, C.: Active learning of expressive linkage rules using genetic programming. Web semantics: science, services and agents on the WorldWideWeb **23**, 2–15 (2013)
11. Heerschap, N., Ortega, S., Priem, A., Offermans, M.: Innovation of tourism statistics through the use of new big data sources. Technical Paper, Statistics Netherlands (2014)
12. Hepp, M.: GoodRelations: an ontology for describing products and services offers on the web. In: Proceedings of the 16th International Conference on Knowledge Engineering and Knowledge Management (EKAW2008), Acitrezza, Italy. vol. 5268. Springer LNCS, 29 Sept–3 Oct 2008, pp. 332–347
13. RELAIS 3.0 User Guide. http://www.istat.it/it/strumenti/metodi-e-strumenti-it/strumenti-di-elaborazione/relais (2015)
14. Tuoto T., Cibella N., Fortini M., Scannapieco M., Tosco L., (2007) RELAIS: Don't Get Lost in a Record Linkage Project, Proc. of the Federal Committee on Statistical Methodologies (FCSM: Research Conference. Arlington, VA, USA (2007)
15. Tuoto, T., Gould, P., Seyb, A., Cibella, N., Scannapieco, N., Scanu, M.: Data Linking: A Common Project for Official Statistics in Proceedings of Conference of European Statistics Stakeholders Rome 24, 25 Nov 2014

Part II
Recent Debates in Statistics and Statistical Algorithms

An Algorithm for Finding Projections with Extreme Kurtosis

Cinzia Franceschini and Nicola Loperfido

Abstract Projection pursuit is a multivariate statistical technique aimed at finding interesting low-dimensional data projections. A projection pursuit index is a function which associates a data projection to a real value measuring its interestingness: the higher the index, the more interesting the projection. Consequently, projection pursuit looks for the data projection which maximizes the projection pursuit index. The absolute value of the fourth standardized cumulant is a prominent projection pursuit index. In the general case, a projection achieving either minimal or maximal kurtosis poses computational difficulties. We address them by an algorithm which converges to the global optimum, whose computational advantages are illustrated with air pollution data.

Keywords Fourth moment · Kurtosis · Projection pursuit · Tensor

1 Introduction

Projection pursuit is a multivariate statistical technique aimed at finding interesting low-dimensional data projections. It deals with three major challenges of multivariate analysis: the curse of dimensionality, the presence of irrelevant features and the limitations of visual perception. As such, projection pursuit is particularly useful when data are high-dimensional, data features are unclear and the approach is exploratory.

A projection pursuit index is a function which associates a data projection to a real value measuring its interestingness: the higher the index, the more interesting

C. Franceschini
Dipartimento di Economia (DEC), Università degli Studi
"G. d'Annunzio" di Chieti-Pescara, Pescara, Italy
e-mail: cinziafranceschini@msn.com

N. Loperfido (✉)
Dipartimento di Economia, Società e Politica (DESP), Università degli Studi
di Urbino "Carlo Bo", Urbino, Italy
e-mail: nicola.loperfido@uniurb.it

C. Perna et al. (eds.), *Studies in Theoretical and Applied Statistics*,
Springer Proceedings in Mathematics & Statistics 227,
https://doi.org/10.1007/978-3-319-73906-9_6

the projection. Consequently, projection pursuit looks for the data projection which maximizes the projection pursuit index. The absolute value of the fourth standardized cumulant is a valid projection pursuit index, according to the criteria stated in [7], and leads to kurtosis-based projection pursuit. Its statistical applications include multivariate normality testing, cluster analysis, outlier detection and independent component analysis. The prominent role of kurtosis as a projection pursuit index has been emphasized by several authors, as for example [6, 16, 17, 20]. Caussinus and Ruiz-Gazen [3] provide the most recent review of literature on projection pursuit.

The directional kurtosis $\kappa_{4,d}^{D}(x)$ of a d—dimensional random vector x is the maximum value attainable by $\gamma_2^2\left(c^T x\right)$, where c is a nonnull, d—dimensional real vector and $\gamma_2(Y)$ is the fourth standardized cumulant of the random variable Y. More formally, let μ and Σ the mean and the variance of x, and let R_0^d the set of all real, d-dimensional, nonnull vectors. The directional kurtosis of x is then

$$\kappa_{4,d}^{D}(x) = \max_{c\in R_0^d} \gamma_2^2\left(c^T x\right) = \max_{c\in R_0^d}\left\{\frac{E\left[\left(c^T x - c^T \mu\right)^4\right]}{\left(c^T \Sigma c\right)^2} - 3\right\}^2. \tag{1}$$

Clearly, $\gamma_2^2\left(c^T x\right)$ attains its maximum if $c^T x$ has either the largest or smallest fourth standardized moment among all linear projections of x. We shall focus on the former projections, that is

$$\beta_{2,d}^{D}(x) = \max_{c\in R_0^d} \beta_2\left(c^T x\right) = \max_{c\in R_0^d} \frac{E\left[\left(c^T x - c^T \mu\right)^4\right]}{\left(c^T \Sigma c\right)^2}, \tag{2}$$

where $\beta_2(Y)$ is the fourth standardized moment of the random variable Y. Projections maximizing kurtosis might be conveniently expressed using fourth moment matrices. Let $x = \left(X_1, \ldots, X_d\right)^T$ be a real, d-dimensional random vector satisfying $E\left(X_i^4\right) < +\infty$, for $i = 1, \ldots, d$. The fourth moment matrix (henceforth fourth moment, for short) of x is the $d^2 \times d^2$ matrix $M_{4,x} = E\left(x \otimes x^T \otimes x \otimes x^T\right)$, where "$\otimes$" denotes the Kronecker product. It conveniently arranges all the fourth-order moments $\mu_{ijhk} = E\left(X_i X_j X_h X_k\right)$ of x, for $i, j, h, k = 1, \ldots, d$. If the variance Σ of x is positive definite, its fourth standardized moment $M_{4,z}$ is the fourth moment of $z = \Sigma^{-1/2}(x - \mu)$, where μ is the expectation of x and $\Sigma^{-1/2}$ is the symmetric, positive definite square root of the concentration matrix Σ^{-1}. Basic properties of fourth moment matrices imply

$$\beta_{2,d}^{D}(x) = \max_{c\in R_0^d} \frac{\left(c^T \otimes c^T\right) M_{4,x-\mu}\left(c \otimes c\right)}{\left(c^T \Sigma c\right)^2} = \max_{c\in S^d}\left(c^T \otimes c^T\right) M_{4,z}\left(c \otimes c\right), \tag{3}$$

where S^d is the set of d-dimensional real vectors of unit length. For computational purposes it is sometimes convenient to arrange the fourth-order moments in a different way. The cokurtosis of a random vector x with mean μ and finite fourth-order moments is $cok(x) = E\left[(x-\mu)\otimes(x-\mu)^T\otimes(x-\mu)^T\otimes(x-\mu)^T\right]$ (see, for example, [10]). Similarly, the standardized cokurtosis of x is the cokurtosis of z: $cok(z) = E\left(z\otimes z^T\otimes z^T\otimes z^T\right)$. As a direct implication of results in [19] on tensor eigenvectors, the d-dimensional real vector of unit length v satisfies $\beta_2\left(v^T x\right) = \beta^D_{2,d}(x)$ if and only if it also satisfies $cok(z)(v\otimes v\otimes v) = \lambda v$ for the largest real value λ.

In the general case, kurtosis-based projection pursuit needs to be carried out numerically. There are several iterative algorithms for kurtosis optimization: gradient methods [8, Chap. 3], the higher-order power method [4], the symmetric higher-order power method [5], the modified Newton's method [17], the fixed-point algorithm [8], the quasi-power method [6]. No one of these methods guarantees achievement of the global optimum, due to the presence of several local maxima [15].

As a direct consequence, the choice of the initial value of the projecting direction is of paramount importance [5, 12], but there is no general consensus on how it should be done. For example [5] use the dominant left singular vector of the standardized cokurtosis [9, p. 186], choose between several starting direction. Kofidis and Regalia [12] use the dominant eigenvector of the matrix whose vectorization is the dominant eigenvector of the fourth standardized moment. Zarzoso and Comon [21] exploit the polynomial character of the objective function. To the best of our knowledge, nobody never compared relative merits of the above proposal.

In the general case, projections achieving either minimal or maximal kurtosis pose computational difficulties, as remarked by [6, 11, 18]. In this paper, we address these difficulties by an algorithm converging to the global optimum.

2 Initialization

In this section we shall investigate the performance of an algorithm for initializing the search for projections with extreme kurtosis, and evaluate its performance via simulations. The algorithm was proposed by [12], and theoretically motivated by [13, 14].

Consider a $n\times d$ data matrix X and the corresponding sample covariance matrix S, which is assumed to be of full rank. The algorithm is defined as follows. First, standardize the data, obtaining the matrix $Z = H_n X S^{-1/2}$, where H_n is the $n\times n$ centring matrix, and $S^{-1/2}$ is the symmetric square root of S^{-1}. Second, obtain the fourth standardized moment $M_{4,Z} = \sum_{i=1}^{d} z_i\otimes z_i^T\otimes z_i\otimes z_i^T$, where z_i^T is the i-th row of Z. Third, find the dominant eigenvector c of $M_{4,Z}$, let the matrix A be such that $vec(A) = c$ and $G = (A + A^T)/2$. The latter operation is motivated by numerical problems, which might lead to asymmetric G matrices. Fourth, when interest lies in projections with maximal (minimal) kurtosis, compute $XS^{-1/2}g$, where g is the eigenvector associated with the largest (smallest) eigenvalue of G.

In order to assess the accuracy of the proposed initialization, it is necessary to compare them with the projections which achieve extreme kurtosis. As shown by

[16, 17], projections with extreme kurtosis have many desirable properties when dealing with two-component normal mixtures. Hence we focused on this class of distributions in order to assess the practical relevance of the proposed approximation. We simulated 10,000 samples of size $n = 50, 100, 150, 200, 250, 300$ from the two-component mixture $\pi_1 N_2\left[(10,10)^T, diag(1,1)\right] + \left(1-\pi_1\right) N_2\left[(0,0)^T, diag(1,1)\right]$, where $\pi_1 = 0.1, 0.3, 0.5$. The same model has been used by [16, 17] to illustrate the use of directions with extreme kurtosis for multivariate outlier detection. We confined the simulations to the bivariate case in order to use the analytical results in the next section for obtaining computationally simple projections either maximizing or minimizing kurtosis.

For each simulated sample, we computed the correlation between the projection minimizing (maximizing) kurtosis and the projection obtained by the initialization method, together with the relative difference between the corresponding kurtoses. Simulations' results are reported in Table 1 and suggest that the proposed initializations (*a*) are very satisfactory, the correlations being always greater than 0.92 (0.97) for projections minimizing (maximizing) kurtosis; (*b*) are more accurate for projections maximizing kurtosis, than for projections minimizing it; (*c*) are only mildly affected by increases in sample sizes and components' weights. We performed another simulation study to compare the performance of the proposed method with another method, based upon the kurtosis matrix $K_4(Z) = n^{-1}\sum_{i=1}^{d} z_i^T z_i z_i z_i^T$, as proposed by [18]. We simulated 10,000 samples of size $n = 50, 100, 150, 200, 250$ from multivariate normal-gamma distributions [1] of dimensions $d = 3, 6$ and shape parameters $g = 0.5, 1.5, 5$. For each sample we computed the absolute correlations between the data projections obtained with our method and the method in [18]. Table 2 reports the simulation's results, which may be summarized as follows. For all sizes and dimensions the proposed method outperforms the one proposed by [18]. The difference tends to increase with the sample size and to decrease with the multivariate kurtosis of the sampled distribution.

Table 1 The first row reports weights of the first mixture's components. The first column under each weight reports the numbers of observations in each simulated sample. The second (third) column under each weight reports the average correlations between projections minimizing (maximizing) kurtosis and their initial values

Weight = 0.1			Weight = 0.3			Weight = 0.5		
Num	Min	Max	Num	Min	Max	Num	Min	Max
50	0.9351	0.9912	50	0.9362	0.9777	50	0.957	0.9763
100	0.9472	0.9978	100	0.9558	0.9754	100	0.9735	0.9766
150	0.9529	0.9991	150	0.9671	0.9738	150	0.9803	0.9758
200	0.9566	0.9996	200	0.9744	0.9733	200	0.9843	0.9763
250	0.9590	0.9999	250	0.9783	0.9720	250	0.9868	0.9761
300	0.9611	0.9999	300	0.9822	0.9714	300	0.9886	0.9764

Table 2 The columns contain the integer part of the absolute correlations, multiplied by 100, between projections maximizing kurtosis and data projections onto the directions of the dominant eigenvectors of sample G and K matrices for $n = 50, 100, 150, 200, 250$, $d = 3, 6$ and $g = 0.5, 1.5, 5$

	d = 3		d = 3		d = 3		d = 6		d = 6		d = 6	
Size	Shape: 0.5		Shape: 1.5		Shape: 5		Shape: 0.5		Shape: 1.5		Shape: 5	
	K	G	K	G	K	G	K	G	K	G	K	G
50	26	94	42	72	47	63	17	82	29	53	32	45
100	20	98	36	82	42	70	12	93	25	65	29	52
150	18	99	31	87	40	76	9	96	22	72	28	57
200	16	99	28	91	37	80	8	97	20	78	26	63
250	15	99	24	93	35	83	7	98	18	82	25	65

3 Iteration

This section describes the iteration step of an algorithm for kurtosis optimization. We shall focus on finding projections of random vectors which achieve maximal kurtosis. The algorithm might be trivially modified when minimal kurtosis is sought or when only a data matrix is available.

The higher-order power method (HOPM) for a symmetric, fourth-order tensor might be summarized as follows [4, 5] . Let $W_0 = v_0^T x$ be our initial guess for the projection of x achieving maximal kurtosis, where v_0 is a d-dimensional real vector. Compute $v_{i+1} = x_i / \left\| x_i \right\|$, where $x_i = cok(z) \left(v_i \otimes v_i \otimes v_i \right)$, where $cok(z)$ is the cokurtosis of z. Alternatively, the iteration step might be repeated until the distance $\left\| v_i - v_{i+1} \right\|$ between v_i, and v_{i+1} becomes negligible. Unfortunately, as remarked by [5], the HOPM might not converge to the global maximum, that is the direction maximizing kurtosis.

In order to overcome this problem, we propose an iteration step which builds upon convenient analytical properties of kurtosis maximization for bivariate random vectors. We look for a linear combination $a_1 X_1 + a_2 X_2$ of the random vector $x = \left(X_1, X_2 \right)^T$ with either minimal or maximal kurtosis, as measured by its fourth standardized moment $\beta_2 \left(a_1 X_1 + a_2 X_2 \right)$. We shall follow the approach outlined in [5] for the eigenvectors of third-order, symmetric and binary tensors. Our presentation will be more detailed and less technical, in order to show the method's potential among statisticians with little background in tensor algebra.

Kurtosis optimization does not depend neither on location nor scale. Hence, without loss of generality, we shall focus on the standardized random vector $z = \left(Z_1, Z_2 \right)^T = \Sigma^{-1/2} (x - \mu)$. We shall obtain a simple analytical form for the projection $v_1 Z_1 + v_2 Z_2$ with extreme kurtosis, where $v = \left(v_1, v_2 \right)^T$ is a real, bivariate vector of unit norm: $v \in \mathfrak{R}^2$, $v^T v = v_1^2 + v_2^2 = 1$. Clearly, the vector $a = \left(a_1, a_2 \right)^T$ is a simple linear function of v: $a = \Sigma^{-1/2} v$.

The assumptions $E(Z_1) = E(Z_2) = E(Z_1 Z_2) = 0$ and $v^T v = E(Z_1^2) = E(Z_2^2) = 1$ imply $\beta_2(v^T z) = \alpha_{40} v_1^4 + 4\alpha_{31} v_1^3 v_2 + 6\alpha_{22} v_1^2 v_2^2 + 4\alpha_{13} v_1 v_2^3 + \alpha_{04} v_2^4$, where $\alpha_{ij} = E\left(Z_1^i Z_2^j\right)$, for $i, j = 0, 1, 2, 3, 4$ and $i + j = 4$. Then we have

$$\beta_2(v_1 Z_1 + v_2 Z_2) = \begin{pmatrix} v_1^2 \\ v_1 v_2 \\ v_1 v_2 \\ v_2^2 \end{pmatrix}^T \begin{pmatrix} \alpha_{40} & \alpha_{31} & \alpha_{31} & \alpha_{22} \\ \alpha_{31} & \alpha_{22} & \alpha_{22} & \alpha_{13} \\ \alpha_{31} & \alpha_{22} & \alpha_{22} & \alpha_{13} \\ \alpha_{22} & \alpha_{13} & \alpha_{13} & \alpha_{04} \end{pmatrix} \begin{pmatrix} v_1^2 \\ v_1 v_2 \\ v_1 v_2 \\ v_2^2 \end{pmatrix}, \tag{4}$$

which may be more compactly represented as $(v^T \otimes v^T) M_{4,z} (v \otimes v)$, where "$M_{4,z}$" denotes the fourth moment of z (that is the fourth standardized moment of x) respectively. The kurtosis of $v^T z$ is either a maximum or a minimum if v is a solution of $\partial\left[\beta_2(v^T z) - \lambda(v^T v - 1)\right]/\partial v = (0, 0)^T$ corresponding to the maximum or minimum value of λ, respectively. Standard differentiation techniques lead to the system of equations

$$\begin{cases} \alpha_{40} v_1^3 + 3\alpha_{31} v_1^2 v_2 + 3\alpha_{22} v_1 v_2^2 + \alpha_{13} v_2^3 = \lambda v_1 \\ \alpha_{31} v_1^3 + 3\alpha_{22} v_1^2 v_2 + 3\alpha_{13} v_1 v_2^2 + \alpha_{04} v_2^3 = \lambda v_2 \end{cases} \tag{5}$$

The vector v is proportional to $(1, 0)^T$ if and only if α_{31} equals zero. In order to rule out this case, we shall assume that both α_{31} and v_2 differ from zero. First, eliminate λ by subtracting the first equation, multiplied by v_2, from the second equation, multiplied by v_1:

$$\alpha_{31} v_1^4 + (3\alpha_{22} - \alpha_{40}) v_1^3 v_2 + 3(\alpha_{13} - \alpha_{31}) v_1^2 v_2^2 + (\alpha_{04} - 3\alpha_{22}) v_1 v_2^3 - \alpha_{13} v_2^4 = 0. \tag{6}$$

Next, divide each side of the above equation by v_2^4 and let $t = v_1/v_2$:

$$\alpha_{31} t^4 + (3\alpha_{22} - \alpha_{40}) t^3 + 3(\alpha_{13} - \alpha_{31}) t^2 + (\alpha_{04} - 3\alpha_{22}) t - \alpha_{13} = 0. \tag{7}$$

The above polynomial has four roots in the complex space, of which at least two are real, corresponding to either miminum or maximum kurtosis. It follows from the kurtosis of a linear combination being never smaller than one (by ordinary properties of kurtosis) and never greater than the dominant eigenvalue of $M_{4,z}$ [14]. The identities $t = v_1/v_2$ and $v_1^2 + v_2^2 = 1$ imply that v equals either $(t_0, 1)^T/\sqrt{1 + t_0^2}$ or $(-t_0, -1)^T/\sqrt{1 + t_0^2}$, where t_0 is an appropriate real root of the above polynomial.

The results obtained for the bivariate case are useful when addressing the problem of kurtosis maximization for random vectors of any dimension. Let $Y_0 = a_0^T x$ be our initial guess for the projection of x achieving maximal kurtosis, where a_0 is a d-dimensional real vector. Replace the first component of a_0 with zero, name the resulting vector a_1 and let $Y_1 = a_1^T x$ be another projection of x. Since $(X_1, Y_1)^T$ is a bivariate random vector, the scalar

$$c_0 = \arg\max_{h \in R_0} \beta_2 \left(hX_1 + Y_1 \right) \tag{8}$$

admits an analytical, easily computable representation. Now repeat the two following steps for $i = 2, \ldots, d$. First compute the scalar

$$c_i = \arg\max_{h \in R_0} \beta_2 \left(hX_i + Y_{i-1} \right). \tag{9}$$

Then replace the i-th and the $(i-1)$-th components of a_{i-1} with zero and c_i, to obtain the vector a_i and the projection $Y_i = a_i^T x$. By construction, $\beta_2 \left(Y_0 \right) \leq \beta_2 \left(Y_1 \right) \leq \ldots \leq \beta_2 \left(Y_d \right)$. Repeat the two steps for the required number of times or until a chosen condition is met.

4 Application

In this section we shall use real data to illustrate both kurtosis-based projection pursuit and the limitations of HOPM which motivated the proposed iterative method. Due to space constraints, we shall focus on kurtosis maximization.

The data belong to the same dataset in [2], and provide an evaluation of PM_{10} (particulate matter having an aerodynamic equivalent diameter of up 10 μ) concentrations recorded in 259 monitoring stations scattered throughout Italy during year 2006. The variables are the average of daily values (AVE), maximum of daily values (MAX) and number of daily exceedances (EXC). Table 3 below reports the summary statistics of the three variables. All variables are positively skewed, AVE and MAX are platikurtic, while EXC is leptokurtic. In neither case kurtosis is significantly different from 3, that is its epected value under normality (the corresponding p-values are always greater than 0.2). Figures 1a, b and c contain the histograms of the three variables. Their visual inspection is consistent with the shape measures in Table 3: the three histograms are clearly unimodal and positively skewed, with no apparent peakedness.

Table 3 Summary statistics (mean, standard deviation, skewness and kurtosis of the average of daily values (AVE) maximum of daily values (MAX) and number of daily exceedances (EXC))

	AVE	MAX	EXC
Mean	125.0579	35.3514	62.3398
Deviation	49.6028	11.4721	49.4461
Skewness	0.6702	0.4155	0.8321
Kurtosis	2.8925	2.7795	3.3852

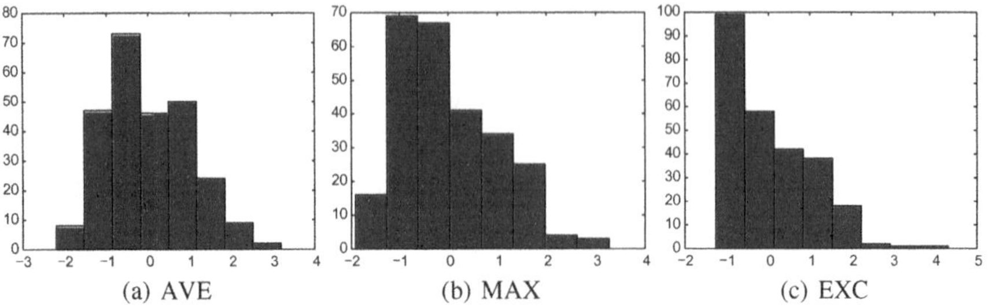

Fig. 1 Histogram of the average of daily values (AVE), Histogram of the maximum of daily values (MAX), Histogram of the number of daily exceedances (EXC)

We first found the projection maximizing kurtosis by direct search, which is a computationally feasible method for tridimensional data:

$$\beta_2 (0.8723 \cdot AVE + 0.1833 \cdot MAX - 0.4533 \cdot EXC) = 10.9917.$$

Remarkably, this kurtosis is almost three times the largest kurtosis among those of the original variables. Then we used the proposed initialization method and found

$$\beta_2 (0.8744 \cdot AVE + 0.1821 \cdot MAX - 0.4498 \cdot EXC) = 10.9887.$$

Both the kurtosis and the weights are very similar to the previous ones, and the correlation between the two projections is about 0.99. Finally, we used the iterative procedure described in Sect. 3, which needed only an iteration to converge to its maximum

$$\beta_2 (0.8721 \cdot AVE + 0.1800 \cdot MAX - 0.4498 \cdot EXC) = 10.9913.$$

The difference between the first and the third kurtoses is negligible and due to rounding errors. Also, the correlation between the corresponding projections is virtually one. However, the direct search method required 4279.910456 s for maximizing kurtosis, that is slightly more than 71 min. Our method completed the same task in 0.057836 s, that is about 74581 times faster. On the other hand, each iteration of the HOPM, starting from the proposed initial value, leads to projections with decreasing kurtoses: 10.6308, 10.4076, 10.2949, 10.2400, 10.2136, 10.2008, 10.1947, 10.1918, 10.1904, 10.1897.

The histogram of the projection with maximal kurtosis (Fig. 2) clearly hints that the excess kurtosis is mainly due to two outlying observations. All the projected data but two are positive, and the negative values are way smaller than the remaining ones. Once the negative values are removed, the kurtosis of the projected data drops to 3.4056, which is consistent with normokurticity (the p-value is slightly less than 0.1).

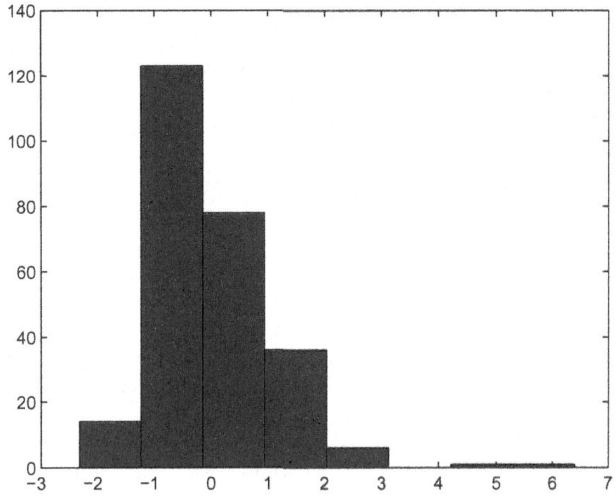

Fig. 2 Histogram of the projection maximizing kurtosis

This example is consistent with the empirical and theoretical results in [16], showing the prominent role of kurtosis-based projection pursuit for outliers detection. The algorithm for kurtosis maximization would benefit from the proposed starting values and iteration steps. On the contrary, HOPM is not fully reliable as an iterative method for kurtosis optimization.

Acknowledgements The authors would like to thank an anonymous referee for her/his comments which greatly helped in improving the present paper.

References

1. Adcock, C.J., Shutes, K.: On the multivariate extended skew-normal, normal-exponential, and normal-gamma distributions. J. Stat. Theory Pract. **6**, 636–664 (2012)
2. Bartoletti, S., Loperfido, N.: Modelling air pollution data by the skew-normal distribution. Stoch. Environ. Res. Risk Assess. **24**, 513–517 (2010)
3. Caussinus, H., Ruiz-Gazen, A.: Exploratory Projection Pursuit. In: Gerard Govaert, G. (ed.) Data Analysis, pp. 76–92. Wiley (2009)
4. de Lathauwer, L., Comon, P., De Moor, B., Vandewalle, J.: Higher-order power method Application in independent component analysis. In: Proceedings of the International Symposium on Nonlinear Theory and Its Applications (NOLTA'95), Las Vegas, NV, pp. 91–96 (1995)
5. de Lathauwer L., de Moor, B., Vandewalle, J.: On the best rank-1 and rank-(R_1,R_2,...R_N) approximation of high-order tensors. SIAM J. Matrix Ana. Appl. **21**, 1324–1342 (2000)
6. Hou, S., Wentzell, P.D.: Fast and simple methods for the optimization of kurtosis used as a projection pursuit index. Anal. Chim. Acta **704**, 1–15 (2011)
7. Huber, P.J.: Projection pursuit (with discussion). Ann. Stat. **13**, 435–525 (1985)
8. Hyvarinen, A., Oja, E.: A fast fixed point-algorithm for independent component analysis. Neural Comput. **9**, 1483–1492 (1997)
9. Hyvarinen, A., Karhunen, J., Oja, E.: Independent Compon. Anal. Wiley, New York (2001)
10. Jondeau, E., Rockinger, M.: Optimal portfolio allocation under higher moments. Eur. Finan. Manag. **12**, 29–55 (2006)

11. Kent, J.T.: Discussion on the paper by Tyler, Critchley, Dumbgen & Oja: invariant co-ordinate selection. J. R. Stat. Soc. Ser. B **71**, 575–576 (2009)
12. Kofidis, E., Regalia, P.A.: On the best rank-1 approximation of higher-order supersymmetric tensors. SIAM J. Matrix Anal. Appl. **23**, 863–884 (2002)
13. Loperfido, N.: Spectral analysis of the fourth moment matrix. Linear Algebra Appl. **435**, 1837–1844 (2011)
14. Loperfido, N.: A new Kurtosis matrix, with statistical applications. Linear Algebra Appl. **512**, 1–17 (2017)
15. Paajarvi, P., Leblanc, J.P.: Skewness maximization for impulsive sources in blind deconvolution. In: Proceedings of the 6th Nordic Signal Processing Symposium—NORSIG 2004, 9–11 June, Espoo, Finland (2004)
16. Peña, D., Prieto, F.J.: Multivariate outlier detection and robust covariance estimation (with discussion). Technometrics **43**, 286–310 (2001)
17. Peña, D., Prieto, F.J.: Cluster identification using projections. J. Am. Stat. Assoc. **96**, 1433–1445 (2001)
18. Peña, D., Prieto, F.J., Viladomat, J.: Eigenvectors of a kurtosis matrix as interesting directions to reveal cluster structure. J. Multiv. Anal. **101**, 1995–2007 (2010)
19. Qi, L.: Rank and eigenvalues of a supersymmetric tensor, the multivariate homogeneous polynomial and the algebraic hypersurface it defines. J. Symb. Comput. **41**, 1309–1327 (2006)
20. Ruiz-Gazen A., Marie-Sainte S.L., Berro, A.: Detecting multivariate outliers using projection pursuit with Particle Swarm Optimization. In: Proceedings of COMPSTAT, pp. 89–98 (2010)
21. Zarzoso, V., Comon, P.: Robust independent component analysis by iterative maximization of the Kurtosis contrast with algebraic optimal step size. IEEE Trans. Neural Netw. **21**, 248–261 (2010)

Maxima Units Search (MUS) Algorithm: Methodology and Applications

Leonardo Egidi, Roberta Pappadà, Francesco Pauli and Nicola Torelli

Abstract An algorithm for extracting identity submatrices of small rank and pivotal units from large and sparse matrices is proposed. The procedure has already been satisfactorily applied for solving the label switching problem in Bayesian mixture models. Here we introduce it on its own and explore possible applications in different contexts.

Keywords Identity matrix · Pivotal unit · Label switching

1 Introduction

Identifying and extracting identity matrices of small rank with given features from a larger, possibly sparse, matrix could appear just of theoretical interest. However, investigating the structure of a given sparse matrix is not only of theoretical appeal but can be useful for a wide variety of practical problems and for statistics.

This kind of matrix appears in clustering ensembles methods, which combine data partitions of the same dataset in order to obtain good data partitions even when the clusters are not compact and well separated. See, for instance, [1] where multiple partitions of the same data (an ensemble) are generated changing the number

L. Egidi (✉)
Dipartimento di Scienze Statistiche, Università degli Studi di Padova,
Via Cesare Battisti 241, 35121 Padova, Italy
e-mail: egidi@stat.unipd.it

R. Pappadà · F. Pauli · N. Torelli
Dipartimento di Scienze Economiche, Aziendali, Matematiche e Statistiche,
'Bruno de Finetti', Università degli Studi di Trieste, Via Tigor 22, 34124 Trieste, Italy
e-mail: rpappada@units.it

F. Pauli
e-mail: francesco.pauli@deams.units.it

N. Torelli
e-mail: nicola.torelli@deams.units.it

C. Perna et al. (eds.), *Studies in Theoretical and Applied Statistics*,
Springer Proceedings in Mathematics & Statistics 227,
https://doi.org/10.1007/978-3-319-73906-9_7

of clusters and using random cluster initializations within the K-means algorithm. Another situation where the global number of zeros of a matrix has a relevant role is in analysing the structure of a matrix of factor loadings; [4] introduces and formulates a statistical index in order to assess how good is the solution based on a factor analysis.

Matrices with a similar structure and for which the sparseness has to be taken into account appear in the so-called cost's optimization theory. [5] builds the well-known Hungarian method, which uses the zeros matrix elements for finding an optimal assignment for a given cost matrix; [6] presents a generalization of such algorithm and an application to a transportation problem.

In this paper we discuss the so-called Maxima Units Search algorithm (hereafter MUS). It has been introduced in [2] and used in the context of the label switching problem [3, 8]. In Bayesian estimation of finite mixture models label switching arises since the likelihood is invariant to permutations of the mixture components. The MUS procedure has proved to be useful in detecting some specific units—one for each mixture component—called pivots, from a large and sparse similarity matrix representing an estimate of the probability that pairs of units belong to the same group. The MUS algorithm is then more generally aimed at identifying for a given partition of the data those units that are not connected with a large number of units selected from the other groups.

A formal description of the MUS algorithm is provided and discussed. In fact, we argue that the proposed approach is of a broader interest and can be used for different purposes especially when the considered matrix presents a non-trivial number of zeros.

In Sect. 2 we introduce the notation, the algorithm and the main quantities of interest. A simulation study conducted for exploring the sensitivity of the algorithm to the choice of some parameters is presented in Sect. 3. Possible applications are illustrated in Sect. 4: in the first example we report the pivotal identification mentioned above, which represents the initial motivation for the procedure. Finally the method is employed to study a small dataset concerning tennis players' abilities. Section 5 concludes.

2 The Methodology

Let us consider a symmetric square matrix C of dimensions $N \times N$ containing a non-negligible number of zeros and suppose that each row's—or equivalently column's—index represents a statistical unit. Moreover, let us suppose that such N units either naturally belong to K different groups or have been preliminarily clustered into them, for instance via a suitable clustering technique.

For some practical purposes an example of which will be given in Sect. 4, we may be interested in detecting those units—one for each group—whose corresponding rows have more zeros than the other units. We preliminarily refer to these units as the *maxima* units. More precisely, the underlying idea is to choose as maxima those units $j_1, ..., j_K$ such that the $K \times K$ submatrix of C, $S_{j_1,...j_K}$ with only the $j_1, ..., j_K$ rows and columns has few, possibly zero, non-zero elements off the diagonal (that is, the submatrix $S_{j_1,...,j_K}$ is identical or nearly identical). Note that an identity submatrix of the given dimension may not exist. From a computational point of view, the issue is non-trivial and involves a global search row by row; as N, K and the number of zeros within C increase, the procedure becomes computationally demanding.

Before introducing mathematical details, let us denote with $i_1, ..., i_K$ a set of K maxima units and with $S_{i_1,...,i_K}$ the submatrix of C containing only the rows and columns corresponding to the maxima. The main steps of the algorithm are summarized below.

(i) For every group k, $k = 1, ..., K$ find the *candidate maxima* units $j_k^1, .., j_k^{\bar{m}}$ within matrix C, i.e. the units in group k with the greater number of zeros corresponding to the units of the other $K-1$ groups, where $\bar{m}$ is a *precision parameter* fixed in advance. Let $\mathscr{P}_k^h$, $h = 1, ..., \bar{m}$, $k = 1, ..., K$ be the entire subset of units belonging to the remaining $K-1$ groups which have a zero in j_k^h, that is

$$\mathscr{P}_k^h = \{j_l, \ l \neq k : C_{(j_k^h, \, j_l)} = 0\}, \quad h = 1, ..., \bar{m}, \ k = 1, ..., K$$

where $C_{(j_k^h, \, j_l)}$ is the element (j_k^h, j_l) of the matrix. We collect a total number of $\bar{m}K$ candidate maxima, $\bar{m}$ for every group.

(ii) For each of these $\bar{m}K$ units, count the number of distinct identity submatrices of C which contain them, constructed by taking a given candidate h and $K-1$ elements of the corresponding set $\mathscr{P}_k^h$. Let us denote this quantity with

$$M_{j_k^h} = \#\{S_{j_1,...j_{k-1}, \, j_k^h, \, j_{k+1},...j_K} \mid j_i \in \mathscr{P}_k^h, \ i = 1, ..., k-1, k+1, ..., K\}. \quad (1)$$

(iii) For each group k, $k = 1, ..., K$, select the unit which yields the greatest number of identity matrices of rank K. In mathematical terms

$$i_k = \underset{j_k^h \in \{j_k^1, ... j_k^{\bar{m}}\}}{\operatorname{argmax}} M_{j_k^h}, \quad h = 1, ...\bar{m}, \ k = 1, ..., K. \quad (2)$$

The steps of the described algorithm are illustrated via a numerical example in Fig. 1. The choice of $\bar{m}$ is crucial in terms of the algorithm performance. This parameter is a sort of benchmark for the size of the K subsets where the algorithm searches for the K maxima units: the greater is this value, the larger is the set of possible candidates involved in Eq. (2). Conversely, a bigger value enhances the possibility to build a larger set $\mathscr{P}_k^h$ and obtain a more accurate result. In Sect. 3 we deal with this issue and we consider different choices for the precision parameter $\bar{m}$.

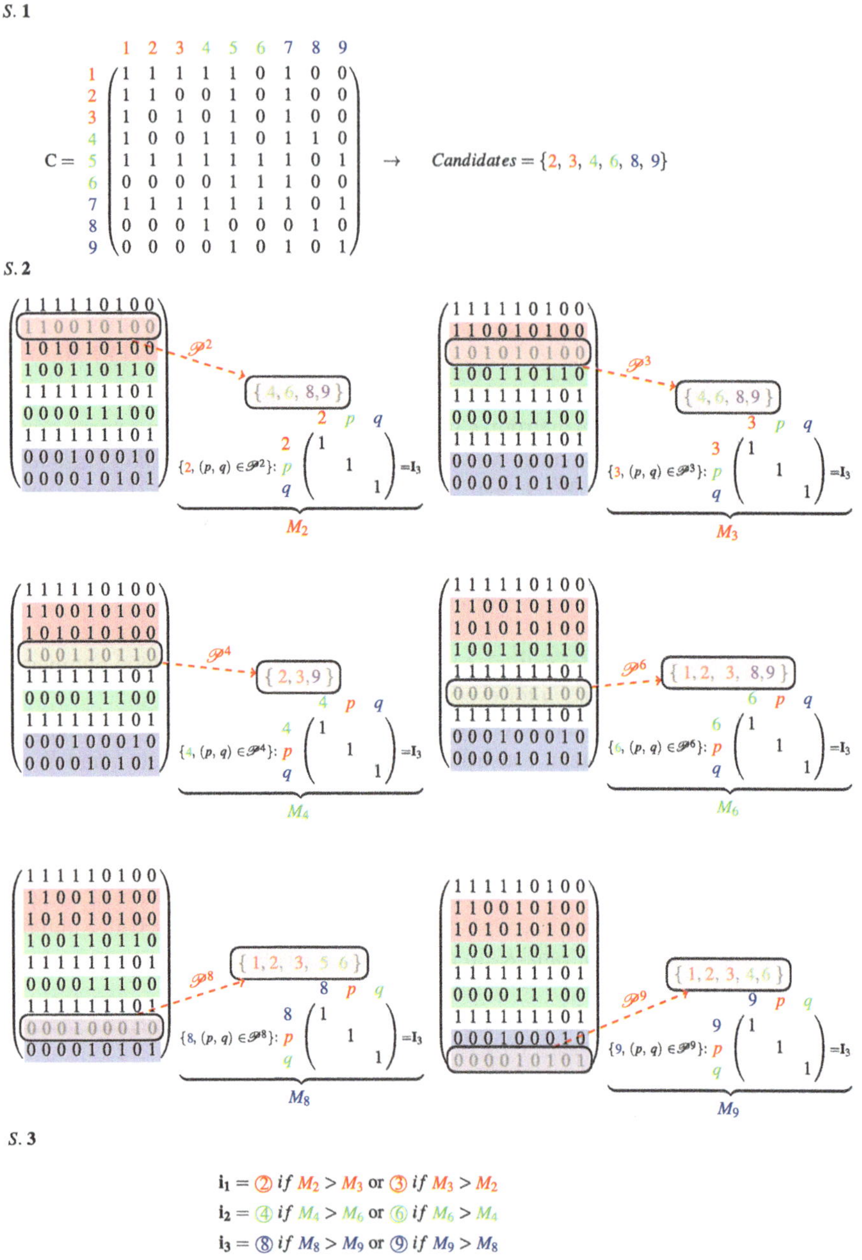

Fig. 1 Graphical scheme of the MUS algorithm for $K = 3$ and precision parameter $\bar{m} = 2$. *S.* **1** Chooses the candidate maxima, the two units for each group with the greatest number of zeros. *S.* **2** Identifies for each candidate the subsets $\mathscr{P}$ of units which belong to a different group (than the candidate) and have a zero in correspondence of it; then builds all the identity matrices of rank three which contain the candidates. *S.* **3** Detects the maxima as the three units—one for each group—that appear the greatest number of times in an identity matrix

3 Simulation Study

The task of this section is to investigate the performance of the MUS algorithm and its sensitivity to the choice of N and $\bar{m}$, for a fixed K, which is determined by some clustering technique or a given grouping of the units. To this aim, we simulate a symmetric $N \times N$ matrix C where the element (i,j) is drawn from a Bernoulli distribution with parameter p. As mentioned in Sect. 2, the i-th row's index, $i, i = 1, ..., N$, is associated to a statistical unit of interest and each unit is here randomly assigned to group k, $k = 1, \ldots, K$, with probability $1/K$. We consider three different values of p, i.e. $p = 0.8, 0.5, 0.2$.

Tables 1, 2 and 3 display the maxima units and the corresponding CPU times in seconds (in brackets) according to the considered scenarios. As expected, the procedure is sensitive to the choices of input weights, both in terms of units selection and computational times.

The first issue one may immediately notice is that, regardless of the weights used for generating data, the computational burden rises dramatically when $K > 3$. Especially when $N = 1000$, the CPU time is huge if compared to the time spent in the same framework—same $\bar{m}$ and weights— for $K = 2$ or $K = 3$. As the probability p decreases (from 0.8 to 0.2) the number of zeros becomes larger and, consequently, the CPU time required keeps growing regardless of the values of N and $\bar{m}$. A second remark is that, by fixing N and K, the choice of the precision parameter $\bar{m}$ does not seem to affect significantly the performance of the procedure: as $\bar{m}$ increases, there is limited variation in the units detection and the difference in the required time between $\bar{m} = 1$ and $\bar{m} = 20$ remains relatively small, as can be seen from Tables 1, 2 and 3. This is suggesting that even the choice of a small precision parameter—e.g. $\bar{m} = 5$—may be accurate enough for detecting the maxima.

4 Applications

4.1 Identification of Pivotal Units

As broadly explained in [2], the identification of some pivotal units in a Bayesian mixture model with a fixed number of groups may be helpful when dealing with the label switching problem [3, 8].

Let N be the number of observations generated from the mixture model. Consider, for instance, the probability of two units being in the same group. Such quantity may be estimated from the MCMC sample and denoted as $\hat{c}_{ij}$. For details, see [2]. The $N \times N$ matrix C with elements $\hat{c}_{ij}$ can be seen as a similarity matrix between units. Now, such matrix can be considered as input for some suitable clustering techniques, in order to obtain a partition of the N observations into K groups. From such partition, we may be interested in identifying exactly K pivotal units—*pivots*—which are (pairwise) separated with (posterior) probability one (that is, the posterior probability of

Table 1 MUS algorithm's maxima and computational times (in brackets) according to $K = 2, 3, 4$, $N = 100, 500, 1000$, $\bar{m} = 1, 5, 10, 20$. Bernoulli data 0, 1 generated with weights $p = 0.8$

		$K = 2$	$K = 3$	$K = 4$
$N = 100$	$\bar{m} = 1$	43, 14 (< **0.1**)	− − − (< **0.1**)	18, 34, 41, 88 (< **0.1**)
	$\bar{m} = 5$	16, 14(< **0.1**)	10, 49, 96 (< **0.1**)	37, 78, 17, 69 (**0.16**)
	$\bar{m} = 10$	16, 14 (< **0.1**)	10, 49, 96 (< **0.1**)	37, 78, 65, 69 (**0.25**)
	$\bar{m} = 20$	16, 14 (< **0.1**)	10, 49, 96 (< **0.1**)	37, 78, 65, 69 (**0.43**)
$N = 500$	$\bar{m} = 1$	− − (**0.56**)	− − − (**0.63**)	27, 44, 59, 263, (**2.3**)
	$\bar{m} = 5$	183, 125 (**0.66**)	346, 373, 500 (**0.68**)	394, 44, 59, 263, (**9.7**)
	$\bar{m} = 10$	183, 125 (**0.64**)	399, 373, 500 (**0.94**)	394, 44, 59, 263, (**17.6**)
	$\bar{m} = 20$	183, 125 (**0.72**)	399, 373, 500 (**1.22**)	394, 44, 59, 263, (**32.71**)
$N = 1000$	$\bar{m} = 1$	− − (**2.32**)	− − − (**2.40**)	350, 825, 916, 204 (**10.9**)
	$\bar{m} = 5$	654, 94 (**2.49**)	909, 499, 868 (**3.02**)	381, 849, 684, 204 (**44.0**)
	$\bar{m} = 10$	654, 94 (**2.62**)	909, 499, 868 (**3.52**)	381, 849, 684, 488 (**81.7**)
	$\bar{m} = 20$	654, 96 (**2.99**)	909, 382, 868 (**4.62**)	381, 849, 748, 488 (**152.82**)

Table 2 MUS algorithm's maxima and computational times (in brackets) according to $K = 2, 3, 4$, $N = 100, 500, 1000$, $\bar{m} = 1, 5, 10, 20$. Bernoulli data 0, 1 generated with weights $p = 0.5$

		$K = 2$	$K = 3$	$K = 4$
$N = 100$	$\bar{m} = 1$	48, 86 (< **0.1**)	− − − (< **0.1**)	32, 62, 38, 55 (**0.44**)
	$\bar{m} = 5$	48, 61 (< **0.1**)	15, 33, 5, (**0.12**)	50, 62, 89, 55 (**1.99**)
	$\bar{m} = 10$	48, 61 (< **0.1**)	15, 62, 5 (**0.14**)	50, 62, 90, 55 (**3.36**)
	$\bar{m} = 20$	48, 61 (**0.10**)	15, 62, 5 (**0.13**)	50, 62, 90, 55 (**1328.73**)
$N = 500$	$\bar{m} = 1$	− − (**1.61**)	− − − (**1.67**)	10, 11, 90, 488 (**56.1**)
	$\bar{m} = 5$	294, 242 (**1.40**)	203, 213, 272 (**2.31**)	273, 242, 292, 383 (**159.8**)
	$\bar{m} = 10$	294, 242 (**1.64**)	203, 213, 272 (**3.31**)	273, 232, 482, 383 (**311.28**)
	$\bar{m} = 20$	66, 242 (**1.78**)	203, 213, 272 (**5.49**)	273, 232, 29, 383 (**582.38**)
$N = 1000$	$\bar{m} = 1$	− − (6.64)	− − − (**7.16**)	123, 964, 813, 238 (**246.28**)
	$\bar{m} = 5$	94, 405 (**6.81**)	67, 995, 688, (**9.78**)	267, 964, 813, 241 (**1208.27**)
	$\bar{m} = 10$	94, 405 (**7.24**)	67, 995, 688, (**12.61**)	267, 964, 813, 241 (**2326.64**)
	$\bar{m} = 20$	398, 405 (**8.23**)	67, 995, 688, (**9.58**)	267, 964, 813, 241 (**4548.47**)

Table 3 MUS algorithm's maxima and computational times (in brackets) according to $K = 2, 3, 4,\ \ N = 100, 500, 1000,\ \bar{m} = 1, 5, 10, 20$. Bernoulli data 0, 1 generated with weights $p = 0.2$

		$K = 2$	$K = 3$	$K = 4$
$N = 100$	$\bar{m} = 1$	42, 32 (< **0.1**)	58, 40, 63 (**0.10**)	24, 68, 34, 89 (**3.75**)
	$\bar{m} = 5$	42, 32 (< **0.1**)	81, 54, 63 (**0.21**)	24, 68, 48, 58 (**8.85**)
	$\bar{m} = 10$	42, 86 (**0.14**)	87, 54, 63 (**0.25**)	24, 68, 48, 58 (**16.8**)
	$\bar{m} = 20$	42, 86 (**0.11**)	87, 54, 63 (**0.43**)	24, 68, 48, 58 (**1985.01**)
$N = 500$	$\bar{m} = 1$	– – (**2.40**)	– – – (**2.74**)	371, 28, 122, 60 (**189.94**)
	$\bar{m} = 5$	326, 288 (**2.39**)	290, 393, 316 (**4.51**)	370, 38, 202, 404 (**949.4**)
	$\bar{m} = 10$	326, 288 (**2.65**)	290, 393, 316 (**7.20**)	370, 413, 202, 196 (**1882.93**)
	$\bar{m} = 20$	284, 288 (**3.06**)	375, 395, 316 (**10.66**)	370, 38, 202, 404 (**3685.61**)
$N = 1000$	$\bar{m} = 1$	555, 892, (**11.30**)	– – – (**12.25**)	427, 452, 218, 631 (**1608.11**)
	$\bar{m} = 5$	434, 892 (**11.28**)	222, 921, 275 (**19.42**)	427, 493, 218, 839 (**8098.54**)
	$\bar{m} = 10$	434, 892 (**11.27**)	387, 921, 255 (**26.72**)	427, 493, 218, 839 (**16629.86**)
	$\bar{m} = 20$	434, 892 (**12.47**)	387, 921, 255 (**45.08**)	427, 493, 218, 839 (**32230.8**)

any two of them being in the same group is zero). In fact, as discussed in [2], the identification of such units allows us to provide a valid solution to the occurrence of label switches.

Following the procedure described in Sect. 2, one can find units $i_1, ..., i_K$ such that the submatrix S of C, with only the rows and columns corresponding to such units, is the identity matrix. It is still worth noticing that the availability of K perfectly separated units is crucial to the procedure, and it can not always be guaranteed.

Practically, our interest is in finding units which should be 'as far as possible' one from each other according to a well defined distance measure. The more separated they are, the better they represent the group they belong to.

Figure 2 shows the pivotal identification of $K = 4$ units for a sample of $N = 1000$ bivariate data generated according to a nested Gaussian mixture of mixtures with K groups and fixed means. Pivots (red) have been detected through the MUS algorithm and through another alternative method, which aims at searching the most distant units among the members that are farthest apart. We may graphically notice that separation is made more efficient by the MUS algorithm, for which the red points appear quite distant from each other. Moreover, in this specific example the pivotal search is made difficult due to the overlapping of the K groups.

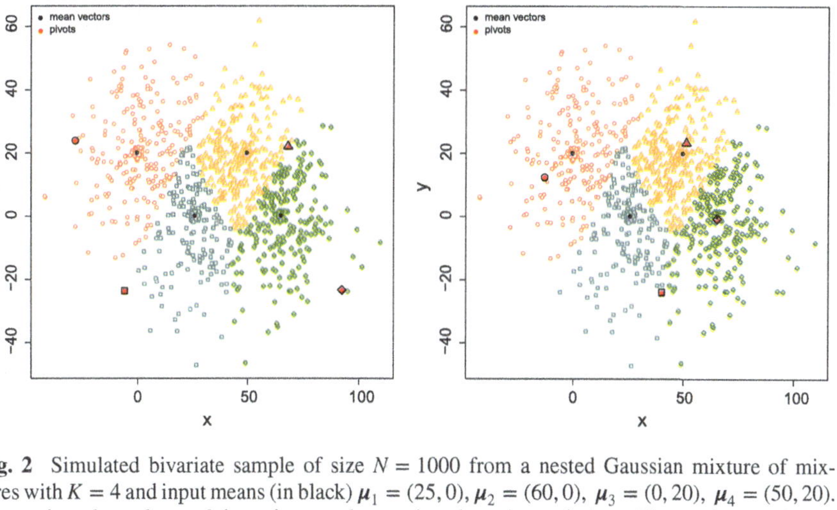

Fig. 2 Simulated bivariate sample of size $N = 1000$ from a nested Gaussian mixture of mixtures with $K = 4$ and input means (in black) $\boldsymbol{\mu}_1 = (25, 0)$, $\boldsymbol{\mu}_2 = (60, 0)$, $\boldsymbol{\mu}_3 = (0, 20)$, $\boldsymbol{\mu}_4 = (50, 20)$. Groups have been detected through an agglomerative clustering technique. Pivots—i.e. maxima—found by MUS algorithm (Left) with $\bar{m} = 5$ are shown in red and seem well separated in the bi-dimensional space. Pivots found by method $\min_{\bar{i}}(\min_{j \notin \mathcal{X}_k} C_{(\bar{i}j)})$ are less distant each other (Right)

4.2 Tennis Singular Features

As a simple example we apply our algorithm to a case study regarding tennis players. We collect $N = 8$ game's features (hereafter *GF*) for $T = 25$ players from the Wimbledon Tournament 2016,[1] and we assign the following values

$$\begin{cases} GF_{i,t} = 1, & \text{if player } t \text{ has GF}i, \\ GF_{i,t} = 0, & \text{if player } t \text{ has } \textit{not} \text{ the GF } i. \end{cases}$$

Game's features belong to $K = 2$ groups, which somehow refer to the attack and the defence skills for each player. We denote with the label 1 the first group and with 2 the second group: 'First Serve Receiving Points' (2), 'Second Serve Receiving Points' (2), 'Break Points Won' (1), 'Serve Speed' (1), 'Aces'(1), 'First Serve Points' (1), 'Second Serve Points' (1), 'Break Points Conversion' (2).

We decide to assign a specific game feature to a given player if this player belongs to the first five positions of that game's feature rank reported by the Wimbledon's website. Hence, let us enumerate the above-mentioned features from one to eight. Our 0–1 dataset has as many records as players. The problem setting is summarized below.

	GF							
Player	1	2	3	4	5	6	7	8
Federer	1	1	0	0	0	0	0	0
Murray	1	1	1	0	0	0	0	0

Note that Federer is assigned game's feature one ('First Serve Receiving Points') and two ('Second Serve Receiving Points') only, Murray is assigned game's feature one, two and three ('Break Points Won') and so on. We define the $N \times N$ symmetric matrix, C, in which the generic element $C_{(i,j)}$ is the number of players that have both features i and j.

$$C = \begin{pmatrix} 1 & 5 & 3 & 1 & 3 & 0 & 0 & 0 \\ 5 & 1 & 2 & 1 & 2 & 0 & 0 & 0 \\ 3 & 2 & 1 & 0 & 1 & 0 & 0 & 0 \\ 1 & 1 & 0 & 1 & 2 & 0 & 0 & 0 \\ 3 & 2 & 1 & 2 & 1 & 0 & 0 & 0 \\ 0 & 0 & 0 & 0 & 0 & 1 & 0 & 0 \\ 0 & 0 & 0 & 0 & 0 & 0 & 1 & 0 \\ 0 & 0 & 0 & 0 & 0 & 0 & 0 & 1 \end{pmatrix}$$

We would like to extract the two 'most distant' game's features for the two groups, i.e. the two features in correspondence of which the matrix C is zero more often, and

[1] http://www.wimbledon.com/en_GB/scores/extrastats/index.html.

for which S_{i_1,i_2} is an identity matrix. We can notice that rows 6 and 7 of C are full of zeros: this means, for instance, that according to our short dataset $C_{(6,1)} = 0$, i.e. the 'First Serve Points' ($k = 1$) and 'First Serve Receiving Points' ($k = 2$) do not coexist to any player. Are they the most distant features between the two groups? To answer this question, we run the MUS algorithm by fixing $\bar{m} = 3$ and we find the candidate maxima $j_1^1 = 6$, $j_1^2 = 7$, $j_2^1 = 8$, $j_2^2 = 1$ and maxima $i_1 = 6$, $i_2 = 8$. Hence we conclude that 'First Serve Points' (six) and 'Break Points Conversion' (eight) are quite unlikely to belong to the same player.

5 Conclusions

A procedure for detecting small identity submatrices from a $N \times N$ matrix has been proposed. It has been initially considered for application to the pivotal approach in label switching problem in the analysis of Bayesian mixture models. The proposed method is discussed in detail and employed for different practical problems.

Its efficiency and its sensitivity to parameter choices is investigated through a simulation study, which shows that for a small number of groups the procedure is quite fast. Moreover, even for small values of the precision parameter $\bar{m}$ the procedure appears quite stable in terms of units indexes, suggesting that a higher value of $\bar{m}$ is often not required. This is also confirmed by the results in Sect. 4.

Further issues for future research are related to the optimization of the proposed algorithm and the definition of suitable indicators for detecting both diagnostic problems inherent to the procedure and goodness of units choice.

References

1. Ana, L.N.F., Jain, A.K.: Data clustering using evidence accumulation. In: 2002 Proceedings 16th International Conference on Pattern Recognition, vol. **4**, pp. 276–280 (2002)
2. Egidi, L., Pappadà, R., Pauli, F., Torelli, N.: Relabelling in Bayesian mixture models by pivotal units. arXiv:1501.05478v2 (2015)
3. Jasra, A., Holmes, C.C., Stephens, D.A.: Markov chain Monte Carlo methods and the label switching problem in Bayesian mixture modeling. Stat. Sci. 50–67 (2005)
4. Kaiser, Henry F.: An index of factorial simplicity. Psychometrika **39**(1), 31–36 (1974)
5. Kuhn, H.W.: The Hungarian method for the assignment problem. Nav. Res. Logist. Q. **2**(1–2), 83–97 (1955)
6. Munkres, J.: Algorithms for the assignment and transportation problems. J. Soc. Ind. Appl. Math. **5**(1), 32–38 (1957)
7. Puolamäki, K., Kaski, S.: Bayesian solutions to the label switching problem. In: Advances in Intelligent Data Analysis VIII, pp. 381–392. Springer (2009)
8. Stephens, M.: Dealing with label switching in mixture models. J. R. Stat. Soc. Ser. B (Stat. Methodo.) **62**(4), 795–809 (2000)

DESPOTA: An Algorithm to Detect the Partition in the Extended Hierarchy of a Dendrogram

Davide Passaretti and Domenico Vistocco

Abstract DESPOTA is a method proposed to seek the best partition among the ones hosted in a dendrogram. The algorithm visits nodes from the tree root toward the leaves. At each node, it tests the null hypothesis that the two descending branches sustain only one cluster of units through a permutation test approach. At the end of the procedure, a partition of the data into clusters is returned. This paper focuses on the interpretation of the test statistic using a data–driven approach, exploiting a real dataset to show the details of the test statistic and the algorithm in action. The working principle of DESPOTA is shown in the light of the Lance–Williams recurrence formula, which embeds all types of agglomeration methods.

1 *Hierarchical Agglomerative Clustering* and the Issue of the Cut

Cluster analysis [3, 6, 8] is an unsupervised technique in which there is no clear indication of the way units should be joined and of the number of clusters that should be finally constituted. In *Hierarchical Agglomerative Clustering*, the grouping is executed incorporating all the dissimilarity levels in a bottom-up process, i.e. in a way that units that are merged cannot be separated anymore. The whole aggregation process is exhaustively embedded in the dendrogram. This depicts all the stages at which units join starting from the leaves up to the root node. Given this (reversed) tree-like structure, one is required to select a method for cutting the tree in order to identify the most adequate number of clusters.

The issue concerning the best cut is a crucial one in literature. Several rules of decision have been conceived to substantially reduce the subjectivity of the final choice. A few of them were proposed to work directly on the dendrogram, and others

D. Passaretti · D. Vistocco (✉)
Dip.to di Economia e Giurisprudenza – Università degli Studi di Cassino e del Lazio Meridionale, Via S. Angelo S.N. – Località Folcara, Cassino (FR), Italy
e-mail: passarettidav@gmail.com

D. Vistocco
e-mail: vistocco@unicas.it

C. Perna et al. (eds.), *Studies in Theoretical and Applied Statistics*, Springer Proceedings in Mathematics & Statistics 227,
https://doi.org/10.1007/978-3-319-73906-9_8

to operate on the results of other cluster methods [7, 12, 13, 18]. The predominant approach for cutting the dendrogram only focuses on all possible horizontal cuts. Among these cuts, the one corresponding to a big jump between two subsequent branches is usually advised in literature. Despite being reasonable, this approach is still partly discretionary and, above all, it does not permit exploration of the entire (*extended*) hierarchy of the partitions housed in the dendrogram. Such extended hierarchy is obtainable by cutting the tree in all possible ways. Considering also non horizontal cuts should provide more valid solutions especially when clusters with different internal consistencies have to be detected.

Bruzzese and Vistocco [1] introduced DESPOTA (DEndrogram Slicing through a PermutatiOn Test Approach), a method that seeks the best partition in the extended hierarchy. The main aim of this paper is to show DESPOTA in action on a real dataset, highlighting the operating principle of its test statistic. In particular, the dataset exploited is the "Wholesale Customer dataset", available on the UCI Machine Learning Repository [11]. It contains the annual spending in monetary units of clients of a wholesale distributor. The clients' purchases are classified according to six product categories: fresh, milk, grocery, frozen, detergents and paper, delicatessen.

The paper is organized as follows. Next section introduces DESPOTA: in particular, it displays the notation adopted, the derivation of the test statistic (Sect. 2.1), the permutation distribution (Sect. 2.2), and the algorithm in action on the *wholesale customer dataset* (Sect. 2.3). Section 3 evaluates the partition detected by DESPOTA via the Silhouette index. The final section contains the concluding remarks and future avenues for research.

2 DEndrogram Slicing through a PermutatiOn Test Approach

DESPOTA exploits permutation tests [5, 14] to detect the final partition on a dendrogram. The algorithm visits nodes from the tree root toward the leaves. At each node, it tests the *null hypothesis* that the two descending branches sustain only one cluster of units. A rejection of H_0 suggests that the split generating the two branches really exists at population level. In case the *null hypothesis* is not rejected, a cluster is detected on the dendrogram and no further sub-branches referring to the node will be visited. The algorithm stops when H_0 can no longer be rejected. The resulting clusters constitute the final partition of the data.

The following subsections are intended to provide a basic understanding and interpretation of the test statistic and of its distribution. The algorithm will be illustrated directly on a working example. The interested reader is referred to the original paper [1] for more technical details.

2.1 *The Test Statistic as a Ratio of Two Costs*

The present section firstly introduces the notation used in Bruzzese and Vistocco [1] to describe the distinctive traits characterizing the dendrogram at each node. The test statistic is then derived.

Let us start by supposing the process is at node k, where k is an integer number that varies from 1 (at the root node) to $N - 1$ (at the node where the two most similar leaves join). The following notation, highlighted in Fig. 1 for the case of $k = 1$, is adopted:

- L_k and R_k define the left and right branch joining at node k;
- $L_k \cup R_k$ identifies node k, because it is where the two branches merge;
- $h(L_k)$, $h(R_k)$ and $h(L_k \cup R_k)$ indicate the dissimilarity levels related to the left branch, the right branch and the node, respectively.

That said, a hierarchical agglomerative cluster analysis is performed on the *wholesale customer dataset*, using a Ward's linkage [19] along with an Euclidean distance, a common choice for this type of data. The resulting dendrogram is depicted in Fig. 1. The algorithm starts operating on the highest node of the tree ($k = 1$), and considers the following two distances:

- $|h(L_1) - h(R_1)|$ is the absolute difference in the dissimilarity level between the two branches descending from the node. It is interpreted as the minimum cost required for merging the two clusters. That is because the dissimilarity measure used in the agglomeration process has to rise at least by this amount in order to join the two branches.

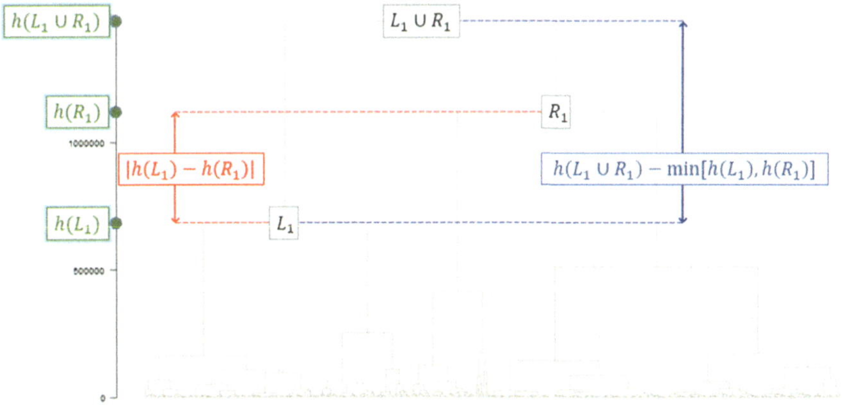

Fig. 1 Main notation and comparison of the two involved distances for the case of $k = 1$ (root node). $|h(L_1) - h(R_1)|$ is the minimum cost required for joining L_1 with R_1, while $h(L_1 \cup R_1) - \min[h(L_1), h(R_1)]$ is the cost actually incurred in the merging process. The actual cost looks much larger than the minimum cost, which might suggest that L_1 and R_1 are not the same cluster

- $h(L_1 \cup R_1) - \min[h(L_1), h(R_1)]$ is the difference in dissimilarity level between the node and the branch being at the lowest height. Therefore, it can be interpreted as the actual cost incurred for merging the two clusters.

The interpretation of these two distances in terms of merging costs makes it reasonable to test whether there is a significant discrepancy between them. Such a discrepancy becomes the determinant for making the decision at the specific node. In particular, if the actual cost is much higher than the minimum cost, this leads to be in favour of the presence of two different clusters. This is due to the fact that a sizeable jump in dissimilarity level is unlikely to occur when making the assumption that there is one homogeneous class of units. Conversely, similar costs indicate that the two branches have joined with reasonably little effort, which suggests that the assumption of the presence of only one cluster holds.
Following this line of reasoning, the test statistic is obtained through the ratio of the minimum cost to the actual cost. Hence, for a general node k, it is defined as:

$$rc_k = \frac{|h(L_k) - h(R_k)|}{h(L_k \cup R_k) - \min[h(L_k), h(R_k)]}$$

which ranges from 0 to 1: the closer to 0 it is, the larger the difference between the actual cost incurred for merging the two clusters and the minimum cost being required. The *null hypothesis* for this test statistic mirrors the above assumption of one homogeneous class, supposing the two costs be non significantly different. Hence, an observed value of rc_k close to 1 would lead to accept H_0.

In order to achieve the sampling distribution of the test statistic, a permutation test approach is exploited, as shown in the next section. The procedure is set in an unifying framework exploiting the Lance–Williams formula, which embeds all types of agglomeration methods used in hierarchical clustering.

2.2 The Permutation Distribution of the Test Statistic Under H_0

In order to introduce the permutation test approach, let us suppose the algorithm is operating at node k, and the observed value of test statistic has been computed. This number alone says almost nothing about the decision to be made at that node. Therefore, the actual aim is to appraise how extreme rc_k is under the *null hypothesis*. This is why a distribution for the test statistic is needed.

In order to obtain it, one can consider that the units referring to a node constitute one homogeneous class under H_0. Thus, since the assumption of *exchangeability* [5] holds, they can be rearranged without substantially alterating the aggregation process under H_0. That is done by randomly assigning these units to one branch or the other, preserving the cardinalities of the two initial clusters. Starting from the two new classes of units, their respective dendrograms are generated, and then joined using the *Lance-Williams* recurrence formula [9, 10]. Such a formula encloses all types of linkage and succinctly defines the update of dissimilarities at each step of the agglomeration process. In particular, let us suppose there are three units (A, B, C), of which A and B merge. Through such a union, the cluster $A \cup B$ is constituted. The distance between $A \cup B$ and unit C is determined by the following formula [9, 10]:

$$\delta_{A\cup B,C} = \alpha_A\, \delta_{A,C} + \alpha_B\, \delta_{B,C} + \beta\, \delta_{A,B} + \gamma\, |\delta_{A,C} - \delta_{B,C}|$$

where $\alpha_A, \alpha_B, \beta, \gamma$ are coefficients characterizing the type of linkage, and $\delta_{i,j}$ indicates the dissimilarity between any two clusters i and j.

The formula is suitably exploited to determine the height at which the node merging the two simulated trees has to be placed. Referring to the actual case, let us define the simulated dendrogram with highest node as $A \cup B$, because the clusters merging at the latest step were A (containing n_A units) and B (containing n_B units). The other simulated tree is C and contains n_C units. Considering that the linkage chosen is Ward's and the metric is the Euclidean norm ${}_2d_{i,j}$, the *Lance-Williams* formula becomes [6]:

$${}_2d_{A\cup B,C} = \frac{n_A + n_C}{n_A + n_B + n_C}\, {}_2d_{A,C} + \frac{n_B + n_C}{n_A + n_B + n_C}\, {}_2d_{B,C} + \frac{-n_C}{n_A + n_B + n_C}\, {}_2d_{A,B}$$

since $\alpha_A = \frac{n_A+n_C}{n_A+n_B+n_C}$, $\alpha_B = \frac{n_B+n_C}{n_A+n_B+n_C}$, $\beta = \frac{-n_C}{n_A+n_B+n_C}$, $\gamma = 0$. Merging the two simulated trees results in a brand new dendrogram on which the test statistic can be calculated.

The above procedure is replicated M times in order to approximate through M random samples the *permutation distribution* of the test statistic, i.e. the distribution that contemplates all possible rearrangements of the units in the two branches, preserving the original cardinalities. The Monte Carlo p-value is then computed as the proportion of times the observed value of the test statistic is greater than the simulated ones:

$$p = \frac{\#(rc_k^m \leq rc_k)}{M}$$

where m is any of the M samples on which the test statistic rc_k^m has been computed.

2.3 The Algorithm in Action

Here, DESPOTA is shown in action on the *wholesale customer dataset* and the steps leading to the detection of the final partition are illustrated. Figure 2 highlights DESPOTA's solution on the same dendrogram depicted in Fig. 1. To achieve the solution, a significance threshold (α) of 1% has been used and 999 random samples have been drawn from the permutation distribution of the test statistic at each visited node. The figure also pinpoints such visited nodes by enumerating them from

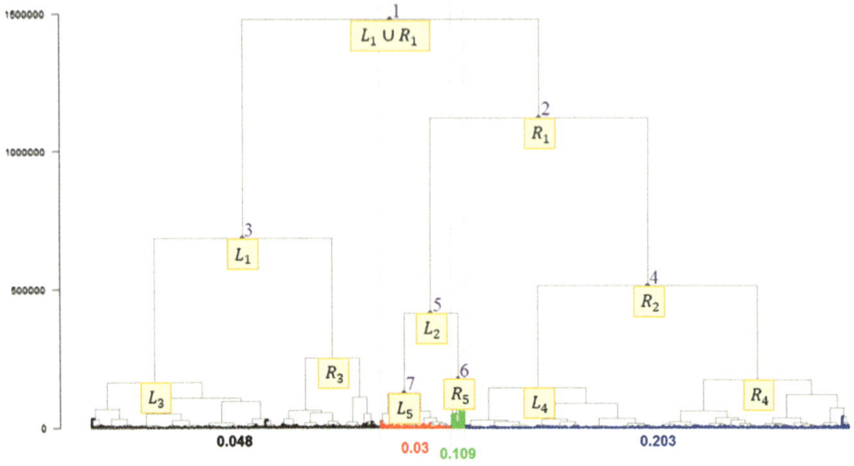

Fig. 2 Partition found by DESPOTA at $\alpha=1\%$ on the wholesale customer dataset. Euclidean distance and Ward's method have been used to generate the dendrogram. The nodes visited by the algorithm are enumerated and the corresponding notation reported. On the bottom, there are the p-values corresponding to the detected clusters

the root downward. Table 1 details the steps needed to accomplish the final partition consisting of four clusters. In order to illustrate the algorithm, it is useful to introduce the two following sets, which are edited at each step:

- *AggregationLevelsToVisit* contains the dissimilarity levels corresponding to the nodes that need to be visited. It is initialized at $h(L_1 \cup R_1)$ so to retrace downward the tree starting from the root.
- *DetectedClusters* includes the clusters that are gradually found. It is clearly empty at the first step.

The algorithm starts at $h(L_1 \cup R_1)$, which is the only dissimilarity level included in *AggregationLevelsToVisit*. Therefore, at the first step, the two branches involved in the test are L_1 and R_1. The *null hypothesis* states $L_1 \equiv R_1$, in the sense that they both correspond to the same homogeneous class $\mathscr{C}_1$. In order to test H_0, the algorithm computes the observed value of rc_1, and all the simulated values corresponding to the trees grown up through M random rearrangements of the units in L_1 and R_1. The permutation distribution of the test statistic is approximated through these simulated values of rc_1. In this case, the observed value is lower than the critical value of the approximate permutation distribution: the *null hypothesis* is rejected. As a consequence of that, *AggregationLevelsToVisit* now contains $h(R_1)$ and $h(L_1)$, the former needing earlier investigation, inasmuch as greater. Hence, what is tested at node 2 is $H_0 : L_2 \equiv R_2 \equiv \mathscr{C}_2$, which, indeed, involves the two branches, L_2 and R_2, descending from R_1. The *null hypothesis* is rejected again. Therefore, $h(R_2)$ and $h(L_2)$ are inserted in *AggregationLevelsToVisit*, while *DetectedClusters* still contains nothing. At step 3, the algorithm tests $h(L_1)$ and fails to reject the *null hypothesis* that $L_3 \equiv R_3 \equiv \mathscr{C}_3$.

This entails that no further sub-branches descending from node 3 are visited: *DetectedClusters* now includes L_1. Conversely, *AggregationLevelsToVisit* contains $h(R_2)$ and $h(L_2)$. The algorithm restarts from $h(R_2)$ and continues, as described in Table 1, until *AggregationLevelsToVisit* is empty, i.e. H_0 can no longer be rejected (step 7). That leads *DetectedClusters* to contain the final partition composed by clusters L_1, R_2, R_5 and L_5.

It is interesting to show in detail what actually leads to an acceptance or a rejection of H_0. To this end, Fig. 3 depicts, in panels *(a)* and *(b)*, the distributions obtained at node 4 and 5, respectively. The area α=1% and the one associated with the observed value of rc_k are shown using two different pattern fills. The values of the two corresponding quantiles are reported on the bottom of each panel. It is evident that the observed value rc_4 is greater (i.e. less extreme) than the critical value of its reference distribution, whereas rc_5 corresponds to a quantile in the critical region of the permutation distribution at node 5. Hence, the *null hypothesis* is accepted in the former case and rejected in the latter.

The p-values associated with the detected clusters are reported below each cluster in Fig. 2 and are all greater than the significance threshold $\alpha = 1\%$. Indeed, it is worth saying that the detection of a cluster always follows from a fail in rejecting H_0, except when the cluster is made of only one unit. In such quite unusual circumstance, not occurring in the current example, the singleton is detected by rejecting H_0 in a test involving a leaf on one side and a class of units on the other side.

Table 1 Compact representation of the process leading to the solution shown in Fig. 2

Node	*AggregationLevelsToVisit*	H_0	Decision	*DetectedClusters*
1	$h(L_1 \cup R_1)$	$L_1 \equiv R_1 \equiv \mathscr{C}_1$	Reject	–
2	$h(R_1)$, $h(L_1)$	$L_2 \equiv R_2 \equiv \mathscr{C}_2$	Reject	–
3	$h(L_1)$, $h(R_2)$, $h(L_2)$	$L_3 \equiv R_3 \equiv \mathscr{C}_3$	Accept	L_1
4	$h(R_2)$, $h(L_2)$	$L_4 \equiv R_4 \equiv \mathscr{C}_4$	Accept	L_1, R_2
5	$h(L_2)$	$L_5 \equiv R_5 \equiv \mathscr{C}_5$	Reject	L_1, R_2
6	$h(R_5)$, $h(L_5)$	$L_6 \equiv R_6 \equiv \mathscr{C}_6$	Accept	L_1, R_2, R_5
7	$h(L_5)$	$L_7 \equiv R_7 \equiv \mathscr{C}_7$	Accept	L_1, R_2, R_5, L_5

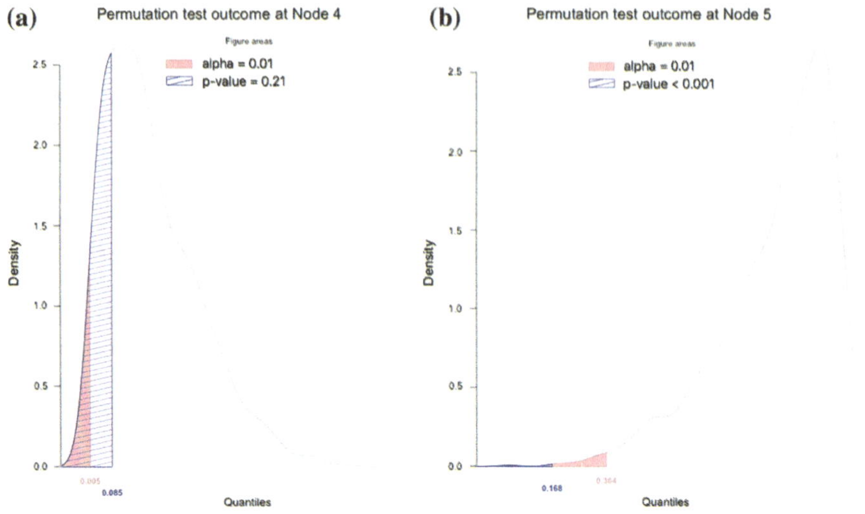

Fig. 3 Comparison of the approximate permutation distributions obtained at nodes 4, in panel (**a**), and at node 5, in panel (**b**). The *null hypothesis* is rejected only in (**b**), where the observed value rc_5 is in the left tail of the reference distribution of the test statistic and over the critical value

3 Evaluating the Detected Partition

Hierarchical agglomerative clustering is usually controversial for various reasons. One, very obvious, refers to the initial choice of the combination of methods for measuring the dissimilarity between units (metric) and between groups of units (linkage): such a combination affects the way units are merged at each step [8].

Another remarkable aspect, related to any kind of cluster analysis, involves the natural structure of the data which can or cannot underlie the presence of actual clusters [2]. This may be a relevant issue when needing an interpretation of what units within each cluster have in common.

That said, the evaluation of the detected partition is a crucial point for validating cluster results. For this purpose, several possibilities are available in literature. Among these, a widely used method is *silhouette* [17]. It aims to evaluate on average how well units are matched to their own clusters. To this end, it computes the distances of each unit from both its own cluster and its neighbouring cluster, the latter being on average the closer cluster not including such a unit. The resulting indices all range from -1 to 1. The closer to 1 a value is, the better the unit is matched to the cluster in which it belongs.

The indices are then usually plotted in order to visually inspect the quality of each detected cluster. Here, such a graphical representation of *silhouette* is not reported. Only the overall corresponding index is computed (i.e. the mean of the indices calculated for all units). The goal is to compare the four-cluster solution provided by DESPOTA to the partition made of the same number of clusters deriving from the

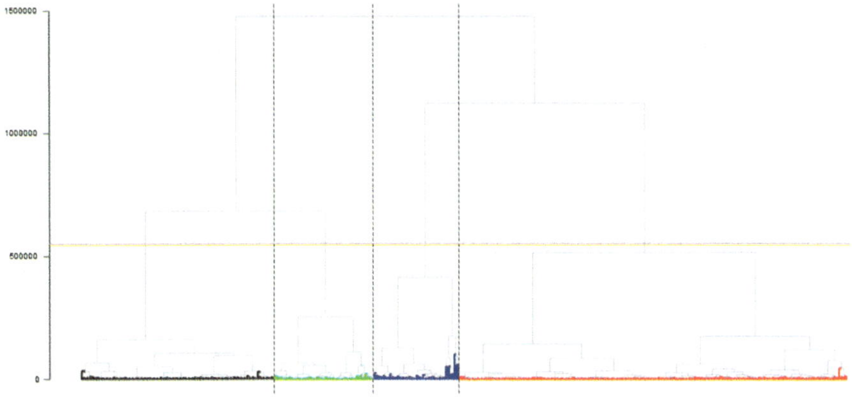

Fig. 4 Partition consisting of four clusters, coming from the horizontal cut

horizontal cut of the dendrogram. Such a horizontal cut defines a partition in which L_1 is split into two clusters, whereas R_5 and L_5 are the same cluster (see Fig. 4). From the point of view of the above index, DESPOTA seems to behave generally better, as it takes on a value of 0.320 which is greater than 0.249, resulting from horizontal cut. That does not mean this is the best partition obtainable on such a dendrogram, since the used index is only one of the many available and does not indicate an undisputable truth. Moreover, even sticking to *silhouette*, a value of 0.320 does not always assert an evidence of such a natural clustering structure in the data: solutions related to different numbers of clusters might take on a greater value.

However, the present example reveals that a standard horizontal cut does not always turn out to be the most appropriate approach for detecting a partition, as already shown in the original paper using various datasets [1].

4 Concluding Remarks and Future Avenues

DESPOTA is an algorithm that only requires as inputs a dendrogram and the significance level to be adopted for the testing procedure. In the permutation procedure, on which it is based, DESPOTA exploits the same methods used to merge units for growing up the dendrogram, i.e. metric and linkage criterion.

The significance level is the main issue concerning the whole process: one cannot avoid taking into consideration the probabilities of the errors that can be made at each node from the root toward the leaves. Therefore, a fixed significance threshold may not be a plausible choice for this algorithm: an approach exploiting multiple testing [16] should be recommended in order to automatically lower α when moving down on the dendrogram. Since such approach has not been implemented yet, the solutions provided by DESPOTA are unlikely to correspond to the actual levels of uncertainty declared in theory. For the actual dataset, a Bonferroni correction has been roughly used to take into account the number of comparisons carried out at the visited nodes. As a result of this, only the p-value associated with the last cut (level 5 in Fig. 2) becomes visibly closer to—but still lower than—the α threshold. However, it is worth remarking that for this dataset the algorithm returns a small number of clusters. A more problematic impact there would have been in case of more clusters detected.

Another, less serious issue, concerns the computational complexity of the algorithm in accomplishing its task. To limit that, the optimization of the existing R [15] code and/or the use of compiled code may be useful. A possible improvement might also come from a step-by-step estimation of the number of samples to be drawn from the permutation distribution of the test statistic. This could provide a reliable p-value at each node without the need of a large number of iterations [4].

That said, DESPOTA seems to be a valid approach to solve the problem of finding a partition on a dendrogram. At each node, it investigates whether the actual cluster structure of the data is plausible under H_0, considering the different possible structures coming from rearrangements of units. That makes it possible to achieve levels of uncertainty for the detected partition in probabilistic terms.

References

1. Bruzzese, D., Vistocco, D.: DESPOTA: DEndrogram slicing through a pemutation test approach. J. Classif. **32**(2), 285–304 (2015)
2. Cormack, R.M.: A review of classification. J, R. Stat. Soc. Ser. A (General) **134**(3), 321–367 (1971)
3. Everitt, B., Landau, M., Leese, M.: Cluster Analysis, 4th edn. Arnold, London (2001)
4. Gandy, A.: Sequential implementation of monte carlo tests with uniformly bounded resampling risk. J. Am. Stat. Assoc. **104**(88), 1504–1511 (2009)
5. Good, P.I.: Permutations Tests for Testing Hypotheses. Springer, New York (1994)
6. Gordon, A.D.: Classification, 2nd edn. Chapman & Hall/CRC Press (1999)
7. Gurrutxaga, I., Albisua, I., Arbelaitz, O., Martìn, J.I., Muguerza, J., Pèrez, J.M., Perona, I.: SEP/COP: an efficient method to find the best partition in hierarchical clustering based on a new cluster validity index. Pattern Recogn. **43**(10), 3364–3373 (2010)
8. Kaufman, L., Rousseeuw, P.J.: Finding groups in data. In: An Introduction to Cluster Analysis. Wiley. New York (1990)
9. Lance, G.N., Williams, W.T.: A generalised sorting strategy for computer classifications. Nature **212**, 218 (1966b)
10. Lance, G.N., Williams, W.T.: A general theory of classificatory sorting strategies. 1. Hierarchical systems. Comput. J. **9**(4), 373–380 (1967)
11. Lichman, M.: UCI machine learning repository. Irvine, CA: University of California, School of Information and Computer Science. http://archive.ics.uci.edu/ml (2013)
12. Milligan, G.W.: A Monte Carlo study of thirty internal criterion measures for cluster analysis. Psychometrika **42**, 187–199 (1981)
13. Milligan, G.W., Cooper, M.C.: An examination of procedures for determining the number of clusters in a dataset. Psychometrika **52**(2), 159–179 (1985)
14. Pesarin, F., Salmaso, L.: Permutation tests for complex data. In: Theory, Applications and Software. Wiley, Chichester, UK (2010)

15. R Core Team: R: A Language and Environment for Statistical Computing. R Foundation for Statistical Computing, Vienna, Austria. https://www.R-project.org/ (2015)
16. Romano, J.P., Wolf, M.: Control of generalized error rates in multiple testing. Ann. Stat. **35**(4), 1378–1408 (2007)
17. Rousseeuw, P.J.: Silhouettes: a graphical aid to the interpretation and validation of cluster analysis. J. Comput. Appl. Math. **20**, 53–65 (1986)
18. Tibshirani, R., Walther, G., Hastie, T.: Estimating the number of clusters in a data set via the gap statistic. J. R. Stat. Soc. B **83**(2), 411–423 (2001)
19. Ward Jr., J.H.: Hierarchical grouping to optimize an objective function. J. Am. Stat. Assoc. **58**, 236–244 (1963)

The p-value Case, a Review of the Debate: Issues and Plausible Remedies

Francesco Pauli

Abstract We review the recent debate on the lack of reliability of scientific results and its connections to the statistical methodologies at the core of the discovery paradigm. Null hypotheses statistical testing, in particular, has often been related to, if not blamed for, the present situation. We argue that a loose relation exists: although NHST, if properly used, could not be seen as a cause, some common misuses may mask or even favour bad practices leading to the lack of reliability. We discuss various proposals which have been put forward to deal with these issues.

Keywords Null hypotheses statistical testing · p-value · Reproducibility

1 Introduction

A discussion on the role of the hypothesis statistical testing method in jeopardizing the reliability of scientific results is underway in the recent literature across many disciplines [18, 39]. It has been argued that a worrying portion of published scientific results, within various disciplines, are actually false discoveries [25]. This state of things has been related to the widespread use—or abuse—of p-values to measure evidence and corroborate new theories [7, 16, 34], to the point that a journal in psychology "banned" p-values [48] (although not in a very clear-cut way, for instance, they are allowed in submissions [49]). That of banning p-values altogether is not a novel idea nor it is exclusive of psychology [42]. According to a recent survey of 1576 researchers made by Nature [1], more than 90% have heard of a 'crisis of reproducibility'. Most of them think that the crisis is in fact in place and has not been overemphasized. Statistics is seen both as part of the problem and as a mean to improve the situation.

It is worth to note that the false discovery rate (FDR) across science (or a discipline) is not a clear-cut concept: a reference population of findings should be identified and a criterion of falsehood defined. In empirical evaluations a (non random)

F. Pauli (✉)
DEAMS, University of Trieste, Trieste, Italy
e-mail: francesco.pauli@deams.units.it

C. Perna et al. (eds.), *Studies in Theoretical and Applied Statistics*,
Springer Proceedings in Mathematics & Statistics 227,
https://doi.org/10.1007/978-3-319-73906-9_9

sample of results is usually considered and falsehood is often equated to lack of replication, which is a different, although related, concept. Notwithstanding how difficult or ambiguous it may be to precisely define the notion, however, the error rate of scientific results is a relevant concept of general interest, as the number of attempts which have been made to quantify it reveals.

Null hypotheses statistical testing (NHST) has a central role in the paradigm which is commonly employed to confirm new scientific theories (Sect. 2), and a long-running controversy on its use is in place. Whilst it may or may not be the culprit of the lack of reliability (Sect. 3), it is relevant to discuss whether alternatives to NHST may lead to a more reliable procedure to confirm scientific results (Sect. 4).

2 Scientific Discoveries and Statistical Testing

Null hypotheses statistical testing (NHST) is a standard topic in academic curricula of various disciplines and a standard tool to analyse data in many scientific fields.

Controversies concerning NHST started since the proposal of significance testing (and p-values) by Fisher and the alternative—and incompatible—procedure for hypotheses testing by Neyman and Pearson. Fisher proposed to measure the strength of evidence of a given observation against a hypothesis on the probabilistic mechanism which generated it with the probability, conditional on that hypothesis, of obtaining a sample at least as extreme as the observed one (p-value) [12]. Neyman and Pearson argued that "no test based upon a theory of probability can by itself provide any valuable evidence of the truth or falsehood of a hypothesis" and propose instead a procedure to choose between two alternative hypotheses on the data generating mechanism keeping under control the (conditional) probabilities of making the wrong choice [37].

NHST plays a central role in the procedure—or mindless ritual as some scholars provocatively called it [15]—which is used to corroborate scientific theories. The procedure goes as follows: a theory is posited according to which a relationship is in place between two quantities; in order to corroborate the theory a null hypothesis of absence of relationship is statistically tested using a sample; a confirmation is claimed whenever the null hypothesis is rejected at a specified level, which is usually 5%. Instances of the use of such a procedure abound across disciplines, for practical examples see [4] in medicine, [2] in psychology, [41] in economics, [20] in zoology. The exact p-value is generally taken as a measure of the evidence against the null hypothesis and possibly also as a measure of the evidence in favor of the alternative hypothesis. Also, acceptance of the null is often taken as evidence of the absence of the posited effect.

It is commonly maintained that, of the two procedures, the Fisherian p-value is the more apt to the described task, while the Neyman-Pearson procedure is more apt to problems which are more naturally cast in a decision framework. It is also to be noted, however, that the actual interpretation given to NHST in applications is sometimes a combination of the two. In fact the p-value is used to draw the conclusion but an alternative hypothesis is also considered (which also helps to clarify what an "extreme result" is in the definition of the p-value) and/or an error probability is attached to the p-value based conclusions [3]. In what follows we refer to NHST having this somewhat imprecise interpretation in mind.

This description does not encompass all uses of NHST in the scientific literature, but it represents the most problematic use and is widespread, and debated, across disciplines. In psychology the debate was already not new in 1994 [7, 28] (it dates back to 1955 according to [44]), and is lively as of today. On the other hand, the use of NHST is increasing, due to academic inertia according to some [44]. In medicine these issues are discussed since the rise of evidence based medicine [16] and still [19]. The practice is also widespread in economics/econometrics [35, 52], although some scholars disagree [23] on the extent of the problem.

While it is sometimes argued that NHST is not used in hard sciences like physics [35, 44], this is not the case: NHST has its place in high energy physics [40], in cosmology [9], in atmospheric sciences [38]. In these contexts, however, it is regarded as "only part of discovering a new phenomenon", the actual degree of belief depending on substantial considerations [8]. A p-value (or, more often, the Z-score, which is the $(1 - p)$-quantile of the standard normal distribution) is used as a measure of surprise, which suggests further investigation of the alternatives, in particular on whether they better explain observations. A peculiarity of some hard sciences is that different thresholds for rejection are customary: threshold values commonly used are $Z = 5$ ($p = 2.87 \times 10^{-7}$) and $Z = 1.64$ ($p = 0.05$). The first is used for 'discovery', that is when the alternative hypothesis includes a sought signal and the null is a 'background only' hypothesis; the second is used when the null is a signal.

3 NHST (p-Value), Good, Bad or Neutral?

The debate on the reliability of scientific results is intertwined with the debate on the suitability of the p-value as a measure of evidence. We argue that there is a relation between the use of p-values and the reliability crisis, albeit loose. In fact, some misuses of the p-value are susceptible to exacerbate some issues of the discovery paradigm outlined in Sect. 2.

The concerns which have been raised upon p-values can be categorized in three classes: one related to interpretation; one to the relationship with the size of the effect; the latter related to the role of the alternative.

Misinterpretations of the p-value take different forms, which in some cases are equivalent. The more trivial, yet common, misinterpretation is to relate the p-value to the probability of the null being true. This amounts at wishful thinking, since the probability of the null is what the researcher actually wants. It is barely worth mentioning that such an interpretation is logically wrong (as the p-value is a probability conditional on the null being true) and potentially strongly misleading, as a given p-value is compatible with any value for the probability of the null being true.

Another common misinterpretation (seen even in "serious use of statistics" [3]) is that the *p*-value is the probability of wrongly rejecting the null (or the probability of the result being due to chance [19]). That is, the (Fisherian) *p*-value, which is conditional to the sample, is mistaken for the (Neyman-Pearson) significance level, which is a long run error probability. The coexistence of the two approaches, whose logical incompatibility is often under-appreciated by non-statisticians users of statistics, is probably to be blamed for this [3].

A second class of issues arises from the fact that the *p*-value is a function of both the estimate of the effect size and the size of the sample; a low *p*-value, indicating a statistically significant effect, is not necessarily associated to a substantially significant effect, and vice versa. In spite of this, in many fields it is common practice to choose models—for example selecting the covariates in a regression analysis—based on the significance of coefficients, that is, only those coefficients which are significantly different from zero at a specified level are reported, implicitly assuming that the others are zero. Within econometrics this practice has been labeled "sign econometrics", interpreting the sign of a significant coefficient regardless of its size, and "star econometrics", ranking importance of variables according to their significance level ignoring their relative sizes [35, 52]. It is contended that the key question in scientific inquiry, establishing "How large is large" (to be substantively relevant), can not be answered by *p*-values [21, 30, 35, 52]. It has also been said that the *p*-value alone may be only a measure of how large is the sample, since in many settings the null hypothesis is a nil hypothesis (of absence of any effect) and this is (almost) surely false due to what Meehl [36] calls the *crud factor*—the fact that in many situations the effect is not precisely zero, but the actual scientific hypothesis of interest is the effect being so low to be irrelevant rather than it being exactly zero.

An obvious solution is to complement the information given by the *p*-value with the estimate of the effect: if the *p*-value leads to rejection, the estimate is reported. It has been noted, however, that coupling NHST and the size estimate is an issue in certain circumstances: if the true effect size is non zero but such that, given the sample size, the test has low power, then the estimate conditional on the *p*-value being lower than the significance threshold is upward biased [24].

A number of authors phrase their critics of the *p*-value saying that it does not convey a valid measure of the evidence against the null or in favor of the alternative. Different meanings may be attached to this, in many cases the same issues outlined above lie at the root of it. For instance, the already mentioned dependence on the sample size reflects a failure of conveying evidence against the null. The fact that the same *p*-value may correspond to very different probabilities of the null if a Bayesian analysis is performed on the same data is also seen as evidence that it does not convey all information [50]. Finally, from a formal point of view the fact that the *p*-value does not measure evidence in favor of the alternative is almost obvious since by using the *p*-value we do not consider an alternative, and the data may be unlikely given the null but even more unlikely given a specific alternative. It is to be noted that this plays well with the fact that the alternative hypothesis is usually phrased vaguely as merely the direction of the effect, if at all.

Rather than being genuine pitfalls of the p-value, the above are instances of misuse of it. In fact, in the search for the causes of the alleged low reliability of scientific results, it has been suggested that the p-value *per se* is not problematic, rather, it is the use which is made that is questionable, prompting the recent statement by the American Statistical Association on p-value use [51].

Roots of the reliability crisis may lie upstream the p-value. It has been pointed out that from an epistemological point of view, the procedure outlined in Sect. 2 may not be suitable to corroborate scientific theories [13, 36]. Despite this, its use is widespread, probably due to its simplicity, apparent objectivity and perceived compelling nature as a measure of evidence. In fact, the p-value alone is a compelling measure of evidence only if misinterpreted through wishful thinking ("it [NHST] does not tell us what we want to know, and we so much want to know what we want to know that, out of desperation, we nevertheless believe that it does!" [7]). The diffusion of its misinterpretations may also be seen as a hint that the simplicity of the p-value is only illusory.

The objectivity of the p-value may be a fallacy as well. In fact, the p-value may be seen as an objective measure of a theory (whatever it measures) if the theory and the test to be performed are specified in advance of collecting the data. If this is not the case, the value which is obtained as a result of the testing procedure may be driven not only objectively by the data but by the subjective judgment of the experimenter through the conscious (or even unconscious) processes of p-hacking [45] or the "garden of forking paths" (GOFP) [14]. The former, p-hacking, refers to the fact that each single researcher or team uses the same data to probe different theories, thus the final p-value is the minimum of a set of p-values obtained from a number of (related) tests. The latter, GOFP, refers to the fact that a single theory (even pre-specified) may be tested, but the details of the data analysis may be driven by the data through mechanisms such as the selection of the relevant variables or the inclusion/exclusion of observations, thus introducing a bias in the testing procedure. Evidence of such phenomena can be found by analyzing the frequency distribution of samples of p-values (p-curves, [46]): both phenomena are expected to lead to a relatively high frequency in a (left) neighborhood of the common 5% threshold. This feature has in fact been observed in different disciplines [5, 22, 32]. A further confirmation of the effect of the researcher degrees of freedom on the likelihood of finding significant results comes from a natural experiment in large NHLBI clinical trials, where it has been noted that, upon the introduction of pre-registration, the share of experiments leading to significant results dropped [26].

4 What Then?

Various changes of the procedure have been proposed to make it able to deal with scientific questions. We may broadly distinguish them based on whether the paradigm itself is left unchanged but the p-value is substituted by an alternative summary of compliance of data to theory (Sect. 4.2) or the paradigm is changed altogether (Sect. 4.1).

4.1 Changing the Paradigm: The Hard Way

Gelman [13] advocates a total change of methodology in which the focus is on estimation rather than testing. He argues that the correct interpretation of the p-value makes it (almost) irrelevant to the purpose for which it is used in science (it remains useful in indicating lack of fit of a model for the purpose of deciding how to improve it). Moreover, from a practical point of view, the use of the p-value may mask p-hacking or GOFP. Any conclusion concerning scientific discoveries should rather be derived from the implications of the estimated model. This approach does not, in general, offer a clear-cut (yes/no) answer and requires more expertise in data analysis than what is needed to use NHST.

A different approach is to remedy the limitations of the p-value by complementing it with some other measure related to the reliability of conclusions. The basic idea can be traced back to Meehl [36], who suggested that the strength of an experiment in corroborating a theory can be measured by the precision with which experimental results can be predicted by the theory. More recently, Mayo and Spanos [33] proposed the severity, which is defined as follows. Suppose that $\{t(Y) > t(y_c)\}$ is a Neyman-Pearson rejection region for $H_0 : \theta \leq \theta_0$. If y_0 is observed and H_0 is accepted, this is evidence against $\theta > \theta_1 (> \theta_0)$ and the strength of this evidence is measured by $P(t(Y) > t(y_0)|\theta = \theta_1)$. A similar notion is defined in case of rejection of the null. Loosely speaking, the evidence against a hypothesis is measured by the probability that the test statistic would have shown less agreement with the null had the hypothesis been false. The severity is related to the power (they are equal if the sample is at the boundary of the rejection region) but is a different concept (it is a function of the observed data, thus being a measure of power given the observed sample). It can be said that it "retains aspects of, and also differs from, both Fisherian and Neyman-Pearsonians accounts" [33], in particular it explicitly allows for the alternative hypothesis but also retains the post-data interpretation of the p-value.

A similar approach is used in physics, where a p-value is often complemented by the "median p-value" (the p-value one would get if the observed value of the test statistic is the median of the sample distribution in the alternative hypothesis) or the expected significance level, both measures of the p-value one would get under specific alternatives.

Focusing on avoiding bias phenomena such as p-hacking and GOFP, it has been proposed to apply the principles of blinded analysis [31]. This method was introduced in particle physics [27] and entails adding noise to data and/or masking labels so that the researcher who performs the data analysis can not anticipate the substantive conclusions of his inferences. The main difficulty is to hide enough information to avoid bias but still allowing a meaningful analysis.

Finally, Bayesian tools may be used [17]. Although a well developed technique, Bayesian analysis has never been widely adopted in applications, likely due to the fact that, with respect to NHST, it is less simple to use and not perceived as objective.

4.2 Changing the Paradigm: The Soft Way

The proposals reviewed in Sect. 4.1 imply a major change of paradigm and, most important, they do not share two of the main perceived advantages of NHST: ease of use and objectivity [50]. Although both "advantages" may be fallacious, their explicit absence may render the suggested alternatives less appealing to potential users and prevent their adoption. An alternative approach is to change the least of the paradigm, substituting the p-value with some other synthetic measure which does not share its pitfalls but keeps the (purported) advantages. We review below the main proposed substitutes.

Substituting NHST with confidence intervals [10] is (at least in standard situations) a change in the way in which the results are communicated rather than a change of method. However, it may still be a relevant change since it is plausible that confidence intervals be less prone to misinterpretations (and some empirical evidence confirming this is available [11]).

Scholars from different fields [6, 50] suggest using model selection criteria: the null and alternative hypotheses correspond to two different models, the null hypothesis is then "rejected" if the model corresponding to the alternative is preferred. This is an appealing strategy because of its simplicity of implementation and "objectivity". A number of options is available for the model comparison criterion: AIC and BIC are the ones which are more often put forward. Besides the link with the likelihood ratio, it should be remembered that AIC is related to cross validation, while BIC is the Bayes factor with suitable priors. We note that using BIC may be one way to introduce the Bayes factor as a substitute of NHST without paying the price of the complications of the Bayesian approach. Beside AIC and BIC, other similar criteria may be considered depending on the models under consideration (Mallows C_p, GCV, UBRE score), standard cross validation (leave-one-out, K-fold, fixed samples) may also be used. Also, using the likelihood alone has been suggested [43] (mainly on the grounds that it does not depend on the sample space (that is, on experimenter intention)).

A further model selection method which is suitable for the task is the lasso method, at least whenever the models can be framed in a (generalized) linear model specification and the null hypothesis is that a coefficient is equal to zero. In that case one may accept the null hypothesis if the lasso estimate of the coefficient is null, the penalization weight being chosen somehow, for instance by cross validation.

Finally, the minimal change which has been suggested is to lower the conventional threshold for significance. It has been noted that the 5% threshold was introduced when fewer hypotheses were being tested so it makes sense to change it today.

A lower threshold is usually employed in hard sciences, which appear less affected—although not immune—by the reliability crisis.

One advantage of the above procedures—which admittedly would probably be seen as a disadvantage by many—is that they offer an automatic choice. This may allow to compare their performances by means of a simulation study to assess, under various scenarios, the false discovery rate they would imply if used as a substitute for NHST/p-value.

5 Discussion

A number of issues have been raised in the literature concerning the use of NHST and the p-value since the introduction of such tools by Neyman-Pearson and Fisher. The debate on whether they are useful or harmful for assessing scientific hypotheses is particularly vivid today and coupled with the debate on the lack of reproducibility and high false discovery rate of scientific results in many disciplines.

In fact, the misuse and misinterpretation of NHST are the reasons why it is often singled out as a major weakness. On the contrary, it can be argued that there are relevant possible reasons for the high FDR/lack of reproducibility which lie upstream the use of NHST.

First, there is a big leap in inferring from the falsification of a null nil hypothesis a confirmation of a specific alternative, particularly when the alternative does not imply a precise prediction of what would have been observed had it been true (i.e., the alternative predicts a positive effect rather than an effect of a given size) [36].

Second, a high number of scientific hypotheses is probed. Each single researcher or team tends to use the same data to probe different theories, thus leading to a multiple testing situation which may be explicit or, more subtly, due to the degrees of freedom in specifying the data processing step and the model. This may be phrased saying that exploratory studies are then treated as confirmatory ones (where by the latter we mean experiments with pre-specified hypotheses and methods) thus creating unrealistic expectations on the reliability of the result (on the probability of it being a false discovery). Moreover, this also happens "science-wide" meaning that, at least in some disciplines, lots of labs and researchers means a high number of hypotheses being tested leading to an uncontrollable multiple testing situation associated to a search for small effects (having the "main ones", the low hanging fruits, already been found) [47].

Based on the above considerations, it is reasonable to think that the "soft" changes to the present paradigm, where basically the p-value is substituted by some other measure of concordance/discordance between theory and data would hardly be a solution [29]. Also, it is probably unrealistic to try to devise a synthetic measure of evidence for or against a scientific theory. A "hard" change of paradigm is more promising, however no generally accepted alternative has been identified as of today. Moreover, it is to be noted that most, if not all, promising changes do not give a clear-cut answer to the posited question (of whether a given theory is true), a circumstance which is likely to make it hard for them to become generally accepted.

Acknowledgements This work was supported by Univesity of Trieste within the FRA project "Politiche strutturali e riforme. Analisi degli indicatori e valutazione degli effetti".

References

1. Baker, M.: Is there a reproducibility crisis? Nature **533**, 452–454 (2016)
2. Beall, A.T., Tracy, J.L.: Women are more likely to wear red or pink at peak fertility. Psychol. Sci. **24**, 1837–1841 (2013)
3. Berger, J.O.: Could Fisher, Jeffreys and Neyman have agreed on testing? Stat. Sci. **18**(1), 1–12 (2003)
4. Boland, M.R., Shahn, Z., Madigan, D., Hripcsak, G., Tatonetti, N.P.: Birth month affects lifetime disease risk: a phenome-wide method. J. Am. Med. Inform. Assoc. ocv046 (2015)
5. Brodeur, A., Lé, M., Sangnier, M., Zylberberg, Y.: Star wars: the empirics strike back. Am. Econ. J. Appl. Econ. **8**(1), 1–32 (2016)
6. Burnham, K., Anderson, D.: P values are only an index to evidence: 20th-vs. 21st-century statistical science. Ecology **95**(3), 627–630 (2014)
7. Cohen, J.: The earth is round ($p < 0.05$). Am. Psychol. **49**, 997–1003 (1994)
8. Cowan, G., Cranmer, K., Gross, E., Vitells, O.: Asymptotic formulae for likelihood-based tests of new physics. Eur. Phys. J. C **71**(2), 1–19 (2011)
9. Cowen, R.: Big bang finding challenged. Nature **510**(7503), 20 (2014)
10. Cumming, G.: The new statistics why and how. Psychol. Sci. **25**, 7–29 (2013)
11. Fidler, F., Loftus, G.R.: Why figures with error bars should replace p values: some conceptual arguments and empirical demonstrations. J. Psychol. **217**(1), 27–37 (2009)
12. Fisher, R.A., et al.: Statistical methods for research workers. In: Statistical Methods for Research Workers, 10th. edn. (1946)
13. Gelman, A.: Commentary: P values and statistical practice. Epidemiology **24**(1), 69–72 (2013)
14. Gelman, A., Loken, E.: The statistical crisis in science. Am. Sci. **102**, 460–465 (2014)
15. Gigerenzer, G.: Mindless statistics. J. Socio-Econ. **33**(5), 587–606 (2004)
16. Goodman, S.N.: Toward evidence-based medical statistics. 1: the p value fallacy. Ann. Intern. Med. **130**(12), 995–1004 (1999)
17. Goodman, S.N.: Toward evidence-based medical statistics. 2: the bayes factor. Ann. Intern. Med. **130**(12), 1005–1013 (1999)
18. Goodman, S.N.: Aligning statistical and scientific reasoning. Science **352**, 1180–1181 (2016)
19. Greenland, S., Poole, C.: Living with p values: resurrecting a bayesian perspective on frequentist statistics. Epidemiology **24**(1), 62–68 (2013)
20. Hart, et al.: Dogs are sensitive to small variations of the Earth's magnetic field. Front. Zool. **10**, 80 (2013)
21. Hauer, E.: The harm done by tests of significance. Accident Analysis & Prevention **36**(3), 495–500 (2004)
22. Head, M.L., Holman, L., Lanfear, R., Kahn, A.T., Jennions, M.D.: The extent and consequences of p-hacking in science. PLoS Biol. **13**(3), e1002,106 (2015)
23. Hoover, K.D., Siegler, M.V.: Sound and fury: Mccloskey and significance testing in economics. J. Econ. Method. **15**(1), 1–37 (2008)
24. Ioannidis, J.P.: Contradicted and initially stronger effects in highly cited clinical research. Jama **294**(2), 218–228 (2005)
25. Ioannidis, J.P.: Why most published research findings are false. PLoS Med. **2**(8), e124 (2005)
26. Kaplan, R.M., Irvin, V.L.: Likelihood of null effects of large nhlbi clinical trials has increased over time. PloS one **10**(8), e0132,382 (2015)

27. Klein, J.R., Roodman, A.: Blind analysis in nuclear and particle physics. Ann. Rev. Nucl. Part. Sci. **55**(1), 141–163 (2005)
28. Krantz, D.H.: The null hypothesis testing controversy in psychology. J. Am. Stat. Assoc. **94**(448), 1372–1381 (1999)
29. Leek, J.T., Peng, R.D.: Statistics: P-values are just the tip of the iceberg. Nature **520**(7549) (2015)
30. Lovell, D.: Biological importance and statistical significance. J. Agric. Food Chem. **61**(35), 8340–8348 (2013)
31. MacCoun, R., Perlmutter, S.: Blind analysis: hide results to seek the truth. Nature **526**(7572), 187–189 (2015)
32. Masicampo, E.J., Lalande, D.R.: A peculiar prevalence of p-values just below.05. Q. J. Exp. Psychol. **65**(11), 2271–2279 (2012)
33. Mayo, D.G., Spanos, A.: Severe testing as a basic concept in a neymanpearson philosophy of induction. Br. J. Philos. Sci. **57**(2), 323–357 (2006)
34. McCloskey, D.: The insignificance of statistical significance. Sci. Am. **272**, 32–33 (1995)
35. McCloskey, D.N., Ziliak, S.T.: The standard error of regressions. J. Econ. Lit. **34**(1), 97–114 (1996)
36. Meehl, P.: The problem is epistemology, not statistics: replace significance tests by confidence intervals and quantify accuracy of risky numerical predictions. In: What if there were no significance tests, pp. 393–425. Psychology press (2013)
37. Neyman, J., Pearson, E.S.: On the problem of the most efficient tests of statistical hypotheses. Philos. Trans. R. Soc. Lon. Ser. A **231**, 289–337 (1933)
38. Nicholls, N.: Commentary and analysis: the insignificance of significance testing. Bull. Am. Meteorol. Soc. **82**(5), 981–986 (2001)
39. Nuzzo, R.: Scientific method: statistical errors. Nature **506**(7487), 150–152 (2014)
40. Reich, E.S.: Timing glitches dog neutrino claim. Nature **483**(7387), 17 (2012)
41. Rogoff, K., Reinhart, C.: Growth in a time of debt. Am. Econ. Rev. **100**, 573–578 (2010)
42. Rothman, K.J.: Writing for epidemiology. Epidemiology **9**(3), 333–337 (1998)
43. Royall, R.: Statistical Evidence: A Likelihood Paradigm (Chapman & Hall/CRC Monographs on Statistics & Applied Probability). Chapman and Hall/CRC (1997)
44. Schmidt, F., Hunter, J.: Eight common but false objections to the discontinuation of significance testing in the analysis of research data. In: S.A.S.J. Harlow L.L. (ed.) What if There were no Significance Tests?, pp. 37–64. Psychology Press (1997)
45. Simmons, J.P., Nelson, L.D., Simonsohn, U.: False-Positive psychology-undisclosed flexibility in data collection and analysis allows presenting anything as significant. Psychol. Sci. **22**(11), 1359–1366 (2011)
46. Simonsohn, U., Nelson, L.D., Simmons, J.P.: P-curve: a key to the file-drawer. J. Exp. Psychol. Gen. **143**(2), 534–547 (2014)
47. Sterne, J.A.C., Smith, G.D., Cox, D.R.: Sifting the evidence-what's wrong with significance tests? Phys. Ther. **81**(8), 1464–1469 (2001)
48. Trafimow, D.: Editorial. Basic Appl. Soc. Psychol. **36**(1), 1–2 (2014)
49. Trafimow, D., Marks, M.: Editorial. Basic Appl. Soc. Psychol. **37**(1), 1–2 (2015)
50. Wagenmakers, E.J.J.: A practical solution to the pervasive problems of p values. Psychon. Bull. Rev. **14**(5), 779–804 (2007)
51. Wasserstein, R.L., Lazar, N.A.: The ASA's statement on p-values: context, process, and purpose. Am. Stat. **70**(2), 129–133 (2016)
52. Ziliak, S., McCloskey, D.: Size matters: the standard error of regressions in the american economic review. J. Socio-Econ. **33**(5), 527–546 (2004)

Part III
Statistical Models and Methods for Network Data, Ordinal and Symbolic Data

A Dynamic Discrete-Choice Model for Movement Flows

Johan Koskinen, Tim Müller and Thomas Grund

Abstract We consider data where we have individuals affiliated with at most one organisational unit and where the interest is in modelling changes to these affiliations over time. This could be the case of people working for organisations or people living in neighbourhoods. We draw on dynamic models for social networks to propose an actor-oriented model for how these affiliations change over time. These models specifically take into account constraints of the system and allow for the system to be observed at discrete time-points. Constraints stem from the fact that for example not everybody can have the same job or live in the same neighbourhood, something which induces dependencies among the decisions marginally. The model encompasses two modelling components: a model for determining the termination of an affiliation; and a discrete-choice model for determining the new affiliation. For estimation we employ a Bayesian data-augmentation algorithm, that augments the observed states with unobserved sequences of transitions. We apply the proposed methods to a dataset of house-moves in Stockholm and illustrate how we may infer the mechanisms that sustain and perpetuate segregation on the housing market.

Keywords Social networks · Stochastic actor-oriented models · Allocation models · Matching models

J. Koskinen (✉)
Social Statistics Discipline Area, University of Manchester
Manchester, Manchester M13 9PL, England
e-mail: johan.koskinen@manchester.ac.uk

J. Koskinen
Institute of Analytical Sociology, University of Linköping,
Norra Grytsgatan 10, 60174 Norrköping, Sweden

T. Müller
Berlin Institute for Integration and Migration Research, Humboldt-Universität zu Berlin, Unter den Linden 6, 10099 Berlin, Germany
e-mail: t.mueller@sowi.hu-berlin.de

T. Grund
School of Sociology, Newman Building, Belfield, Dublin 4, Ireland
e-mail: thomas.grund@ucd.ie

C. Perna et al. (eds.), *Studies in Theoretical and Applied Statistics*,
Springer Proceedings in Mathematics & Statistics 227,
https://doi.org/10.1007/978-3-319-73906-9_10

1 Introduction

In the economic, social, and behavioural sciences there is a long tradition of analysing why people end up in certain professions, particular schools, and different neighbourhoods. Standard statistical approaches such as discrete-choice models allow for investigating how different characteristics increase a person's likelihood to end up in one destination rather than another or how characteristics of the final destinations explain who ends up there. It has however been recognised that there are systemic properties of these phenomena that constrain and structure choices and destinations. For example, whereas everybody might want to go to the best school, one school cannot accommodate everyone [1]. When it comes to, say, occupations, typically (excluding individuals with for example multiple part-time jobs) people only have one occupation at any one time and the likelihood that an individual ends up in one job might be a function of their own preference as well as the composition of employees already in that job [3].

To accommodate the systemic dependencies and restrictions associated with an allocation system we formulate a constrained version of the stochastic actor-oriented modelling (SAOM) family for two-mode networks proposed by Koskinen and Edling [10]. The SAOM, originally developed for modelling the evolution of one-mode networks by Snijders [17], models the dynamic network as a discrete Markov chain in continuous time, where changes to the network can be seen as incremental updates where the network changes one tie at a time. The process thus defined is a random walk on a d-cube [2] for d number of tie-variables. A particularly attractive aspect of Snijders' [17] modelling framework is that each incremental change to the network can be modelled as an independent discrete-choice model, conditional on the current state of the network (Snijders [16], additionally provided a model formulation that enjoys similar properties but that relaxes the assumption of choice). Furthermore, even if we only observe the network at discrete points in time, the SAOM provides us with a model for imputing or augmenting our observed data with any changes that might have occurred in-between observations. The extension to two-mode network data or, indeed, any system that affords dissagregating complex dependencies into a series of cumulative changes modelled as discrete-choice models is mostly a choice of appropriate restrictions and determining the decision process.

In the following we proceed by defining the data-structure that motivates our model. We then defined the model in detail and outline the inference scheme. Finally we provide an illustrative example of applying the model to a data set that is a subset of a rich and very complex register data set for inhabitants in Sweden.

2 Data Structure

We will distinguish between two types of social entities, individuals and their locations. Individuals may be alternatively people, organisations, or some other social entity that we are willing to endow with some form of decision-power. The locations will represent collections of individuals, such as organisations, sectors, etc. For the sake of our empirical setting we will refer to these two entities as *individuals* and *neighbourhoods*. Let $\mathcal{I} = \{1, \ldots, n\}$ be a set of individuals and $\mathcal{J} = \{1, \ldots, m\}$ a set of neighbourhoods. A binary variable $Y_{ij}(t)$ indicates whether $i \in \mathcal{I}$ lives in neighbourhood $j \in \mathcal{J}$ at time $t \in \mathbb{R}^+$.

We denote by $v_i(t)$ a $k \times 1$ vector of covariates of i and $w_j(t)$ a vector of covariates for j. In addition we may have dyadic covariates for the individuals, e.g. a_{iu} being one or zero according to whether there is a kinship tie between i and u.

We assume that the matrix $Y = (Y_{ij})_{(i,j) \in \mathcal{I} \times \mathcal{J}}$ is row-regular such that $Y_{i+} = \sum_j Y_{ij} = 1$ for all individuals. This eliminates homeless individuals and individuals registering multiple living addresses, as well as individuals moving to areas that are not in $\mathcal{J}$. For the purposes of defining the model, we need not necessarily need to distinguish between individuals being people or households. However, for the purpose of analysis, individuals may conveniently be collapsed into households. This prevents direct analysis of transition into cohabitation, children transitioning from dependants to independants, etc.

Data could be time-stamped in the sense that we observe the exact time at which every change occurs. In most cases we would however expect that there is some form of discretisation of observation times, for example that only the month or year of a change is known. For this reason we make the tacit assumption that data are observed at discrete points in time $t_0, t_1, \ldots, t_{M+1}$ but in order to account for the fact that changes do not happen simultaneously, we assume that there are unobserved time-points $s_{1,u}, \ldots, s_{K_u,u}$ in-between t_{u-1} and t_u at which changes are made to Y. For the purposes of inference we will assume that $Y_{i\cdot}$ may change at most once in-between t_{u-1} and t_u for any $u = 1, \ldots, M+1$. We use the notational conventions of letting $j(Y_{i\bullet}(t)) = \{j \in \mathcal{J} : Y_{ij}(t) = 1\}$ and $t_i = t_i(\{Y(t)\}_{t \in [s,T]}) = \inf\{t \in [s,T] : Y_{i\bullet}(t) \neq Y_{i\bullet}(s)\}$.

We may construe the process of changes as a process of evolution of a bipartite network [10] subject to the constraints of row-regularity and parsimonious paths.

3 Model Formulation

We follow the schema of Snijders [16, 17] in defining the process in terms of waiting times and conditionally on this a discrete-choice model. When an individual i chooses to move at, say, time t, we assume that i has complete information about the entire state of the system and that i may choose to move to any of the neighbourhoods in $\mathcal{J}$. We assume that i associates with $j \in \mathcal{J}$ a utility $U_i(j; Y(t), t, \theta) = f_i(j; Y(t), t, \theta) + \varepsilon(i,j,t)$, where f_i is an objective function that we aim to model and $\varepsilon(i,j,t)$ are random components that we will assume are i.i.d. extreme value type one distributed for all i, j, and t. We assume that the individuals maximise this function which implies [12] that the conditional probability when a change is made conditional on the current state (at $t = t_i - \delta$) is given by

$$\Pr(Y_{ij}(t_i) = 1 | Y_{i\cdot}(t)) = \frac{e^{f_i(j;Y(t),t,\theta)}}{\sum_{k \in \Omega(\mathscr{J};Y(t))} e^{f_i(k;Y(t),t,\theta)}},$$

where $\Omega(\mathscr{J}; Y(t))$ denotes the option set. We shall limit the formulation of the systematic part of the utility function to being a weighted sum of statistics. The latter shall be functions of Y, v, w, and a.

While the conditional discrete choice model is the main focus, for purposes of specifying a generative model we also need to specify a process for picking who is moving. In the most general form, we assume that the rate at which an individual i moves out of their current neighbourhood is $\lambda_i(Y(t), t, \theta)$, and that the holding-times are conditionally independent conditional on the current state $Y(t)$ and observable covariates. We may construct indicators $A_i(t)$ for each actor, indicating whether at time t, actor i has moved ($A_i(t) = 1$) or not $A_i(t) = 0$. Once $A_i(t) = 1$, the actor is no longer permitted to move. This is to fit the credible constraint that people move at most once in each interval. From standard properties of exponential distributions (see also the definition of rates in SAOMs for social networks [17]), the probability at time t that actor $i \in \mathscr{A}_t = \{i \in \mathscr{I} : A_i(t) = 0\}$ is the first person to move is

$$\frac{\lambda_i(Y(t), t, \theta)}{\sum_{i \in \mathscr{A}_t} \lambda_i(Y(t), t, \theta)}.$$

We make the explicit assumption that the rates of moving out are not confounding the discrete choice model for the destination. Thus we tacitly disallow individuals deciding to move because they have decided where to move to.

For computational purposes, we further make the simplifying assumption that $\lambda_i(Y(t), t, \theta) = \lambda_i(\theta)$ and that these rates may only depend on individual-level covariates and time-constant characteristics of the neighbourhoods. This simplification means that independently for each i, $\Pr(A_i(t) = 1) = \Pr(T_i \leq t) = 1 - e^{-t\lambda_i(\theta)}$ and $\Pr(A_i(t) = 0) = \Pr(T_i > t) = e^{-t\lambda_i(\theta)}$. Assuming that the system is not 'full', i.e. that there is always somewhere to move to, this means that the parameters of the moving out process may be estimated independently of the destination model. In particular, with $\lambda_i = e^{\eta_i}$, for some linear function η_i of fixed covariates, this model formulation implies that $\text{cloglog}(\Pr(A_i(t) = 1)) = \eta_i$, where cloglog$(\cdot)$ is the complimentary log-log link function. Note that this simplification does not imply that the time-ordering does not matter. Individuals with higher rates are likely to move earlier and the order of moves consequently has to be weighted in the course of estimating the destination process.

Should $\lambda_i(\cdot)$ be allowed to depend on time-varying characteristics, then the rates would have to be updated every time anyone moves from a neighbourhood. However, while neighbourhood characteristics can be accounted for by including a 'lagged' measure, e.g. the ethnic composition at the start of a period, this assumption prevents us from investigating phenomena such as 'white flight' [4] in the same detail as the destination processes (e.g. 'white avoidance'). Note that the simplifying assumptions for the decision to move will not systematically bias our analysis of the destination choice as long as the assumption that the processes are not confounded is true.

3.1 Example Statistics

We will focus here on defining elements of the discrete-choice model for the destinations of moves. We can break the effects down into properties of the actors, origins, potential destinations, as well as combinations of these. As the origin of i at the time t_i of a change $j^* = j(Y_{i\bullet}(s))$, $s < t$, is fixed in evaluating the decisions $\Omega(\mathcal{J}; Y(t))$, effects in f_i cannot depend on j^* only. Similarly, effects in f_i cannot depend only on properties of i. However, all interactions between properties of j^* and i on the one hand and properties of $j \in \Omega(\mathcal{J}; Y(t))$ are admissible. Some effects relevant to our empirical example are the following.

Destination Capacity and Popularity A natural restriction on movements is availability of housing—no matter how desirable a neighbourhood, if there is no housing no move is possible. We call this the 'capacity' c_j of j at time t. We may take the capacity into account through the objective function f_i. The sequential nature of the model means that moves will free up accommodation in the sense that the move will increase the utility for others and thereby create vacancy chains [20]. A potential effect in the opposite direction is that popular neighbourhoods may become more popular (a Matthew effect, see [14, 19]). We may combine these effects in a statistic f_i that makes sure that the sizes of the neighbourhoods do not fluctuate wildly over time and also prevents heaping in popular neighbourhoods.

It is clear from data over time, however, that the number of residents remain relatively fixed. This may be reflected by a stricter use of capacity, namely that $\Omega(\mathcal{J}; Y(t)) = \{j \in \mathcal{J} : Y_{+j} < c_j\}$. The utility for j given that $j \in \Omega(\mathcal{J}; Y(t))$ should reflect the increasing utility of availability as reflecting market prices as well as a decreasing utility of availability as reflecting decreased popularity (the Yogi Berra effect, [7]). In addition to the restrictions on Ω, we model the effect of occupancy as a decaying utility of available housing

$$S_{ij}(t) = \exp\{-\alpha \min(c_j - y_{+j}(t), c)\}, \tag{1}$$

where c is some truncation point set for computational convenience.

Prices of Housing and Household Assets To reflect differences in the prices of housing, we may include as a statistic $S_{ij}(t) = w_j(t)$, where $w_j(t)$ is the average log price or another appropriate measure. Instead, or in addition, we may here have a compositional variable $S_{ij,\text{comp}}(t) = \sum_i y_{kj}(t) v_k(t) / y_{+j}$ for some measure of household assets v_k.

Assuming that housing costs w_j and household assets v_k are measured on comparable scales, household-area pricing discrepancy can me modelled using the effect

$$S_{ij,\mathrm{diff}}(t) = (v_I(t) - w_j(t))^2.$$

Instead of w_j we may use average income $S_{ij,\mathrm{comp}}(t)$, which comprises two aspects, the compositional aspect and a pragmatic proxy for cost of housing. Additionally, we may consider income differences between neighbourhoods $(S_{ij^*,\mathrm{comp}}(t) - S_{ij,\mathrm{comp}}(t))^2$.

Ethnic Mixing To model ethnic mixing, let v_i indicate if i belongs to a minority ethnic group or not (the extension to multiple categories is straightforward), and we can model mixing through:

$$S_{ij}(t) = 1 - \left| v_i(t) - \frac{\sum_k y_{kj}(t)v_k(t)}{y_{+j}(t)} \right|. \tag{2}$$

A positive parameter means that individuals tend to associate higher utility with areas with high proportion of inhabitants of the same ethnic group.

Spatial Embedding A plausible factor in moves is the spatial embedding of neighbourhoods. On the one hand, neighbourhoods that are more central may be more attractive, on the other hand there are considerable transaction costs associated with moves over large distances. We focus here only on the latter factor and define a statistic

$$S_{ij,5}(t) = \log ||w_j - w_{j^*}||^2,$$

where $j^* \in \mathcal{J}$ is the current area of i. For longitudinal networks it is common to use distances on a log-scale [9, 15], something which may be motivated through a relation to the attenuated power-law form of spatial decay (see further e.g. [5]).

Kinship and Social Networks The literature suggests that information is not symmetrically disseminated and that considerable advantages may be had through extra-market structures [6]. While networks of acquaintances, work colleagues etc., are difficult to gather data on, we may have register data on kinship networks. A simplistic approach for modelling a preference for moving to neighbourhoods where you have relatives is to include the effect $S_{ij,6}(t) = \sum_k y_{kj}(t)a_{ik}$ that is large for areas where i has others k that are kin.

Dynamic Endogeneity The incremental nature of the model captures two aspects of endogeneity and emergence. Firstly, the move of i from j^* to j implies that any compositional variables for areas are updated by removing the contribution of i to j^* and giving j the additional contribution of i. A compositional change in the ethnic composition of a neighbourhood thus comes about by successive updates to the neighbourhood ethnic composition, for example ethnic minorities moving into a neighbourhood, incrementally increasing the proportion of ethnic minorities.

Second, the conditional nature of the model means that we may include path dependencies. The simplest form of path dependency is the first-order, whereby we model the probability that an individual moves into a neighbourhood j given that the individual currently lives in neighbourhood j^*.

Similar to homophily and influence effects in stochastic actor-oriented models for social network evolution [17], there may be strong effects of segregation without there being any evidence in the changes of the marginal distributions of ethnic composition. If there is no effect of segregation, we expect that the ethnic compositions of neighbourhoods will tend to be evenly distributed across networks, everything else equal. If, however, there is already segregation initially (as reflected in unequal distribution of ethnic proportions), a strong effect for segregation may be needed to *maintain* the distribution of ethnic composition across neighbourhoods (for longitudinal exponential random graph models this has been addressed through joint modelling of initial conditions and dynamics; [9]).

4 Estimation Issues

Given an observed sequence of moves by individuals with known attributes, the likelihood is tractable and inference is straightforward. In the case that data are only observed, say, once every year, the sequence of moves is not observed. Had utilities been independent of past moves and constant over time, the exact sequence might have been of minor importance. Here, the order of moves is potentially important and while it is unobserved, the composition of areas and individual attributes may bias transitions in an informative way. Of particular interest is the case of moves are the restrictions imposed by capacity of areas [20].

Given that the process is equivalent to that of Koskinen and Edling [10] with added constraints, a straightforward modification of the Bayesian inference procedure proposed by Koskinen and Snijders [11] applies. In particular, given that actors may only move once in any given time-interval, this means that the Bayesian data-augmentation scheme can be considerably simplified. To propose updates to the unobserved sequence we only need to shuffle the order of moves. Here, we pick two changes and propose swapping their order provided this does not lead to a violation of the maximal capacity of any neighbourhood.

5 Empirical Illustration

Figure 1 illustrates the structure of the movements on the Stockholm housing market. The 128 nodes represent the neighbourhoods defined by the small area market statistics of Statistics Sweden.

We consider here a simplified model where we only use transitions in the period 1990–1991. In total there were 572,389 actors (households) in the Stockholm area in this period and a total of 41,578 actors were registered at different addresses in 1990 and 1991. The main focus of interest is to explain ethnic segregation on the housing market as measured by the different proportions of actors that were either second or first generation immigrants in neighbourhoods. The overall proportion of

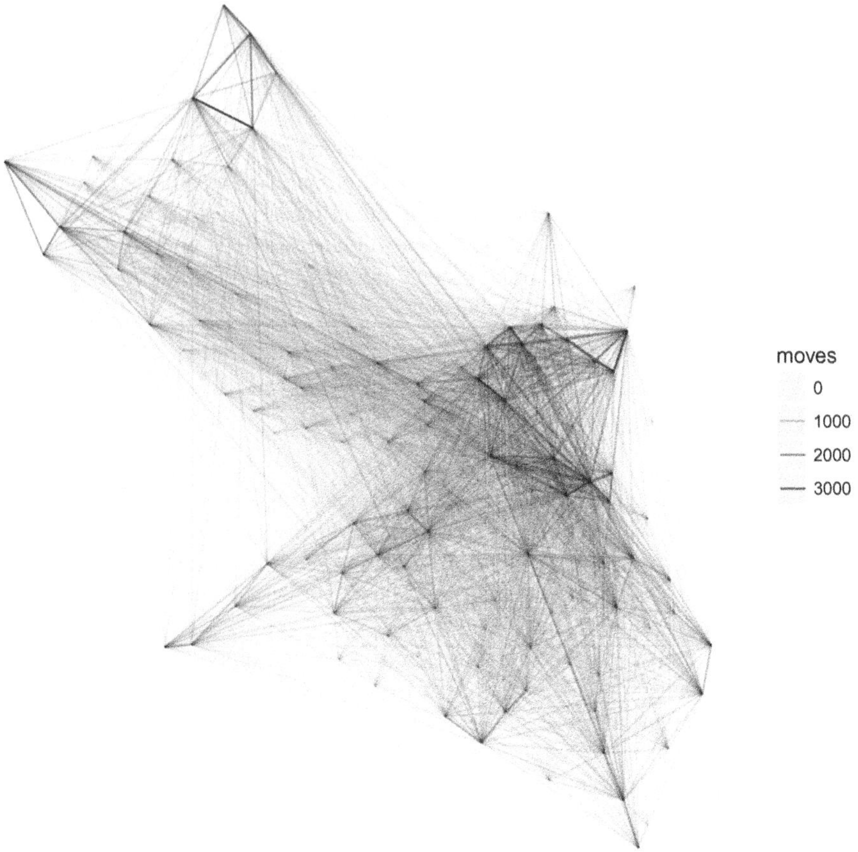

Fig. 1 Movement flows among 128 Stockholm neighbourhoods (based on small area market statistics) aggregated across 1990–2003 (directionality has been suppressed for visibility). In the period 1990–1991 there were 572,389 households across the 128 areas and 41,578 moves. The thickness and saturation of a line is proportional to the number of moves between respective areas (code from *spatial.ly* gratefully acknowledged)

thus defined immigrants in the Stockholm area was 22.4%. We model segregation by including the two effects 'mover-dest immig sim' and 'orig-dest immig diff'. The former is defined as Eq. 2 and the latter as $|\sum_i y_{ij*} v_i(t)/y_{ij} - \sum_i y_{+j} v_i(t)/y_{+j}|$, which measures the dissimilarity in proportion of immigrants in the origin and potential destination neighbourhood of i. To control for income distribution, two effects 'orig-dest income diff', $(S_{ij*,\mathrm{comp}}(t) - S_{ij,\mathrm{comp}}(t))^2$, and 'mover-dest income diff', $(v_j(t) - S_{ij,\mathrm{comp}}(t))^2$, were constructed.

To capture the potential of vacancy chains the capacity c_j of each neighbourhood is set to 100.05% of the total number of inhabitants in 1991. Thus a neighbourhood that increases its population by 0.05% early on in the period is effectively

Table 1 Posterior summaries for SAOM fitted to the movement flows on the Stockholm Housing market between 1990 and 2003

Effect	Parameter mean	Std
capacity	−2.945	0.027
dest income	−5.409	0.508
orig-dest income diff	−2.396	5.089
mover-dest income diff	−10.890	0.744
orig-dest immig diff	−3.500	0.145
mover-dest immig sim	1.292	0.048
orig-dest distance	−128.725	1.423

removed from the choice-set of actors until someone moves out of this neighbourhood. The scaling parameters in Eq. 1 are set to $\alpha = 0.25$ and $c = 19$.

For this example we have not estimated the moving-out process nor have we taken into account actors entering or leaving the Stockholm market. Estimation results from a Markov chain Monte Carlo algorithm based on a modified version of Koskinen and Snijders [11] using 5,000 iterations are provided in Table 1.

The negative capacity effect could be interpreted as neighbourhoods being more popular the more popular they are, or equivalently, the further they are from capacity, the less attractive they are. However, the proxy for capacity is blunt and its interpretation in terms of popularity could be argued to be somewhat circular.

There is no discernible effect of neighbourhood homophily on income level but a clear preference for actors moving to neighbourhoods with a similar average income level to their own. The negative coefficient for distance indicate that actors prefer to move only short distances. Over and above these other effects, there are two clear segregation effects: actors move between neighbourhoods with similar proportions of immigrants (coef: −3.5); and immigrant actors prefer to move to neighbourhoods that have high proportion of immigrants (coef: 1.29) (and conversely, non-immigrants prefer low proportion of immigrants).

The relative effect of vacancy chains is indicated by the proportion of proposed permutations in the course of estimation that would have resulted in a neighbourhood exceeding its capacity. Here, in 30% of proposals for draws from the full conditional posterior of the unobserved sequence of moves, given the rest, an actor was attempting to move to a neighbourhood that had no vacancy.

6 Conclusions and Future Directions

We presented a generative model for modelling allocation of individuals to neighbourhoods while taking systemic dependencies into account. For the example of the Stockholm housing market we found that there is clear evidence of households sorting themselves according to ethnic background, sustaining and reproducing ethnic segregation on the housing market. As one of the key purposes is to generate predictions under various scenarios we need to specify the moving-out process in order to fully leverage the generative capacity of the model. Furthermore, applying the model to data across 13 consecutive years, gives us an opportunity to investigate how the model changes over time. This could be done by drawing on work on change-point analysis for sequential models [13] or by specifying a hierarchical model as in Koskinen et al. [9] and model a smooth change to the parameters as a function of time.

There are obvious connections to demographic micro-simulation techniques [8] but, as pointed out by Snijders and Steglich [18], the inferential framework of the actor-oriented model allows a simulation model that is informed by real data. As the modelling of housing moves demonstrates, the principles of Snijders [17] stochastic actor-oriented model provides a general framework that applies much more widely than to just social networks.

Acknowledgements The research leading to these results has received funding from the European Research Council under the European Union's Seventh Framework Programme (FP7/2007-2013)/ERC grant agreement no 324233, Riksbankens Jubileumsfond (DNR M12-0301:1), and the Swedish Research Council (DNR 445-2013-7681) and (DNR 340-2013-5460).

References

1. Abdulkadiroğlu, A., Sönmez, T.: School choice: a mechanism design approach. Am. Econ. Rev. **93**, 729–747 (2003)
2. Aldous, D.: Minimization algorithms and random walk on the d-cube. Ann. Probab. **11**, 403–413 (1983)
3. Butts, C.T.: Models for generalised location systems. Sociol. Methodol. **37**, 283–348 (2007)
4. Crowder, K., South, S.J.: Spatial dynamics of white flight: the effects of local and extralocal racial conditions on neighborhood out-migration. Am. Sociol. Rev. **73**, 792–812 (2008)
5. Daraganova, G., Pattison, P., Koskinen, J., Mitchell, B., Bill, A., Watts, M., Baum, S.: Networks and geography: modelling community network structures as the outcome of both spatial and network processes. Soc. Netw. **34**, 6–17 (2012)
6. Granovetter, M.S.: The strength of weak ties? Am. J. Sociol. **78**, 1360–1380 (1973)
7. Hedström, P.: Rational imitation. In: Hedström, P., Swedberg, R. (eds.) Social Mechanisms: An Analytical Approach to Social Theory, pp. 306–327. Cambridge University Press, Cambridge (1998)
8. Li, J., O'Donoghue, C.: A survey of dynamic microsimulation models: uses, model structure and methodology. Int. J. Microsimul. **6**, 3–55 (2013)
9. Koskinen, J., Caimo, A., Lomi, A.: Simultaneous modeling of initial conditions and time heterogeneity in dynamic networks: an application to Foreign Direct Investments. Netw. Sci. **3**, 58–77 (2015)
10. Koskinen, J., Edling, C.: Modelling the evolution of a bipartite network—peer referral in interlocking directorates. Soc. Netw. **34**, 309–322 (2012)
11. Koskinen, J.H., Snijders, T.A.B.: Bayesian inference for dynamic social network data. J. Stat. Plan. Infer. **137**, 3930–3938 (2007)
12. Maddala, G.S.: Limited-Dependent and Qualitative Variables in Econometrics. Cambridge University Press, Cambridge (1983)

13. McCormick, T.H., Raftery, A.E., Madigan, D., Burd, R.S.: Dynamic logistic regression and dynamic model averaging for binary classification. Biometrics **68**, 23–30 (2012)
14. Merton, R.K.: The Matthew effect in science. Science **159**, 56–63 (1968)
15. Preciado, P., Snijders, T.A.B., Burk, W.J., Stattin, H., Kerr, M.: Does proximity matter? Distance dependence of adolescent friendships. Soc. Netw. **34**, 18–31 (2012)
16. Snijders, T.A.B.: Statistical methods for network dynamics. In: Luchini, S.R. (ed.) XLIII Scientific Meeting, Italian Statistical Society, pp. 281–296. CLEUP, Padova (2006)
17. Snijders, T.A.B.: The statistical evaluation of social network dynamics. Sociol. Methodol. **31**, 361–395 (2001)
18. Snijders, T.A.B., Steglich, C.E.G.: Representing micro-macro linkages by actor-based dynamic network models. Sociol. Methods Res. **44**, 222–271 (2015)
19. de Solla Price, D.: A general theory of bibliometric and other advantage processes. Am. Soc. Inf. Sci. **27**, 292–306 (1976)
20. White, H.: Chains of Opportunity, System Models of Mobility in Organizations. Harvard University Press, Cambridge (1970)

On the Analysis of Time-Varying Affiliation Networks: The Case of Stage Co-productions

Giancarlo Ragozini, Marco Serino and Daniela D'Ambrosio

Abstract Multiple Correspondence Analysis and Multiple Factor Analysis have proved appropriate for visually analyzing affiliation (two-mode) networks. However, more could be said about the use of these tools within the positional approach of social network analysis, relying upon the ways in which both these factorial methods and blockmodeling can lead to an appraisal of positional equivalences. This paper presents a joint approach that combines all these methods in order to perform a positional analysis of time-varying affiliation networks. We present an application to an affiliation network of theatre companies involved in stage co-productions over four seasons. The study shows how the joint use of Multiple Factor Analysis and blockmodeling helps us understand network positions and the longitudinal affiliation patterns characterizing them.

Keywords Multiple Factor Analysis · Generalized blockmodeling · Positional approach

1 Introduction

Affiliation networks are a special case of two-mode networks [1] and occur whenever a number of social units (i.e. individual or collective agents) are connected to each other through specific activities or settings. These networks consist of two disjoint sets: a set of actors and a set of events (or activities) in which those actors are involved. When a number of actors attend a number of events over different time occasions, we deal with a time-varying affiliation network. This type of network is of interest in a number of fields, such as organizational, economic, sociological, and political studies, but also in the arts [2].

In the present paper we focus on a time-varying affiliation network resulting from the co-productions that theatre companies release by participating in joint projects

---

G. Ragozini (✉) · M. Serino · D. D'Ambrosio
Department of Political Science, University of Naples Federico II, Naples, Italy
e-mail: giragoz@unina.it

C. Perna et al. (eds.), *Studies in Theoretical and Applied Statistics*,
Springer Proceedings in Mathematics & Statistics 227,
https://doi.org/10.1007/978-3-319-73906-9_11

during different seasons. In such a case it is interesting to study roles and positions that can emerge from a thorough analysis of the relational patterns of the network units, also considering their attributes.

In social network analysis (SNA) literature, two main tools have recently been considered to undertake the positional analysis [3] of two-mode networks, namely generalized blockmodeling [4, 5] and Multiple Correspondence Analysis (MCA) and/or Multiple Factor Analysis (MFA) [6, 7]. In this paper we take a further step in this direction by presenting a joint approach that combines factorial methods and generalized blockmodeling for two-mode networks [4, 5]. Thanks to the connection between the principle of distributional equivalence, which pertains to MCA, and that of structural equivalence related to blockmodeling [8], the combination of these two methods is pursued in order to accomplish in a novel way the positional analysis typical of SNA. This joint approach can provide greater insights into the analysis of the relational patterns of both actors and events, along with their attributes and variation over time, and into positional analysis in particular.

2 The Stage Co-production Network

In this paper we deal with a collaboration network in the art world of theater. Actually, collaboration networks are of interest especially in the case of complex artworks such as theater productions [9]. SNA scholars have paid increasing attention to collaboration networks in the art worlds (see e.g. [10, 11]), and co-production networks in the theater sector are a particular, relatively unexplored example of such networks [12, 13]. Collaboration among theater companies takes the form of the co-production when two or more companies jointly release a given theatrical work. These linkages among co-producing organizations give rise to an affiliation network, which is characterized by a set of nodes (the co-producing organizations) and a set of events (or affiliations, i.e., the co-productions). When co-productions occur over different seasons we have to do with a time-varying affiliation network.

Formally, a time-varying affiliation network can be represented by a set of K affiliation networks $\{\mathcal{G}_k\}_{k=1,\ldots,K}$. The k index refers to different time points, and in the following we will generally refer to them as occasions [7]. Each affiliation network $\mathcal{G}_k$ consists of two sets of relationally connected units, and can be represented by a triple $\mathcal{G}_k\left(V_{1k}, V_{2k}, \mathcal{R}_k\right)$ composed of two disjoint sets of nodes—V_{1k} and V_{2k} of cardinality N_k and J_k, respectively—and one set of edges or arcs, $\mathcal{R}_k \subseteq V_{1k} \times V_{2k}$. By definition, $V_{1k} \cap V_{2l} = \emptyset, \forall k$.

For the sake of presentation, in this paper we assume that $V_{1k} = V_1\ \forall k$.[1] The set $V_1 = \{a_1, a_2, \ldots, a_N\}$ represents the set of N actors, whereas $V_{2k} = \{e_{1k}, e_{2k}, \ldots, e_{J_k}\}$ represents the set of J_k events. We are thus considering networks in which the set of actors is fixed over time while the events can be fixed or fleeting occurrences over

[1]The case of nodes in V_{1k} that partially change over k can be treated by considering V_1 as the overall set of nodes that includes all the V_{1k}'s, i.e. $V_1 = \cup_{k=1}^{K} V_{1k}$.

time. The edge $r_{ijk} = (a_i, e_{jk})$, $r_{ijk} \in \mathcal{R}_k$ is an ordered couple and indicates whether or not an actor a_i attends an event e_{jk}.

Each set $V_1 \times V_{2k}$ can be fully represented by a binary *affiliation matrix* $\mathbf{F}_k = (f_{ijk})$, $i = 1, \dots, N$, $j = 1, \dots, J_k, k = 1, \dots, K$, with $f_{ijk} = 1$ if $(a_i, e_{jk}) \in \mathcal{R}_k$ and 0 otherwise. Given $\mathbf{F}_k$, the row and column marginals $f_{i.k} = \sum_{j=1}^{J_k} f_{ijk}$ and $f_{.jk} = \sum_{i=1}^{N} f_{ijk}$ coincide with the degree d_{ik} of the ith actor at occasion k and the size s_{jk} of the jth event at occasion k, respectively, i.e., $f_{i\cdot k} = d_{ik}$ and $f_{\cdot jk} = s_{jk}$. The set of all affiliation matrices $\{\mathbf{F}_k\}_{k=1,\dots,K}$ can be represented through a data table, usually called the *grand table*, $\mathbb{F} = [\mathbf{F}_1 | \dots | \mathbf{F}_K] = (f_{ijk})$, with $f_{ijk} \in \{0, 1\}$, $i = 1, \dots, N, j = 1, \dots, J_k, k = 1, \dots, K$. The grand table $\mathbb{F}$ is built up by stacking the subtables $\{\mathbf{F}_k\}_{k=1,\dots,K}$ side by side. Given $\mathbb{F}$, for each i, $i = 1, \dots, N$, the row marginal $f_{i..} = \sum_{j=1}^{J_k} \sum_{k=1}^{K} f_{ijk}$ coincides with the total degree of the ith actor, and $L = \sum_{i=1}^{N} \sum_{j=1}^{J_k} \sum_{k=1}^{K} f_{ijk}$ is the total number of edges over all the occasions.

We aim to (i) analyze the relational (affiliation) patterns and characteristics of both companies and co-productions over the four-season span; (ii) assess the space of network positions of actors (theater companies) on the basis of their participation in events (co-productions); (iii) identify the models of structure characterizing our network; (iv) explore the variation the relational patterns over time. In order to pursue the previously listed aims, we propose to jointly use Multiple Factor Analysis (MFA) based on Multiple Correspondence Analysis (MCA) for affiliation networks, and generalized blockmodeling.

3 The Joint Use of Multiple Factor Analysis and of Generalized Blockmodeling

Our proposal seeks to combine the main advantages of both factorial and positional analysis methods. Indeed, both methods seek to identify and synthesize the underlying structure of the observed phenomena by looking at them from a global perspective. Both methods allow detecting the structural similarity in the network on the basis of a certain criterion of equivalence between the units. Thus, they both have the same object but examine it with different levels of detail and from different points of view, allowing an in-depth analysis. Their combination rests on the existence of formal connections between them.

MCA [14], like simple Correspondence Analysis (CA), is based on the principle of distributional equivalence, while blockmodeling, in one of its formulation, is based on the principle of structural equivalence [15]. These two forms of equivalence have a different nature. Distributional equivalence is related to the frequency distributions of the row and column profiles, while structural equivalence refers to the links between network units. The connection between these two principles is visible especially in the case of MCA, since the complex system of weights on which it is based makes the resulting distances very similar to the correct Euclidean distances

[6, 8]. These latter measures are compatible with structural equivalence. It follows that graphic representations made by MCA are based on properties that allow reproducing a relational structure akin to the one emerging from a blockmodeling based on structural equivalence. This connection legitimates our choice of combining the two techniques, i.e., our joint approach.

In order to do that, we follow three steps. We first analyze the time-varying two-mode network through MFA. In a second step, starting from the previous results, we derive for each season the generalized blockmodeling solutions. Finally, in a third step, we incorporate the blockmodeling network positions in MFA along with actors' and events' attributes.

3.1 Multiple Factor Analysis for Time-Varying Two-Mode Networks in Brief

Given the relational structure embedded in each $\mathbf{F}_k$, and in the grand matrix $\mathbb{F}$, we analyze the relational patterns by using MFA, which provides a unifying and general framework to deal with multiple-way matrices, like $\mathbb{F}$. MFA is an extension of factorial techniques [16] tailored to handle multiple data tables. This allows to jointly analyze quantitative and qualitative variables, providing displays in which representations of the set of individuals associated to each group of variables are superimposed. By applying MFA to time-varying affiliation networks, we can perform four different analyses [8]: (i) analysis of each $\mathbf{F}_k$, $k = 1, \ldots, K$ through a suitable factorial method (partial analysis); (ii) analysis of $\mathbb{F}$ (global analysis); (iii) analysis of structural changes among occasions; (iv) analysis of actor/event variation over the occasions by projecting the weighted affiliation matrices $\mathbf{F}_k$ on the global factorial plane. To perform MFA, a factorial method has to be chosen for the analysis of both $\mathbf{F}_k$ and $\mathbb{F}$. We choose to use MCA because of its nice properties in the analysis of network data [6, 8].

In order to perform MCA for the partial analyses, and then for the global analysis, we consider actors as observational units and the participation in events as dichotomous categorical variables, and apply the usual CA algorithm—SVD of the doubled normalized and centered profile matrix—to the *multiple indicator matrix* $\mathbf{Z}_k$ derived from $\mathbf{F}_k$ through a full disjunctive coding, as described in [6]. To this end, we consider each event e_{jk} as a dichotomous variable with categories e^+_{jk} and e^-_{jk}, where e^+_{jk} is a dummy variable coding the participation in the event, and e^-_{jk} is a dummy variable coding the non-participation. Hence, each $\mathbf{Z}_k$ matrix is a $N \times 2J_k$ matrix of the form: $\mathbf{Z}_k \equiv \left[\mathbf{F}^+_k, \mathbf{F_k}^-\right]$, where $\mathbf{F}^+_k = (e^+_{jk}) = \mathbf{F}_k$, and $\mathbf{F}^-_k = (e^-_{jk}) = \mathbf{1} - \mathbf{F}^+_k = \mathbf{1} - \mathbf{F}_k$, with $\mathbf{1}$ the $N \times J_k$ all-ones matrix. The indicator matrix $\mathbf{Z}_k$ turns out to be a *doubled matrix*.

For the sake of presentation, in the present paper we focus on the joint representation of actors and events in the common barycentric space, considering the space given by the global analysis of $\mathbb{F}$ and actors' and events' variations over the occasions by projecting the weighted affiliation matrices $\mathbf{F}_k$ on the global factorial plane.

3.2 Generalized Blockmodeling for Affiliation Networks

In the second step, we perform generalized blockmodeling. This class of methods aims to partition network units into equivalent classes, which are called *network positions*, or clusters, as well as the links within and between such equivalent classes [5, 17]. These partitions of links are called *blocks*. Generalized blockmodeling for two-mode networks relies upon the same general theoretical basis concerned with blockmodeling for one-mode networks, but aims to obtain a two-clustering [4, 5]. As "the partitions of rows and columns are not identical", in two-mode blockmodeling the rows and columns of the matrix, i.e. the two clusters, are partitioned simultaneously but in a different way [5, p. 249]. Given the dual nature of the clusters in affiliation networks [18], blocks that connect two different clusters of units (actors and events) have special properties: each block is formed by the ties between a cluster (position) of the first set and a cluster (position) of the second set. The generalized blockmodeling approach [5] to affiliation (two-mode) networks is developed in order to cope with such properties, so as to obtain a two-clustering that reflects the structure of a selected block type among a wider set of blocks that have to do with a relaxation of the equivalence concept, beyond structural and regular equivalences. Generalized blockmodeling allows us to treat of this possibly wide range of equivalence forms, which can be named *generalized equivalences*.

Formally, given the affiliation networks $\{\mathcal{G}_k\}_{k=1,\dots,K}$, we look for a set of a two-clustering $\mathbf{C}_k = (\mathbf{C}_k^1, \mathbf{C}_k^2) \in \mathbf{\Phi}$, with $\mathbf{C}_k^1 = \{C^1_{1_k}, C^1_{2_k}, \dots, C^1_{H_k}\}$ a partition (or clustering) of the set V_1 in H_k clusters, $\mathbf{C}_k^2 = \{C^2_{1_k}, C^2_{2_k}, \dots, C^2_{M_k}\}$ a partition of the set V_{2k} in M_k clusters, $1 \le H_k \le N$ and $1 \le M_k \le J_k$, $\mathbf{\Phi}$ the set of all possible feasible partitions. Two-mode generalized blockmodeling can be formulated as the following optimization problem [5]:

$$P(\mathbf{C}_k^*) = \min_{\mathbf{C}_k \in \Phi} P(\mathbf{C}_k), \tag{1}$$

with $\mathbf{C}_k^*$ the optimal partition, and $P(\mathbf{C}_k)$ the criterion function defined as

$$P(\mathbf{C}_k) = P((\mathbf{C}_k^1, \mathbf{C}_k^2)) = \sum_{C_{i_k} \in \mathbf{C}_k^1, C_{j_k} \in \mathbf{C}_k^2} \min_{B \in \mathcal{B}(C_{i_k}, C_{j_k})} d(R(C_{i_k}, C_{j_k}), B), \tag{2}$$

where $\mathcal{B}(C_{i_k}, C_{j_k})$, with $C_{i_k} \subseteq V_1$ and $C_{j_k} \subseteq V_{2k}$, is the set of all ideal blocks corresponding to the observed blocks $R(C_{i_k}, C_{j_k})$, and the expression $d(R(C_{i_k}, C_{j_k}), B)$ the measure the number of inconsistencies, i.e. the number of not corresponding ties between empirical blocks $R(C_{i_k}, C_{j_k})$ and the corresponding ideal blocks B.

Given the wide range of possible ideal blocks and equivalence forms, it could be very hard to find a unique solution. In our case, we use the results from MFA in order to make hypothesis on blocks to be tested. This *prespecified* blockmodeling can be considered as a sort of confermative analysis [5].

3.3 Incorporate Blockmodel Positions in MCA Representation

In the last step, we integrate all the results in a unique analysis, characterizing actors and events by using also their attributes. This is possible by adding to the indicator matrix $\mathbf{Z}_k$ a set of columns (rows) corresponding to complete disjunctive coding of both qualitative variables used as attributes [6], and actors' and events' clusters (blockmodel positions) arising from the two-mode blockmodeling procedure. These latter are treated as supplementary rows and columns in MFA.

Starting from $\mathbf{Z}_k$, an augmented matrix $\mathbf{Z}_k^+$ is defined as:

$$\mathbf{Z}^+ = \begin{pmatrix} \mathbf{Z}_k & \mathbf{A}_k^a & \mathbf{B}_k^a \\ \mathbf{A}_k^e & 0 & 0 \\ \mathbf{B}_k^e & 0 & 0 \end{pmatrix}, \tag{3}$$

where $\mathbf{A}_k^a$ and $\mathbf{A}_k^e$ are the complete disjunctive coding matrices of attributes (of actors and of events, respectively), and $\mathbf{B}_k^a$ and $\mathbf{B}_k^e$ are the matrices encoding the memberships of actors and events in clusters (positions) given by the partitions resulting from blockmodeling.

Attributes and clusters can be projected as supplementary points on the factorial plane, and their positions can be used to support the interpretation of the network structure emerging by the factorial configuration. In addition, it is possible to explore the relationships between the relational patterns and the corresponding clusters established through the blockmodeling procedure. In this way, we can maximize the informative contribution present in the data and enrich the interpretation of the final solution, for the purposes of both MCA and blockmodeling.

In turn, MCA, among other advantages, adds to the blockmodeling the possibility of representing actors/events' positions, or clusters, in a metric space, allowing us to analyze their associations with the underlying relational structure and to assess their location in a "social space" in terms of distance. In addition, the presence of actors' and events' attributes permits us to reconstruct a "significant information context" by which the actors/events positions can be usefully characterized and interpreted.

4 The Case Study of Stage Co-productions

In the present paper we focus upon the theater system of the Campania region (Italy), where a distinguished theatrical tradition survives side by side with avant-garde productions [19]. In addition, this system resembles the national one for it is composed of all types of major and minor theater institutions acknowledged by the Italian Ministry of Cultural Heritage (MIBACT, in Italian). Our network data consist of a main set including information on the network ties between companies and the co-productions which they participated in, along with the categorical attributes of both. By means of web-based questionnaires and relying upon information available on companies' websites, between 2013 and 2015 we have collected data on the co-productions released in the course of four seasons (2011/2012, 2012/2013, 2013/2014, 2014/2015) by the resident theater companies that were located in the region and actively operating during the period of data gathering. In the affiliation network, each company is an actor, and each co-production is an event. The participation of a company to a co-production at time k gives rise to a collaboration tie, which may also involve other organizations than those considered for the basic data collection, with $i = 1, \dots, 100, j = 1, \dots, J_k, k = 1, \dots, 4$, and $J_1 = 37, J_2 = 45, J_3 = 40, J_4 = 35$ for a total of $J = 157$ events. The 100 organizations consist of 67 theater companies plus other 33 non-theatrical organizations.

The global MFA shown in Fig. 1 displays the joint representation of companies (actors), represented by points, and co-productions (events), represented by vectors. In the map, points' coordinates result from the weighted average of the coordinates of the related points over the four seasons. By this map it is possible to illustrate the (dis)similarity among the patterns of participation of companies in co-productions, and among the attendance patterns of co-productions with respect to the companies

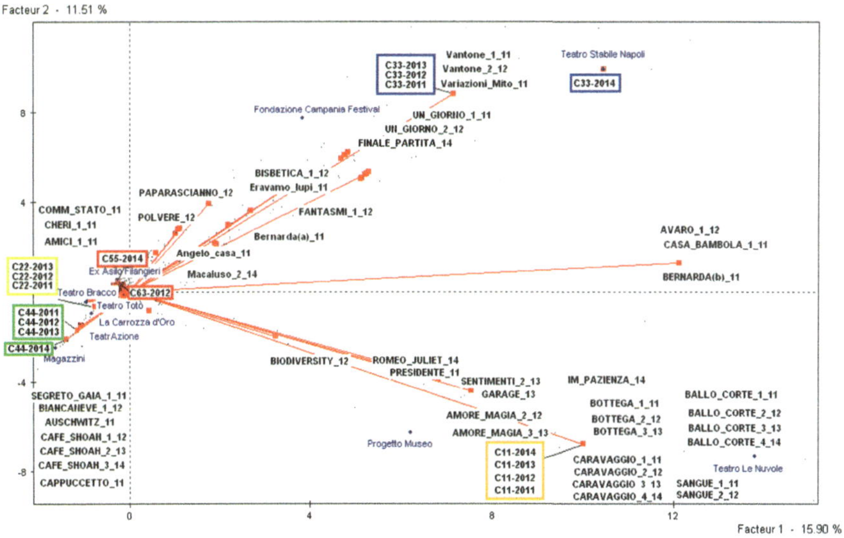

Fig. 1 Global analysis: companies' and co-productions' joint representation. Points represent companies (regular labels), vectors represent co-productions (bold labels ending with the number of times the same co-production occurs). Blockmodeling clusters are superimposed (bold labels in the boxes)

involved. This can be done by looking at the distances between points but also at the angle between vectors. In addition, the map shows the clusters (network positions) to which blockmodeling assigns the companies by reason of the similarity of their relational patterns.

By a visual inspection of the factorial map, five subgroups of companies can be easily identified. At the top right, Teatro Stabile di Napoli (*TSN*) and Fondazione Campania Festival (*FCF*) form a first subgroup, while, below, a second subgroup consists of the companies Teatro Le Nuvole (*TLN*) and Progetto Museo (*PM*). These four companies are the ones with the largest coordinates, and thus give the main contribution to the first and second axes. Further, the points representing them are the farthest ones from the axes origin. Thus, their relational patterns are very rare and the most peculiar in the network: they participate in a large number of co-productions with a given number of partners and this differentiates them from the rest of the network. This peculiarity also emerges from the length of the vectors lying close to each pair of points. The vectors on the top right panel characterize *TSN* and *FCF*, while a bundle of vectors on the bottom right panel highly characterizes the relational patterns of *TLN* and *PM*. Interestingly, the longest vector standing very close to the horizontal axis corresponds to the co-productions that involve both groups and act as a bridge between these two positions.

On the left-hand side of the map, other two subgroups appear: the first one includes Magazzini, TeatrAzione, and La Carrozza d'Oro, while the second one comprises Teatro Bracco and Teatro Totò. At the bottom left of the map the factorial representation provides a plain visual counterpart to the patterns displayed on the other side. The vectors of the co-productions involving Magazzini and *FCF* at the top right are negatively correlated, which clearly expresses the opposite relational patterns of these companies.

A fifth subgroup of companies lies close to the axes origin by virtue of their very sparse participation patterns. In fact, around the origin we find the most common co-production profiles, characterized by very low rates of attendance of companies, as well as those co-productions (and companies) which give the lowest contribution to the first two axes. Overall, these first results drove the prespecification of the blockmodeling [5], i.e. the definition of block types and their location in the blockmodel structure, which has been subject to a fitting procedure whose results are put directly into the map of Fig. 1. For the sake of brevity we do not show this procedure (detailed results are in [13]). We project the resulting clusters as supplementary points on the map following the procedure described in Sect. 3.3.

As the blockmodeling procedure assigns these companies to distinct clusters, we have an evidence of the segmented structure of the network, while the factorial plane displays the mutual exclusion of these clusters by reason of their different relational patterns. In addition, the resulting blockmodels reveal a hierarchical structure emerging from some core-periphery models in which the clusters with peculiar relational

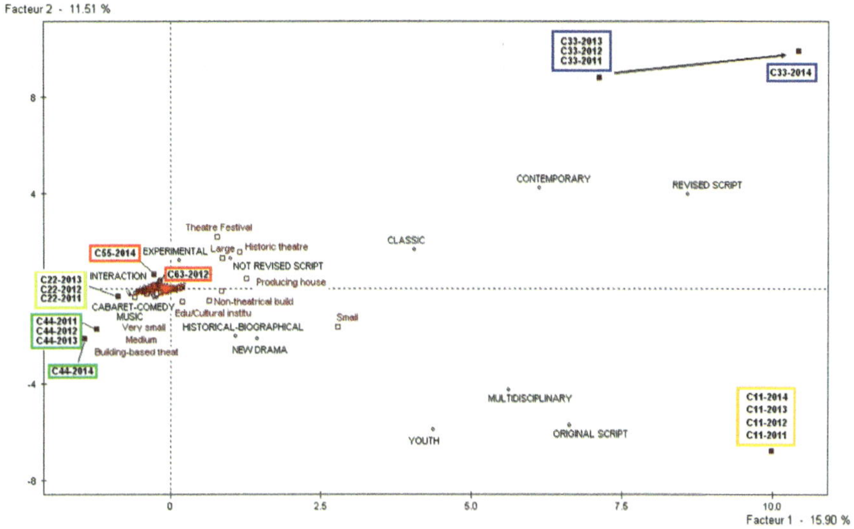

Fig. 2 Global analysis: joint representation of clusters (solid squares) and companies' and co-productions' attributes (empty circles and squares)

patterns (i.e. the clusters[2] (C_3^1, C_3^2) are linked to those having common profiles (like the clusters (C_5^1, C_5^2)). These are some of the benefits of our joint approach.

MFA also allows considering to what extent the clusters are characterized by companies' and co-productions' attributes. The map in Fig. 2 shows that the clusters (C_3^1, C_3^2) differ from (C_1^1, C_1^2) as for the type of theatrical work the corresponding companies mainly focus on. For instance, the former are more inclined to produce classic and contemporary plays; the latter, instead, mainly devote their work to youth theater with a multidisciplinary orientation. The location of these two clusters clearly marks the distinction between them and those lying at the opposite side of the map in line with the difference we observed in the respective relational patterns.

Transitions of companies in the space of relational patterns over the time occasions (partial analysis) are also of interest. In Fig. 2 we have highlighted the changing pattern of participation of companies in the clusters (C_3^1, C_3^2), which turn out to include only *TSN* in the 2014/2015 season. This transition is better understood by looking at the individual trajectories of the companies over time, as shown in Fig. 3, where *FCF* moves towards the axes origin, i.e. the area of non participation, until it no longer belongs to clusters (C_3^1, C_3^2) (in the last time occasion), and *TSN* denotes an odd trajectory mainly due to the complex patterns of its participation in co-productions over the four-time span.

[2]We omit the reference to the occasion, as the clusters are stable during the four seasons.

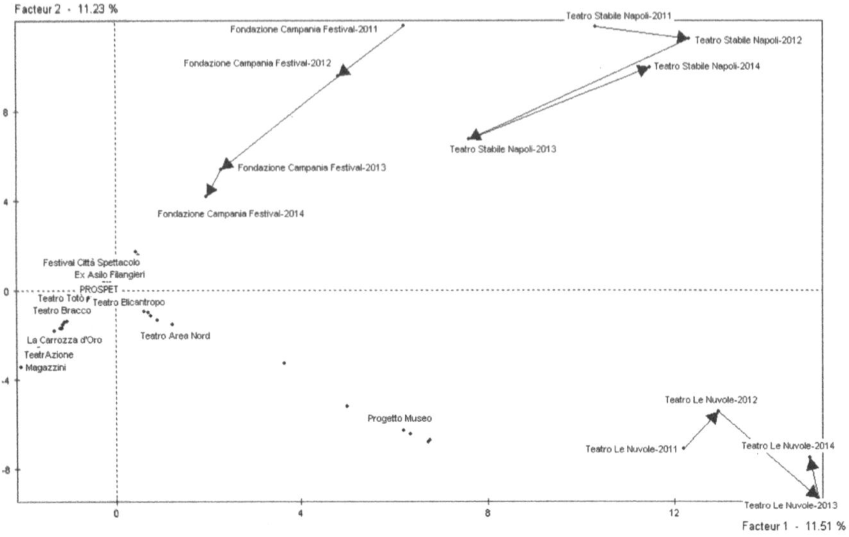

Fig. 3 Partial analysis: trajectories of some companies over time

References

1. Wasserman, S., Faust, K.: Soc. Netw. Anal. Cambridge University Press, New York (1994)
2. Scott, J., Carrington, P.J.: The SAGE Handbook of Social Network Analysis. SAGE publications, London (2011)
3. Doreian, P., Batagelj, V., Ferligoj, A.: Positional Analyses of Sociometric Data. In: Carrington, P.J., Scott, J., Wasserman, S. (eds.) Models and methods in social network analysis, pp. 77–96. Cambridge University Press, Cambridge (2005)
4. Doreian, P., Batagelj, V., Ferligoj, A.: Generalized blockmodeling of two-mode network data. Soc. Netw. **26**(1), 29–53 (2004)
5. Doreian, P., Batagelj, V., Ferligoj, A.: Generalized Blockmodeling. Cambridge University Press, Cambridge (2005)
6. D'Esposito, M.R., De Stefano, D., Ragozini, G.: On the use of multiple correspondence analysis to visually explore affiliation networks. Soc. Netw. **38**, 28–40 (2014)
7. Ragozini, G., De Stefano, D., D'Esposito, M.R.: Multiple factor analysis for time-varying two-mode networks. Netw. Sci. **3**(1), 18–36 (2015)
8. D'Esposito, M.R., De Stefano, D., Ragozini, G.: A comparison of Chi-square metrics for the assessment of relational similarities in affiliation networks. In Vicari, D., Okada, A., Ragozini, G., Weihs, C. (eds.) Analysis and Modeling of Complex Data in Behavioral and Social Sciences, pp. 113–122. Springer, Heidelberg (2014)
9. Becker, H.S.: Art Worlds. University of California Press, Berkeley and Los Angeles (1982)
10. Crossley, N.: The man whose web expanded: network dynamics in Manchester's post/punk music scene 1976–1980. Poetics **37**(1), 24–49 (2009)
11. McAndrew, S., Everett, M.: Music as collective invention: a social network analysis of composers. Cult. Sociol. **9**(1), 56–80 (2015)
12. Serino, M.: Reti di collaborazione tra teatri di produzione in Campania. Sociol. Del lav. **138**, 121–137 (2015)
13. Serino, M., D'Ambrosio, D., Ragozini, G.: Bridging social network analysis and field theory through multidimensional data analysis: the case of the theatrical field. Poetics **62**, 66–80 (2017). https://doi.org/10.1016/j.poetic.2016.12.002

14. Greenacre, M., Blasius, J.: Multiple Correspondence Analysis and Related Methods. CRC Press (2006)
15. Lorrain, F., White, H.C.: Structural equivalence of individuals in social networks. J. Math. Sociol. **1**, 49–80 (1971)
16. Escofier, B., Pagès, J.: Analyse Factorielles Simples et Multiples: objectifs, methodes et interpretation. Dunod, Paris (1998)
17. White, H.C., Boorman, S.A., Breiger, R.L.: Social structure from multiple networks. I. blockmodels of roles and positions. Am. J. Sociol. **81**(4), 730–780 (1976)
18. Breiger, R.L.: The duality of persons and groups. Soc. Forces **53**(2), 181–190 (1974)
19. Serino, M.: Theatre provision and decentralization in a region of Southern Italy. New Theatre Q. **29**(1), 61–75 (2013)

Similarity and Dissimilarity Measures for Mixed Feature-Type Symbolic Data

Manabu Ichino and Kadri Umbleja

Abstract This paper presents some preliminary results for the similarity and dissimilarity measures based on the *Cartesian System Model* (*CSM*) that is a mathematical model to manipulate mixed feature-type symbolic data. We define the notion of concept size for the description of each object in the feature space. By extending the notion to the concept sizes of the *Cartesian join* and the *Cartesian meet* of the descriptions for objects, we can obtain various similarity and dissimilarity measures. We present especially asymmetric and symmetric similarity measures useful for pattern recognition problems.

Keywords Cartesian system model · Symbolic data · Concept size
Pattern recognition

1 Introduction

Many similarity and dissimilarity measures have been extensively developed in various research fields. Johnson [1] treated similarity in clustering problems, and Hubert [2] extended Johnson's hierarchical clustering algorithms. Tversky [3] presented a detail discussion for features of similarity. As a generalization of clustering problems, Michalski and Stepp [4] described conceptual clustering in which each cluster is described by conjunctive logical expression. Jain, Murty, and Flynn [5] presented a wide range of review for data clustering including conceptual clustering.

As noted by Bock and Diday [6], Billard and Diday [7], and Diday and Noirhomme [8], symbolic data analysis aims at extending the classical data models to take into account more complete and complex information. De Carvalho and

M. Ichino (✉)
Tokyo Denki University, Tokyo, Japan
e-mail: ichino@mail.dendai.ac.jp

K. Umbleja
Tallinn University of Technology, Tallinn, Estonia

C. Perna et al. (eds.), *Studies in Theoretical and Applied Statistics*,
Springer Proceedings in Mathematics & Statistics 227,
https://doi.org/10.1007/978-3-319-73906-9_12

De Souza [9] presented clustering methods for mixed feature-type symbolic data with a thorough survey. Ichino and Yaguchi [10] defined generalized Minkowski metrics for mixed feature-type data based on the *Cartesian System Model* (*CSM*) and present dendrogram obtained from the application of standard linkage methods. Guru et al. [11] presented proximity measure and concept of mutual similarity value for clustering interval-valued symbolic data.

The purpose of this paper is to present some preliminary results of the similarity and dissimilarity measures for mixed feature-type symbolic data under the *CSM*. In Sect. 2, we describe the *Cartesian System Model* (*CSM*). In Sect. 3, we present various similarity measures defined with respect to the *concept size* and the *conditional concept size*. We describe the *asymmetric* similarity measure and then *compactness* measure useful for pattern recognition problems with simple examples.

2 The Cartesian System Model [10] and a Generalization

Let D_k be the domain of feature $F_k, k = 1, 2, \ldots, d$. Then, the *feature space* is defined by

$$\boldsymbol{D}^{(d)} = D_1 \times D_2 \times \cdots \times D_d. \tag{1}$$

Since we permit the simultaneous use of various feature types, we use the notation $\boldsymbol{D}^{(d)}$ for the feature space in order to distinguish it from the d-dimensional Euclidean space $\boldsymbol{D}^d$. Each element of $\boldsymbol{D}^{(d)}$ is represented by

$$\boldsymbol{E} = E_1 \times E_2 \times \cdots \times E_d \text{ or } \boldsymbol{E} = (E_1, E_2, \ldots, E_d), \tag{2}$$

where E_k, $k = 1, 2, \ldots, d$, is the feature value taken by the feature F_k. We permit not only continuous and discrete quantitative features (e.g. height, weight, and the number of family members) but also interval values of the form $E_k = [a, b]$. By assuming a proper numerical coding, we can also manage ordinal qualitative features (e.g. academic career, etc.). We call the Cartesian product (2) as an *event*. Any *point* in the feature space is also an event of the reduced form.

2.1 The Cartesian Join Operator

The Cartesian join, $\boldsymbol{A} \boxplus \boldsymbol{B}$, of a pair of events $\boldsymbol{A} = (A_1, A_2, \ldots, A_d)$ and $\boldsymbol{B} = (B_1, B_2, \ldots, B_d)$ in the feature space $\boldsymbol{D}^{(d)}$, is defined by

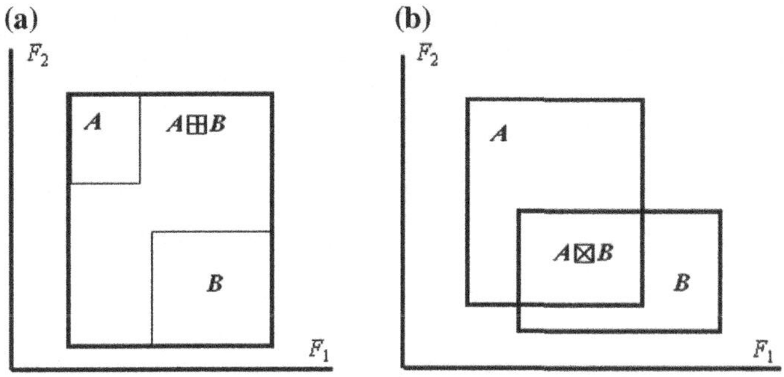

Fig. 1 Cartesian join and meet in the Euclidean plane

$$\boldsymbol{A} \boxplus \boldsymbol{B} = [A_1 \boxplus B_1] \times [A_2 \boxplus B_2] \times \cdots \times [A_d \boxplus B_d], \tag{3}$$

where $[A_k \boxplus B_k]$ is the Cartesian join of feature values $A_k = [A_{kL}, A_{kU}]$ *and* $B_k = [B_{kL}, B_{kU}]$ defined by

$$[A_k \boxplus B_k] = [min(A_{kL}, B_{kL}), max(A_{kU}, B_{kU})]. \tag{4}$$

Figure 1a illustrates the Cartesian join of two interval-valued objects $\boldsymbol{A}$ and $\boldsymbol{B}$ in the Euclidean plane.

2.2 *The Cartesian Meet Operator*

The Cartesian meet, $\boldsymbol{A} \boxtimes \boldsymbol{B}$, of a pair of events $\boldsymbol{A} = (A_1, A_2, \ldots, A_d)$ and $\boldsymbol{B} = (B_1, B_2, \ldots, B_d)$ in the feature space $\boldsymbol{D}^{(d)}$ is defined by

$$\boldsymbol{A} \boxtimes \boldsymbol{B} = [A_1 \boxtimes B_1] \times [A_2 \boxtimes B_2] \times \cdots \times [A_d \boxtimes B_d], \tag{5}$$

where $[A_k \boxtimes B_k]$ is the Cartesian meet of feature values A_k and B_k for feature F_k defined by the intersection

$$[A_k \boxtimes B_k] = A_k \cap B_k. \tag{6}$$

When the intersection (6) takes the empty value $\boldsymbol{\phi}$, for at least one feature, the events $\boldsymbol{A}$ and $\boldsymbol{B}$ have no common part. We say this fact that $\boldsymbol{A}$ and $\boldsymbol{B}$ are disjoint.

If all feature values have no common part simultaneously (see Fig. 1a), we say that $\boldsymbol{A}$ and $\boldsymbol{B}$ are completely disjoint and we denote as

$$\boldsymbol{A} \boxtimes \boldsymbol{B} = \emptyset. \tag{7}$$

Figure 1b illustrates the Cartesian meet of two interval-valued objects $\boldsymbol{A}$ and $\boldsymbol{B}$. We call the triplet $(\boldsymbol{D}^{(d)}, \boxplus, \boxtimes)$ the *Cartesian System Model* (*CSM*).

2.3 A Generalization of the CSM to Histogram Valued Data

We use the Oils' data [10, 12] to illustrate the generalization of our model. In this data, we describe *six* plant oils; Linseed, Perilla, Cotton, Sesame, Camellia, and Olive, and *two* animal fats; Beef and Hog, by *four* interval valued features; Specific Gravity, Freezing Point, Iodine Value, and Saponification Value, and *one* nominal histogram feature; Major Acids in Table 1.

In Table 1, the order of *six* major acids owes to their molecular weights. Each row shows the set of bin probabilities. We attach the unit interval to each nominal bin, and we assume that each object is defined on the whole interval [0, 6]. Then, we can find the cumulative distribution function for each object. Figure 2a is the illustration of eight distribution functions as parallel monotone line graphs. Figure 2b is the corresponding quantile functions represented by 0, 10, 25, 50, 75, 90, and 100% values. For each object, we select here 10 and 90% quantile values denoted by large dots. As the result, we obtain *five* interval-valued Oils' data in Table 2.

We use the Oils' data in Table 2 to illustrate the asymmetric and symmetric (dis) similarity measures in the following sections.

Table 1 Major acids for oils and fats

	Palmitic C16:0	Stearic C18:0	Oleic C18:1	Linoleic C18:2	Linolenic C18:3	Arachic C20:0
Linseed	0.07	0.04	0.19	0.17	0.53	0.00
Perilla	0.13	0.03	0.00	0.16	0.68	0.00
Cotton	0.24	0.02	0.18	0.55	0.00	0.00
Sesame	0.11	0.05	0.41	0.43	0.00	0.00
Camellia	0.08	0.02	0.82	0.07	0.00	0.00
Olive	0.12	0.02	0.74	0.00	0.10	0.01
Beef	0.32	0.17	0.46	0.03	0.01	0.01
Hog	0.27	0.14	0.38	0.17	0.01	0.03

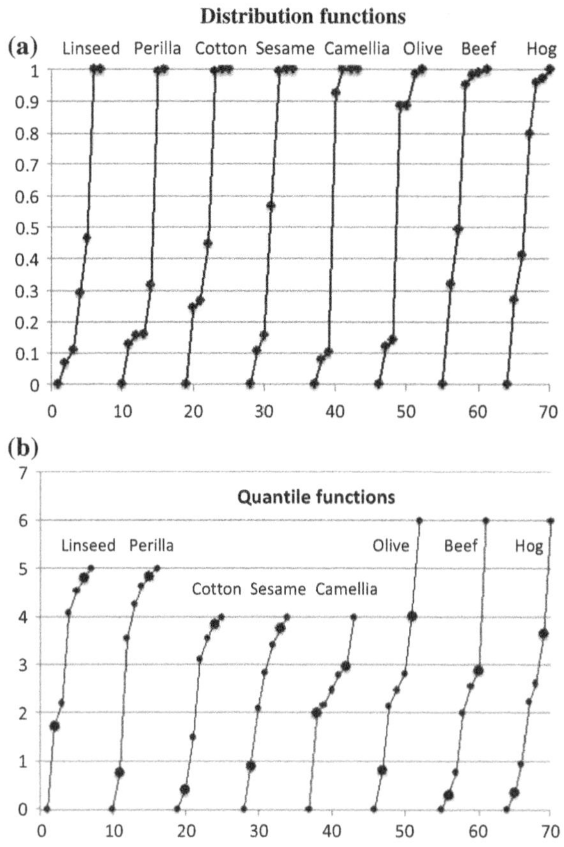

Fig. 2 Distribution functions and quantile functions

Table 2 Oils' data by five interval-valued features

	Specific gravity: F_1	Freezing point: F_2	Iodine value: F_3	Saponification value: F_4	Major acids: F_5
Linseed	[0.930, 0.935]	[− 27, −18]	[170, 204]	[118, 196]	[1.75, 4.81]
Perilla	[0.930, 0.937]	[− 5, −4]	[192, 208]	[188, 197]	[0.77, 4.85]
Cotton	[0.916, 0.918]	[− 6, −1]	[99, 113]	[189, 198]	[0.42, 3.84]
Sesame	[0.920, 0.926]	[− 6, −4]	[104, 116]	[187, 193]	[0.91, 3.77]
Camellia	[0.916, 0.917]	[− 21, −15]	[80, 82]	[189, 193]	[2.00, 2.98]
Olive	[0.914, 0.919]	[0, 6]	[79, 90]	[187, 196]	[0.83, 4.02]
Beef	[0.860, 0.870]	[30, 38]	[40, 48]	[190, 199]	[0.31, 2.89]
Hog	[0.858, 0.864]	[22, 32]	[53, 77]	[190. 202]	[0.37, 3.65]

3 Similarity and Dissimilarity Measures

3.1 Concept Size, Conditional Concept Size, and Several Similarity Measures

Let $U = \{\omega_1, \omega_2, \ldots, \omega_N\}$ be the given set of objects, and let each object ω_k be described by an event $\boldsymbol{E}_k = (E_{k1}, E_{k2}, \ldots, E_{kd})$ in $\boldsymbol{D}^{(d)}$.

Definition 1 We define the concept size $P(E_{kj})$ of ω_k in terms of feature F_j as

$$P(E_{kj}) = |E_{kj}|/|D_{kj}|, j = 1, 2, \ldots, d;\ k = 1, 2, \ldots, N, \tag{8}$$

where $|E_{kj}|$ and $|D_{kj}|$ are the lengths of intervals E_{kj} and D_{kj}.

Definition 2 We define the concept size for the event $\boldsymbol{E}_k$ by the arithmetic mean:

$$P(\boldsymbol{E}_k) = \left(\sum_j P(E_{kj})\right)/d. \tag{9}$$

It is clear that $0 \leq P(\boldsymbol{E}_k) \leq 1$.

For two objects ω_p and ω_q, we can define various symmetric similarity measures based on Definition 2 as follows.

Definition 3

(1) $S_{Jaccard}(\omega_p, \omega_q) = P(\boldsymbol{E}_p \boxtimes \boldsymbol{E}_q)/P(\boldsymbol{E}_p \boxplus \boldsymbol{E}_q)$
(2) $S_{Dice}(\omega_p, \omega_q) = 2P(\boldsymbol{E}_p \boxtimes \boldsymbol{E}_q)/(P(\boldsymbol{E}_p) + P(\boldsymbol{E}_q))$
(3) $S_{Simpson}(\omega_p, \omega_q) = P(\boldsymbol{E}_p \boxtimes \boldsymbol{E}_q)/min(P(\boldsymbol{E}_p), P(\boldsymbol{E}_q))$
(4) $S_{cosine}(\omega_p, \omega_q) = P(\boldsymbol{E}_p \boxtimes \boldsymbol{E}_q)/(P(\boldsymbol{E}_p) \times P(\boldsymbol{E}_q))^{1/2}$

They have desirable properties as similarity measures. However, they always take *zero* whenever the given two events are completely disjoint (see the case in Fig. 1a).

Definition 4 We define the conditional concept size of ω_p under ω_q by

$$P(\boldsymbol{E}_p|\boldsymbol{E}_q) = P(\boldsymbol{E}_p \boxtimes \boldsymbol{E}_q)/P(\boldsymbol{E}_q), (P(\boldsymbol{E}_q) \neq 0). \tag{10}$$

We have the following proposition for the conditional concept size.

Proposition 1

(1) $0 \leq P(\boldsymbol{E}_p|\boldsymbol{E}_q) \leq 1$.
(2) *From* (10), *we have the relations*:

$$P(\boldsymbol{E}_p|\boldsymbol{E}_q)P(\boldsymbol{E}_q) = P(\boldsymbol{E}_p \boxtimes \boldsymbol{E}_q) \text{ and } P(\boldsymbol{E}_q|\boldsymbol{E}_p)P(\boldsymbol{E}_p) = P(\boldsymbol{E}_p \boxtimes \boldsymbol{E}_q).$$

These lead to the Bayes theorem:

$$P(\boldsymbol{E}_q|\boldsymbol{E}_p)P(\boldsymbol{E}_p) = P(\boldsymbol{E}_p|\boldsymbol{E}_q)P(\boldsymbol{E}_q). \tag{11}$$

Definition 5 We can define an asymmetric similarity measure of ω_p to ω_q by

$$S_{Inclusion}(\omega_p|\omega_q) = P(\boldsymbol{E}_p \boxtimes \boldsymbol{E}_q)/P(\boldsymbol{E}_q), (P(\boldsymbol{E}_q) \neq 0). \tag{12}$$

However, this *inclusion* type asymmetric measure is *null* again, whenever two objects are completely disjoint in the feature space.

3.2 Improved Asymmetric and Symmetric Similarity Measures

In order to remove the drawback occurring in completely disjoint cases, we redefine new asymmetric similarity measure as follows.

Definition 6 We define the asymmetric similarity of ω_p to ω_q by

$$\begin{aligned} S_A(\omega_p|\omega_q) &= P(\boldsymbol{E}_q|\boldsymbol{E}_p \boxplus \boldsymbol{E}_q) = P(\boldsymbol{E}_q \boxtimes (\boldsymbol{E}_p \boxplus \boldsymbol{E}_q))/P(\boldsymbol{E}_p \boxplus \boldsymbol{E}_q) \\ &= P(\boldsymbol{E}_q)/P(\boldsymbol{E}_p \boxplus \boldsymbol{E}_q). \end{aligned} \tag{13}$$

In this definition, the similarity of ω_p to ω_q, i.e., $S_A(\omega_p|\omega_q)$, and the similarity of ω_q to ω_p, i.e., $S_A(\omega_q|\omega_p)$, owes to their respective concept size and the size of their Cartesian join. Therefore, we can determine the degree of similarity whether or not their Cartesian meet is completely disjoint.

We have the following two propositions.

Proposition 2

(1) $0 \leq S_A(\omega_p|\omega_q) \leq 1$
(2) *If* $\omega_p = \omega_q, S_A(\omega_p|\omega_q) = S_A(\omega_q|\omega_p) = 1$
(3) $\boldsymbol{E}_p \subseteq \boldsymbol{E}_q$ *implies* $S_A(\omega_p|\omega_q) = 1$
(4) *If* $S_A(\omega_p|\omega_q) = S_A(\omega_q|\omega_p)$, *then* $P(\boldsymbol{E}_p) = P(\boldsymbol{E}_q)$, *however, we may not have* $\omega_p = \omega_q$ (Fig. 5b *illustrates a counter example*).

Proposition 3 ("A larger tree is a better shelter")

$$S_A(\omega_p|\omega_q) \leq S_A(\omega_q|\omega_p) \text{ iff } P(\boldsymbol{E}_q) \leq (\boldsymbol{E}_p).$$

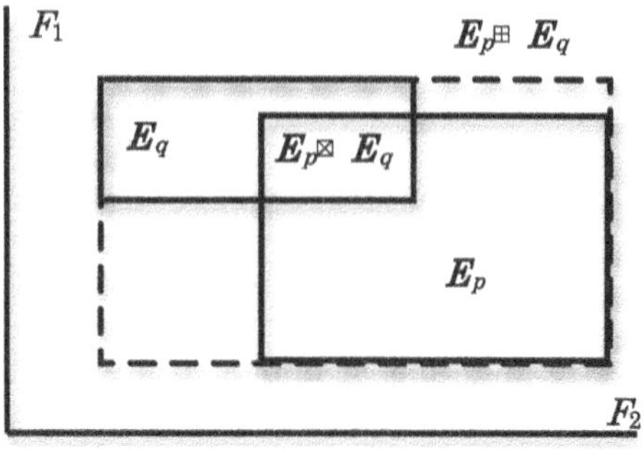

Fig. 3 An example of two different sized concepts

Proof

$$S_A(\omega_p|\omega_q) \le S_A(\omega_q|\omega_p) \text{ iff } P(\boldsymbol{E}_q)/P(\boldsymbol{E}_p \boxplus \boldsymbol{E}_q) \le P(\boldsymbol{E}_p)/P(\boldsymbol{E}_p \boxplus \boldsymbol{E}_q) \text{ iff } P(\boldsymbol{E}_q) \le P(\boldsymbol{E}_p).$$

Q.E.D.

Tversky [3] presented a detail discussion about features of similarity. According to Tversky, let $\boldsymbol{E}_p$ and $\boldsymbol{E}_q$ in Fig. 3 are the descriptions for "tiger" and "leopard", respectively. A larger sized concept "tiger" is more common compared to a smaller sized concept "leopard" with respect to the concept spanned by these two concepts, i.e., the Cartesian join $\boldsymbol{E}_p \boxplus \boldsymbol{E}_q$. Then, the Proposition 3 asserts that "leopard" is more similar to "tiger" than "tiger" is to "leopard".

We should note that our new asymmetric similarity measure works well in wider situations as the measure of membership degree in pattern recognition problems. Figure 4 illustrates the result of clustering of Oils' data by the asymmetric similarity

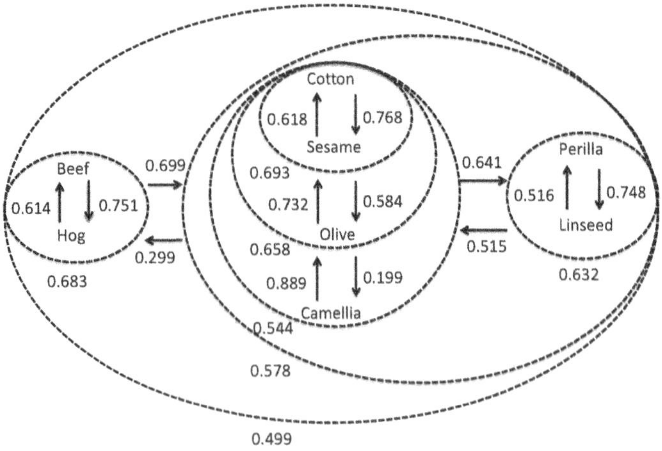

Fig. 4 Clustering result for Oils' data by the asymmetric similarity measure

measure by Definition 6. The numbers attached to each circle by dotted line are the average values of asymmetric similarities between pair of objects and/or clusters.

We can also define the following *min* or *max* type symmetric similarity measures that depend only on the Cartesian join.

Definition 7

$$S_{min}(\omega_p, \omega_q) = min(P(\boldsymbol{E}_p), P(\boldsymbol{E}_q))/P(\boldsymbol{E}_p \boxplus \boldsymbol{E}_q) \tag{14}$$

$$S_{max}(\omega_p, \omega_q) = max(P(\boldsymbol{E}_p), P(\boldsymbol{E}_q))/P(\boldsymbol{E}_p \boxplus \boldsymbol{E}_q). \tag{15}$$

These measures have the following properties.

Proposition 4

(1) $0 \leq S_{min}(\omega_p, \omega_q) \leq S_{max}(\omega_p, \omega_q) \leq 1$
(2) $S_{min}(\omega_p, \omega_q) = S_{max}(\omega_p, \omega_q) = 1$ *if* $\omega_p = \omega_q$
(3) $\boldsymbol{E}_p \subseteq \boldsymbol{E}_q$ implies $S_{min}(\omega_p, \omega_q) \leq 1$ and $S_{max}(\omega_p, \omega_q) = 1$
(4) $S_{min}(\omega_p, \omega_q) = S_{min}(\omega_q, \omega_p)$ *and* $S_{max}(\omega_p, \omega_q) = S_{max}(\omega_q, \omega_p)$.

We should note that the similarity measure $S_{max}(\omega_p, \omega_q)$ *may also be useful as the measure of membership degree in pattern recognition problems.*

3.3 A Measure of Compactness

In Fig. 5, the Cartesian join regions generated by event pairs ($\boldsymbol{E}_p$, $\boldsymbol{E}_q$) in (a) and ($\boldsymbol{E}_p$, $\boldsymbol{E}_q$) in (b) are the same. However, the values of the similarity for these event pairs are different. Therefore, the similarity (or the dissimilarity) of the given objects is a different aspect from the size of the generalized concept (the Cartesian join region) that is obtained by the objects. This fact suggests the measure of compactness for the concept generated by two objects ω_p and ω_q as the following definition.

Definition 8 We define the compactness of the generalized concept by ω_p and ω_q as

$$C(\omega_p, \omega_q) = P(\boldsymbol{E}_p \boxplus \boldsymbol{E}_q). \tag{16}$$

We can obtain the compactness from three different similarity measures with respect to the *whole concept* by the feature space $\boldsymbol{D}^{(d)}$ as follows.

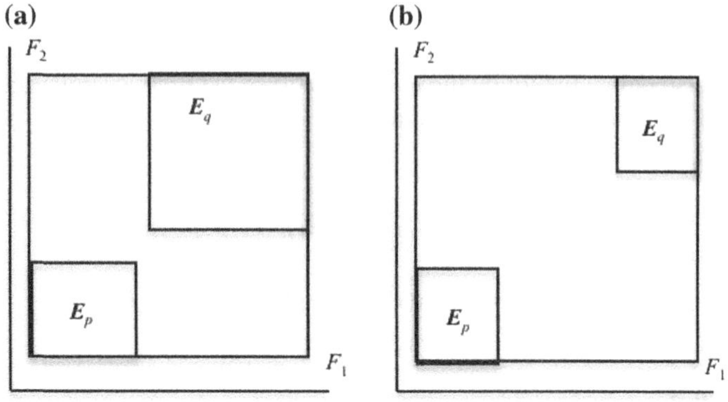

Fig. 5 Minimum concept spanned by two events

Proposition 5

(1) $C(\omega_p, \omega_q) = S_{Inclusion}(\{\omega_p, \omega_q\}|whole\ concept)$,
(2) $C(\omega_p, \omega_q) = S_A(whole\ concept|\{\omega_p, \omega_q\})$, *and*
(3) $C(\omega_p, \omega_q) = S_{min}(\{\omega_p, \omega_q\}, whole\ concept)$,

where $\{\omega_p, \omega_q\}$ *denotes the concept generated by objects* ω_p*and* ω_q.

Proof Definitions 5, 6, and 7 lead to

$$(1)\quad S_{Inclusion}(\{\omega_p, \omega_q\}|whole\ concept) = P(\boldsymbol{E}_p \boxplus \boldsymbol{E}_q|D^{(d)}) = P((\boldsymbol{E}_p \boxplus \boldsymbol{E}_q) \boxtimes D^{(d)})/P(D^{(d)}) = P(\boldsymbol{E}_p \boxplus \boldsymbol{E}_q)$$

$$(2)\quad S_A(whole\ concept|\{\omega_p, \omega_q\}) = P((\boldsymbol{E}_p \boxplus \boldsymbol{E}_q) \boxtimes D^{(d)})|(\boldsymbol{E}_p \boxplus \boldsymbol{E}_q) \boxplus D^{(d)}) \text{ and } = P(\boldsymbol{E}_p \boxplus \boldsymbol{E}_q)/P(D^{(d)}) = P(\boldsymbol{E}_p \boxplus \boldsymbol{E}_q),$$

$$(3)\quad S_{min}(\{\omega_p, \omega_q\}, whole\ concept) = min(P(\boldsymbol{E}_p \boxplus \boldsymbol{E}_q), P(D^{(d)}))/P((\boldsymbol{E}_p \boxplus \boldsymbol{E}_q) \boxplus D^{(d)}) = P(\boldsymbol{E}_p \boxplus \boldsymbol{E}_q)/P(D^{(d)}) = P(\boldsymbol{E}_p \boxplus \boldsymbol{E}_q).$$

Q.E.D.

Since the Cartesian join $\boldsymbol{E}_p \boxplus \boldsymbol{E}_q$ generates the smallest description spanned by the given two events $\boldsymbol{E}_p$ and $\boldsymbol{E}_q$, the compactness $C(\omega_p, \omega_q)$ evaluates quantitatively the size of the generated concept.

The compactness satisfies the following properties.

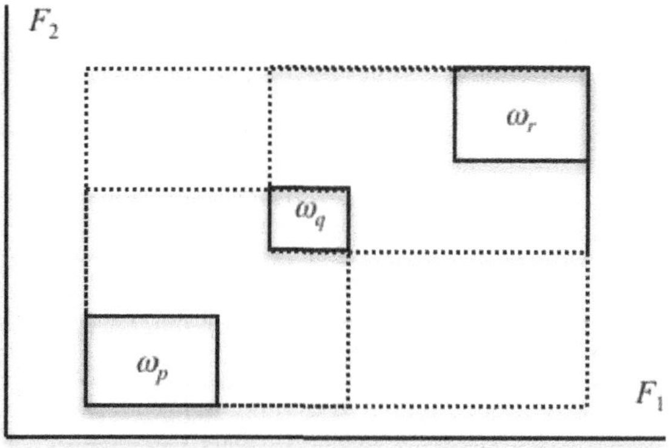

Fig. 6 A counter example for property 6

Proposition 6

(1) $0 \leq C(\omega_p, \omega_q) \leq 1$
(2) $C(\omega_p, \omega_p) = P(\boldsymbol{E}_p) \geq 0$
(3) $C(\omega_p, \omega_p), C(\omega_q, \omega_q) \leq C(\omega_p, \omega_q)$
(4) $C(\omega_p, \omega_q) = 0$ *iff* $\boldsymbol{E}_p \equiv \boldsymbol{E}_q$ *and has null size* $(P(\boldsymbol{E}_q) = 0)$
(5) $C(\omega_p, \omega_q) = C(\omega_q, \omega_p)$
(6) $C(\omega_p, \omega_r) \leq C(\omega_p, \omega_q) + C(\omega_q, \omega_r)$ *may not hold in general* (*see the counter example of* Fig. 6).

In hierarchical clustering, the compactness by Definition 8 works to generate clusters as compact as possible contrast to the whole concept space. Therefore, we may expect that the compactness play not only the role of similarity measure between objects and/or clusters, but also the role of measure of cluster quality.

Figure 7 is the result of hierarchical conceptual clustering for Oils' data by using the compactness. By cutting the dendrogram around 0.5, we obtain three clusters and their mutually disjoint conjunctive logical expressions in Table 3.

3.4 Other Similarity and Dissimilarity Measures

We define a symmetric similarity measure by using the asymmetric similarity measures in Sect. 3.2 as follows.

Definition 9

$$S(\omega_p, \omega_q) = (S_A(\omega_p|\omega_q) + S_A(\omega_q|\omega_p))/2 = (P(\boldsymbol{E}_p) + P(\boldsymbol{E}_q))/(2P(\boldsymbol{E}_p \boxplus \boldsymbol{E}_q)). \quad (17)$$

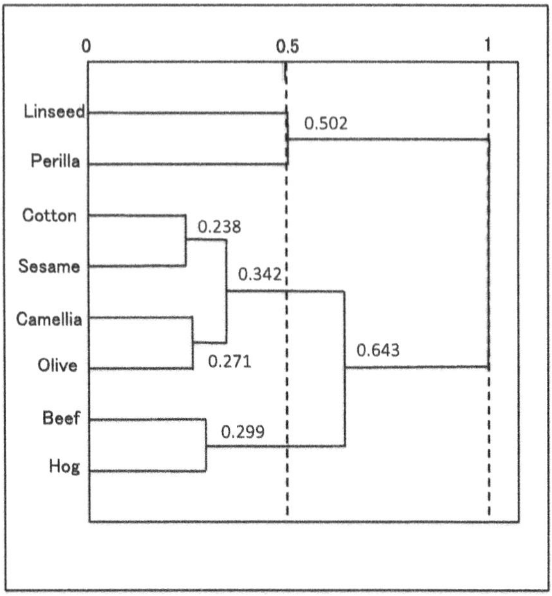

Fig. 7 Clustering of Oils' data by the compactness

Table 3 Mutually disjoint conjunctive logical expressions for three clusters

{Linseed, Perilla}: 0.502	[Specific gravity = [0.930, 0.937]] & [Freezing point = [− 27, −4]] & [Iodine value = [170, 208]]
{Cotton, Sesame, Camellia, Olive}: 0.342	[Specific gravity = [0.914, 0.926]] & [Freezing point = [− 21, 6]] & [Iodine value = [79, 116]]
{Beef, Hog}: 0.299	[Spceific gravity = [0.858, 0.870]] & [Freezing point = [22, 38]] & [Iodine value = [40, 77]]

We have another way to obtain the same measure as follows:

$$\begin{aligned} S(\omega_p, \omega_q) &= (S_{min}(\omega_p, \omega_q) + S_{max}(\omega_p, \omega_q))/2 \\ &= (min(P(\boldsymbol{E}_p), P(\boldsymbol{E}_q))/P(\boldsymbol{E}_p \boxplus \boldsymbol{E}_q) \\ &\quad + max(P(\boldsymbol{E}_p), P(\boldsymbol{E}_q))/P(\boldsymbol{E}_p \boxplus \boldsymbol{E}_q))/2 \\ &= (P(\boldsymbol{E}_p) + P(\boldsymbol{E}_q))/(2P(\boldsymbol{E}_p \boxplus \boldsymbol{E}_q)). \end{aligned} \quad (18)$$

By using Definition 9, we have the following dissimilarity measure.

Definition 10

$$D(\omega_p, \omega_q) = 1 - S(\omega_p, \omega_q) = 1 - (P(\boldsymbol{E}_p) + P(\boldsymbol{E}_q))/(2P(\boldsymbol{E}_p \boxplus \boldsymbol{E}_q)). \quad (19)$$

We can prove the metric properties of this measure by the same way in [10]. We summarize the properties of the similarity measure (17) and the dissimilarity measure (19) in the following proposition.

Proposition 7

(1) $S(\omega_p, \omega_q) \geq 0,\ S(\omega_p, \omega_q) = 1$ *iff* $\omega_p = \omega_q$
(2) $S(\omega_p, \omega_q) = S(\omega_q, \omega_p)$
(3) $S(\omega_p, \omega_q) + S(\omega_q, \omega_r) \leq 1 + S(\omega_p, \omega_r)$
(4) $D(\omega_p, \omega_q) \geq 0, D(\omega_p, \omega_q) = 0$ *iff* $\omega_p = \omega_q$
(5) $D(\omega_p, \omega_q) = D(\omega_q, \omega_p)$
(6) $D(\omega_p, \omega_q) + D(\omega_q, \omega_r) \geq D(\omega_p, \omega_r)$.

These measures are also applicable to analyse mixed feature-type symbolic data in the same manner as in [10].

4 Concluding Remarks

This paper presented some preliminary results for various asymmetric and symmetric (dis)similarity measures based on the *Cartesian System Model (CSM)*. We summarize our results as follows.

(1) By using the concept size defined in the *CSM*, we presented *four* symmetric similarity measures of Definitions 3, and an asymmetric measure of Definition 5 using the conditional concept size. However, these measures do not work well when the given objects are completely disjoint in the feature space.
(2) In order to remove the drawback for the measures in (1) we defined conditional concept size and an asymmetric similarity measure in Definitions 6 and 7, respectively, based only on the Cartesian join operator.
(3) The compactness of Definition 8 owes only to the Cartesian join, and is useful as the measure for hierarchical clustering. In each step of clustering, we merge objects and/or clusters so as to minimize the compactness. This means that the compactness plays the role of similarity measure for objects and/or clusters. On the other hand, in each step of clustering, the generated concept maximizes dissimilarity from the *whole concept*. Therefore, the compactness plays not only the role of similarity measure between objects and/or clusters but also the role of measure for cluster quality contrasting with the whole concept.
(4) In the design of pattern classifiers, we can apply the compactness of Definition 8 to obtain class-conditional conceptual description for each pattern class. Then, we can use asymmetric similarity measure of Definition 6 to determine the class membership degrees for new coming patterns.
(5) The measures in Definitions 9 and 10 are also applicable as the measures of similarity and dissimilarity in standard agglomerative hierarchical clustering.

As a further research, we are interesting in *feature selection* and dimensionality reduction based on *the compactness* and the *geometrical thickness* [13].

Acknowledgements The authors thank anonymous referees for their helpful comments. This work was supported by JSPS KAKENHI (Grants–in–Aid for Scientific Research) Grant Number 25330268.

References

1. Johnson, S.C.: Hierarchical clustering schemes. Psychometrika **32**(3), 241–254 (1967)
2. Hubert, L.: Some extensions of Johnson's hierarchical clustering algorithms. Psychometrika **37**(3) 261–27 L. 4 (1972)
3. Tversky, A.: Features of similarity. Psychol. Rev. **84**(4) (1977)
4. Michalski, R., Stepp, R.: Learning from observation: Conceptual clustering. In: Michalski, R.S., Carbonell, J.G., Mitchel, T.M. (eds.) Machine Learning, An Artificial Intelligence Approach, vol. II, pp. 331–363. TIOGA Publishing Co., Palo Alto (1983)
5. Jain, A.K., Murty, M.N., Flynn, P.J.: Data clustering: a review. ACM Comput. Surv. **31**(3), 264–323 (1999)
6. Bock, H.-H., Diday, E.: Analysis of Symbolic Data. Exploratory Methods for Extracting Statistical Information from Complex Data. Springer, Berlin, Heidelberg (2000)
7. Billard, L., Diday, E.: Symbolic Data Analysis: Conceptual Statistics and Data Mining. Wiley, Chichester (2007)
8. Diday, E., Noirhomme-Fraiture, M.: Symbolic Data Analysis and the SODAS Software. Wiley, Chichester (2008)
9. De Carvalho, F.D.A.T., De Souza, M.C.R.: Unsupervised pattern recognition models for mixed feature-type data. Pattern Recognit. Lett. **31**, 430–443 (2010)
10. Ichino, M., Yaguchi, H.: Generalized Minkowski metrics for mixed feature-type data analysis. IEEE Trans. Syst. Man Cybern. **24**(4), 698–708 (1994)
11. Guru, D.S., Kiranagi, B.B., Nagabhushan, P.: Multivalued type proximity measure and concept of mutual similarity value useful for clustering symbolic patterns. Pattern Recognit. **25**, 1203–1213 (2004)
12. Ichino, M.: The quantile method of symbolic principal component analysis **4**, 184–198 (2011)
13. Ono, Y., Ichino, M.: A new feature selection method based on geometrical thickness. Int. J. Off. Stat. **1**(2), 19–38 (1998)

Dimensionality Reduction Methods for Contingency Tables with Ordinal Variables

Luigi D'Ambra, Pietro Amenta and Antonello D'Ambra

Abstract Several extensions of correspondence analysis have been introduced in literature coping with the possible ordinal structure of the variables. They usually obtain a graphical representation of the interdependence between the rows and columns of a contingency table, by using several tools for the dimensionality reduction of the involved spaces. These tools are able to enrich the interpretation of the graphical planes, providing also additional information, with respect to the usual singular value decomposition. The main aim of this paper is to suggest an unified theoretical framework of several methods of correspondence analysis coping with ordinal variables.

Keywords Ordinal variables · Single and double cumulative correspondence analysis · Orthogonal polynomials · Generalized singular value decomposition

1 Introduction

Correspondence Analysis (CA) is a widely used tool for obtaining a graphical representation of the interdependence between the rows and columns of a contingency table, and it is usually performed by applying a generalized singular value decomposition to the standardised residuals of a two-way contingency table obtaining a

L. D'Ambra (✉)
Department of Economics, Management and Institutions,
University "Federico II" of Naples, Via Cinthia Monte Sant'Angelo, Naples, Italy
e-mail: dambra@unina.it

P. Amenta
Department of Law, Economics, Management and Quantitative Methods,
University of Sannio, Piazza Arechi II, Benevento, Italy
e-mail: amenta@unisannio.it

A. D'Ambra
Department of Economics, University of Campania "L.Vanvitelli",
Corso Gran Priorato di Malta, Capua, Italy
e-mail: antonello.dambra@unicampania.it

C. Perna et al. (eds.), *Studies in Theoretical and Applied Statistics*,
Springer Proceedings in Mathematics & Statistics 227,
https://doi.org/10.1007/978-3-319-73906-9_13

dimensionality reduction of the space. Dimension reduction can be achieved via a suite of interrelated methods for a two-way contingency table. We draw our attention to singular value decomposition (hereafter SVD), generalised singular value decomposition (GSVD), bivariate moment decomposition (BMD), and hybrid decomposition (HMD). After a review of several CA extensions coping with ordinal categorical variables, the last section presents an unified approach of these methods. This framework can aid the user to better understand the methodological differences between these approaches and provides the basic rationale for an easier single software implementation of all of them.

2 Basic Notation

We consider samples from I different populations $(A_1, \ldots, A_I)$, each of which is divided into J categories $(B_1, \ldots, B_J)$. We assume that the samples of sizes $n_1, \ldots, n_I$ from different populations are independent and that each sample follows a multinomial distribution. The probability of having an observation falls in the i-th row and j-th column of the table is denoted π_{ij}. An $I \times J$ contingency table $\mathbf{N}$ of the observations n_{ij} $(i = 1, \ldots, I; j = 1, \ldots, J)$ is considered with $n = \sum_{i=1}^{I} \sum_{j=1}^{J} n_{ij}$. The (i, j)-th element of the relative frequencies matrix $\mathbf{P}$ is defined as $p_{ij} = n_{ij}/n$ such that $\sum_{i=1}^{I} \sum_{j=1}^{J} p_{ij} = 1$. Let's suppose that $\mathbf{N}$ has one ordered set of categories (column) with row and column marginal frequencies given by $p_{i.} = \sum_{j=1}^{J} p_{ij}$ and $p_{.j} = \sum_{i=1}^{I} p_{ij}$, respectively. Finally, let $\mathbf{D}_I$ and $\mathbf{D}_J$ represent the diagonal matrices of row and column marginal $p_{i.}$ and $p_{.j}$, respectively, with $\mathbf{r} = \mathbf{D}_I \mathbf{1}_I$ and $\mathbf{c} = \mathbf{1}_J^T \mathbf{D}_J$.

3 Correspondence Analysis and Its Extensions Based on SVD

3.1 Correspondence Analysis

CA of two categorical variables is usually presented as a multivariate method that decomposes the chi-squared statistic associated with a contingency table into orthogonal factors. Row and column categories are usually displayed in two-dimensional graphical form. This approach has been described from other points of view. For instance, Goodman [10] shows that CA can be performed by applying the Singular Value Decomposition (SVD) on the Pearson's ratios table [10]. That is, for the $I \times J$ correspondence matrix $\mathbf{P}$ then its Pearson ratio α_{ij} is decomposed so that $\alpha_{ij} = p_{ij}/(p_{i.}p_{.j}) = 1 + \sum_{m=1}^{K} \lambda_m a_{im} b_{jm}$ with a_{im} and b_{jm} $\{m = 1, \ldots, K = \min(I, J) - 1\}$ singular vectors associated with the i'th row and j'th column category, respectively, and λ_m is the m'th singular value of the ratio. Moreover, these quantities are such to satisfy the conditions $\sum_j p_{.j} b_{jm} = \sum_i p_{i.} a_{im} = 0$ and $\sum_j p_{.j} b_{jm} b_{jm'} = \sum_i p_{i.} a_{im} a_{im'} = 1$ for $m = m'$, 0 otherwise.

Using the matrix notation, the above least squares estimates are obtained by a generalized singular value decomposition (GSVD) of matrix $\Omega = \mathbf{D}_I^{-1}(\mathbf{P} - \mathbf{D}_I \mathbf{1}_I \mathbf{1}_J^T \mathbf{D}_J)\mathbf{D}_J^{-1} = \mathbf{A}\mathbf{D}_\lambda \mathbf{B}^T$ with $\mathbf{A}^T\mathbf{D}_I\mathbf{A} = \mathbf{I}$, $\mathbf{B}^T\mathbf{D}_J\mathbf{B} = \mathbf{I}$ and where $\mathbf{D}_\lambda$ is a diagonal matrix with all singular values λ_m in descending order. It is well known that

$$\phi^2 = \frac{\chi^2}{n} = trace(\mathbf{D}_I^{-1/2}(\mathbf{P} - \mathbf{D}_I \mathbf{1}_I \mathbf{1}_J^T \mathbf{D}_J)\mathbf{D}_J^{-1}(\mathbf{P} - \mathbf{D}_I \mathbf{1}_I \mathbf{1}_J^T \mathbf{D}_J)^T \mathbf{D}_I^{-1/2}) = \sum_{m=1}^{K} \lambda_m^2$$

where χ^2 is the Pearson's chi squared statistic. See [3] for a bibliographic review.

3.2 Non Symmetrical Correspondence Analysis

In two-way contingency tables, rows and columns often assume an asymmetric role. This aspect is not taken into account by correspondence analysis where it is supposed a symmetric role between the categorical variables. When the variables are asymmetrical related, it has been introduced in literature a new approach named non symmetrical correspondence analysis (NSCA) which aim is to examine predictive relationships between rows and columns of a contingency table, and where it is assumed that columns depend on rows, but not viceversa (see [3, 8] for a comprehensive review). NSCA amounts to the GSVD $\mathbf{D}_I^{-1}(\mathbf{P} - \mathbf{D}_I \mathbf{1}_I \mathbf{1}_J^T \mathbf{D}_J) = \mathbf{A}\mathbf{D}_\lambda \mathbf{B}^T$ with $\mathbf{A}^T\mathbf{D}_I\mathbf{A} = \mathbf{I}$, and $\mathbf{B}^T\mathbf{B} = \mathbf{I}$. This GSVD leads to

$$N_\tau = n \times trace(\mathbf{D}_I^{-1/2}(\mathbf{P} - \mathbf{D}_I \mathbf{1}_I \mathbf{1}_J^T \mathbf{D}_J)(\mathbf{P} - \mathbf{D}_I \mathbf{1}_I \mathbf{1}_J^T \mathbf{D}_J)^T \mathbf{D}_I^{-1/2}) = n \times \sum_{m=1}^{K} \lambda_m^2$$

where N_τ is the numerator of the Goodman-Kruskal's tau index [11]

$$\tau = \frac{\sum_{i=1}^{I} \sum_{j=1}^{J} p_{i.} \left(\frac{p_{ij}}{p_{i.}} - p_{.j}\right)^2}{1 - \sum_{j=1}^{J} p_{.j}^2} = \frac{N_\tau}{1 - \sum_{j=1}^{J} p_{.j}^2}$$

See [13] to assess the statistical significance of dependence.

3.3 The Decomposition of Cumulative Chi Squared Statistic

Beh et al. [2] perform CA when one of the cross-classified variables has an ordered structure. This approach determines graphically how similar cumulative categories are with respect to nominal ones. Let $z_{ik} = \sum_{j=1}^{k} n_{ij}$ be the cumulative frequency of the i-th row category up to the k-th column category providing a way of ensuring that the ordinal structure of the column categories is preserved. Similarly, let $d_k = \sum_{j=1}^{k} n_{.j}/n = \sum_{j=1}^{k} p_{.j}$ be the cumulative relative frequency up to the k-th column category. Moreover, let $\mathbf{W}$ be the $((J-1)\times(J-1))$ diagonal matrix of weights w_j and $\mathbf{M}$ a $((J-1)\times J)$ lower triangular matrix of ones. CA of cumulative frequencies (TA) amounts to the GSVD $\mathbf{D}_I^{-}(\mathbf{P}-\mathbf{D}_I\mathbf{1}_I\mathbf{1}_J^T\mathbf{D}_J)\mathbf{M}^T\mathbf{W}^{\frac{1}{2}} = \mathbf{U}\mathbf{D}_\lambda\mathbf{V}^T$ with $\mathbf{U}^T\mathbf{D}_I\mathbf{U} = \mathbf{I}$ and $\mathbf{V}^T\mathbf{V} = \mathbf{I}$, such that

$$T = n \times trace(\mathbf{D}_I^{-\frac{1}{2}}(\mathbf{P}-\mathbf{D}_I\mathbf{1}_I\mathbf{1}_J^T\mathbf{D}_J)\mathbf{M}^T\mathbf{W}\mathbf{M}(\mathbf{P}-\mathbf{D}_I\mathbf{1}_I\mathbf{1}_J^T\mathbf{D}_J)^T\mathbf{D}_I^{-\frac{1}{2}}) = \sum_{i=1}^{I}\lambda_i^2$$

where $T = \sum_{j=1}^{J-1} w_j \left[\sum_{i=1}^{I} n_{i.}(z_{ij}/n_{i.} - d_j)^2\right]$ is the Taguchi's statistic [19, 20], with $0 \le T \le [n(J-1)]$ and w_j suitable weights. Possible choices for w_j could be to assign constant weights to each term ($w_j = 1/J$) or assume it proportional to the inverse of the conditional expectation of the k-th term under the hypothesis of independence ($w_j = 1/[d_j(1-d_j)]$). T performs better χ^2 when there is an order in the categories on the columns of the contingency table and it is more suitable in studies (such as clinical trials) where the number of categories within a variable is equal to (or larger) than 5 [12]. The T statistic was originally developed for the one-way Anova model for industrial experiments to test the hypothesis of homogeneity against monotonicity in the treatment effects. An $I \times J$ contingency table with row multinomial model with equal row totals ($n_{i.} = K$ observations per level of a factor A with I levels) has been then obtained. For this model, Nair [16] shows that the sum of squares for the factor A is given by $SSA = n\sum_{j=1}^{J-1}\sum_{i=1}^{I}(z_{ij}/K - d_j)^2/(d_j(1-d_j))$ which is T with fixed and equal rows totals. He highlighted also that $T = \sum_{j=1}^{J-1}\chi_j^2$ where χ_j^2 is the Pearson chi-squared for the $I \times 2$ contingency tables obtained by aggregating the first j column categories and the remaining categories $(j+1)$ to J, respectively. Several authors refer the Taguchi's statistic T as the "Cumulative Chi-Squared statistic" (CCS). Moreover, Nair [16] showed that the distribution of T can be approximated using the Satterthwaite's method [18].

Let $\mathbf{d}^T = [d_1, \ldots, d_{J-1}]^T$ be the vector of the cumulative column relative marginal frequencies d_j, and let be the joined $(J-1)\times J$ matrix $\tilde{\mathbf{D}} = [(\mathbf{H} - \mathbf{d}\mathbf{1}_{J-1}^T)|(-\mathbf{d})]$ where $\mathbf{H}$ is a unit lower triangular matrix. In addition, let $\mathbf{y}_i = [n_{i1}, \ldots, n_{iJ}]^T$ be a vector of observed frequencies for the i-th row. Then, the CSS statistic can be written as $T = \sum_{i=1}^{I}\mathbf{y}_i^T\tilde{\mathbf{D}}^T\mathbf{W}\tilde{\mathbf{D}}\mathbf{y}_i/n_{i.} = trace(\mathbf{D}_I^{-1/2}\mathbf{N}\tilde{\mathbf{D}}^T\mathbf{W}\tilde{\mathbf{D}}\mathbf{N}^T\mathbf{D}_I^{-1/2})$. The matrix $\tilde{\mathbf{D}}^T\mathbf{W}\tilde{\mathbf{D}}$ can be then decomposed such that $\tilde{\mathbf{D}}^T\mathbf{W}\tilde{\mathbf{D}} = \mathbf{Q}\mathbf{D}_\lambda\mathbf{Q}^T$. Under the assumption of equiprobable columns ($w_j = J/[j \times (J-j)]^{-1}$ with $j = 1, \ldots, J-1$), the eigenvectors are linked to the j-th degree Chebychev polynomial on the integers $\{1, \ldots, J\}$. The first (linear) and the second (quadratic) component are equivalent to the Wilcoxon statistic for the $2 \times J$ table and to the Mood's test [15], respectively.

3.4 *The Decomposition of Double Cumulative Chi Squared Statistic*

D'Ambra et al. [7] proposed a generalisation of the Taguchi decomposition based on cumulative frequencies for the rows and columns (HDA). This approach introduces two suitable cumulative matrices **R** and **C** to pool the rows and columns of contingency table. Let **R** be a $2(I-1) \times I$ matrix formed by alternating the rows of an $(I-1) \times I$ unit lower triangular matrix with the rows of an $(I-1) \times I$ unit upper triangular matrix (by first removing the row consisting of all ones in both matrices). **C** is a $J \times 2(J-1)$ matrix obtained likewise **R**. Moreover, $\mathbf{P}_R$ and $\mathbf{P}_C$ ($\tilde{\mathbf{D}}_R$ and $\tilde{\mathbf{D}}_C$) are the diagonal matrices with the marginal relative (absolute) frequencies $h_{i.}$ and $h_{.j}$ of the doubly cumulative table $\mathbf{H} = \mathbf{RPC}$ ($\mathbf{H} = \mathbf{RNC}$), respectively. This approach amounts to the SVD of the matrix $\mathbf{D}_R^{-\frac{1}{2}}\mathbf{R}(\mathbf{P} - \mathbf{D}_I\mathbf{1}_I\mathbf{1}_J^T\mathbf{D}_J)\mathbf{C}^T\mathbf{D}_C^{-\frac{1}{2}}$. This approach [12] considers the decomposition of Hirotsu's index $\chi^{**2} = \sum_{i=1}^{I-1}\sum_{j=1}^{J-1}\chi_{ij}^{**2} = n(I-1)(J-1)\sum_{k=1}^{K}\lambda_k^2$. Hirotsu [12] introduced χ^{**2} in order to measure the association between two ordered categorical variables in a two-way contingency table, where χ_{ij}^{**2} is the chi-squared statistic for the 2×2 contingency table obtained by pooling the original table $I \times J$ data at the i-th row and j-th column. He showed also that the null distribution of the statistic χ^{**2} is approximated by $d\chi_v^2$ with $d = d_1 \times d_2$ and $v = (I-1)(J-1)/d$, where $d_1 = 1 + 2(J-1)^{-1}[\sum_{k=1}^{J-2}(\sum_{s=1}^{k}\lambda_s)/\lambda_{k+1}]$ e $d_2 = 1 + 2(I-1)^{-1}[\sum_{=1}^{I-2}(\sum_{s=1}^{k}\gamma_s)/\gamma_{k+1}]$ con $\lambda_s = (\sum_{h=1}^{s} n_{.h})/(\sum_{g=s+1}^{J} n_{.g})$ e $\gamma_s = (\sum_{h=1}^{s} n_{h.})/(\sum_{g=s+1}^{I} n_{g.})$. See [7] for deeper theoretical aspects.

Another approach to deal with the study of the association between ordered categorical variables has been suggested in literature by Cuadras and Cuadras [4] named CA based on double accumulative frequencies. The Cuadras and Cuadras's method (DA) is based on the SVD of the matrix $\mathbf{D}_I^{-\frac{1}{2}}\mathbf{L}(\mathbf{P} - \mathbf{D}_I\mathbf{1}_I\mathbf{1}_J^T\mathbf{D}_J)\mathbf{M}^T\mathbf{W}^{\frac{1}{2}}$ which amounts to $\mathrm{GSVD}[\mathbf{D}_I^{-}\mathbf{L}(\mathbf{P} - \mathbf{P}_I\mathbf{1}_I\mathbf{1}_J^T\mathbf{P}_J)\mathbf{M}^T\mathbf{W}^{\frac{1}{2}}]_{\mathbf{D}_I,\mathbf{I}}$ where **L** is an unit upper triangular matrix. This approach does not seem to lead to the decomposition of any known association index, and matrices **L** and **M** pool the rows and the columns of table in a successive manner such that they do not provide the necessary $2(I-1) \times 2(J-1)$ 2×2 tables to compute Hirotsu's index χ^{**2}.

4 Correspondence Analysis of Ordinal Cross-Classifications Based on the Bivariate Moment Decomposition

Consider the matrix $\Omega = \mathbf{D}_I^{-1/2}(\mathbf{P} - \mathbf{D}_I\mathbf{1}_I\mathbf{1}_J^T\mathbf{D}_J)\mathbf{D}_J^{-1/2}$ of size $I \times J$ with $\mathbf{R} = \Omega\Omega^T$ and $\mathbf{C} = \Omega^T\Omega$. It is well known [3] that the generalised singular value decomposition of the matrix of Pearson contingencies yields results that are equivalent to those obtained by performing an eigen-decomposition on the matrices **R** and **C**. A reliable method of calculating the orthogonal vectors with the Gram–Schmidt process is to use the recurrence formulae of Emerson [9]. Using Emerson's orthogonal polynomials (OPs) it is possible to decompose ϕ^2 in different components, each of which represents a different power of the supposed relationship between row and column (linear, quadratic, ...). The resulting scoring scheme allows then a clear interpretation of these components.

4.1 *Double Ordered Correspondence Analysis*

Double Ordered Correspondence Analysis [1] (DOCA) decomposes the (i, j)-th Pearson ratio α_{ij} so that $\alpha_{ij} = 1 + \sum_{u=1}^{I-1} \sum_{v=1}^{J-1} a_u(i) b_v(j) \tilde{z}_{uv}$. For this method of decomposition, $\tilde{z}_{uv} = \sqrt{n} \sum_{i,j} a_u(i)\, b_v(j)\, p_{ij}$ is the (u, v)th generalised correlation where $\{a_u(i) : u = 1, \ldots, I-1\}$ and $\{b_v(j) : v = 1, \ldots, J-1\}$ are the OPs [9] for the i-th row and j-th column respectively. The bivariate association $\tilde{z}_{uv}$ are collected in $\mathbf{Z} = \mathbf{A}_*^T \mathbf{P} \mathbf{B}_*$ where $\mathbf{A}_*$ contains the $I-1$ non-trivial row OPs and $\mathbf{B}_*$ is the $J \times (J-1)$ matrix of the $J-1$ non-trivial column OPs. The matrix Ω can be then rewritten as $\mathbf{D}_I^{-1/2}(\mathbf{P} - \mathbf{D}_I \mathbf{1}_I \mathbf{1}_J^T \mathbf{D}_J)\mathbf{D}_J^{-1/2} = \mathbf{A}_* \mathbf{Z} \mathbf{B}_*^T$ with $\mathbf{A}_*^T \mathbf{D}_I \mathbf{A}_* = \mathbf{I}$ and $\mathbf{B}_*^T \mathbf{D}_J \mathbf{B}_* = \mathbf{I}$. This kind of decomposition has been named "Bivariate Moment Decomposition".

The elements of $\mathbf{Z}$ (that is, the bivariate associations $\tilde{z}_{uv}$) are independent and asymptotically standard normal [1]. Moreover, Rayner and Best [17] showed that the χ^2 can be decomposed into the sum of squares of the generalized correlations so that $\phi^2 = \chi^2/n = \sum_{u=1}^{I-1} \sum_{v=1}^{J-1} \tilde{z}_{uv}^2$. Observe that χ^2 is partitioned into $(I-1)(J-1)$ terms, where the significance of each term can be compared with the χ^2 with one degree of freedom (dof). Sources of variation for the row and column profiles can be easily obtained. For instance, any difference in the row profiles in terms of their location is computed by $\sum_{v=1}^{J-1} \tilde{z}_{1v}^2$ while the row dispersion component is given by $\sum_{v=1}^{J-1} \tilde{z}_{2v}^2$. The significance of each component can be compared with the χ^2 with $(J-1)$ dof. Similarly, location and dispersion column components can be computed. This approach to correspondence analysis uses then the bivariate moment decomposition to identify linear (location), quadratic (dispersion) and higher order moments.

4.2 *Double Ordered Non Symmetric Correspondence Analysis*

For the analysis of the cross-classification of two ordinal variables, Lombardo et al. [14] propose the BMD decomposition of the τ index by the methodology named Doubly Ordinal Non Symmetric Correspondence Analysis (DONSCA). Thy apply BMD to the centred row profiles such that $\mathbf{D}_I^{-1}(\mathbf{P} - \mathbf{D}_I \mathbf{1}_I \mathbf{1}_J^T \mathbf{D}_J) = \mathbf{A}_* \Lambda \mathbf{B}_*$ where $\Lambda = \mathbf{A}_*^T \mathbf{D}_I \mathbf{P} \mathbf{B}_*$ with $\mathbf{A}_*^T \mathbf{D}_I \mathbf{A}_* = \mathbf{I}_{I-1}$ and $\mathbf{B}_*^T \mathbf{B}_* = \mathbf{I}_{J-1}$. Lombardo et al. [14] showed that the numerator of the Goodman-Kruskal τ index can be decomposed into the sum of squares of the generalized correlations so that $N_\tau = \sum_{u=1}^{I-1}\sum_{v=1}^{J-1}\lambda_{uv}^2$. Each term λ_{uv}^2 shows the quality of the symmetric/asymmetric association of the ordered categorical variables. The linear components for row and column variables can be obtained. The significance overall predicability can be tested by the C statistics given by $C = \sum_{u=1}^{I-1}\sum_{v=1}^{J-1}\tilde{\lambda}_{uv}^2$ where each term $\tilde{\lambda}_{uv}^2 = \lambda_{uv}^2[(n-1)(J-1)/(1-\sum_{j=1}^{J}p_{.j}^2)]^{1/2}$ is a random variable from an asymptotically standard normal distribution [6] and $C \sim \chi^2_{(I-1)(J-1)}$ and each $\tilde{\lambda}_{i.}^2 \sim \chi^2_{(J-1)}$.

5 Correspondence Analysis Based on the Hybrid Moment Decomposition

5.1 *Singly Ordered Correspondence Analysis*

An alternative approach to partitioning χ^2 for a two-way contingency table with one ordered set of categories is given by the Singly Ordered Correspondence Analysis [1] (SOCA). This method combines the approach of orthogonal polynomials for the ordered columns and singular vectors for the unordered rows (named Hybrid Moment Decomposition, HMD), such that $\chi^2 = \sum_{u=1}^{M^*}\sum_{v=1}^{J-1}\mathbf{Z}_{(u)v}^2$ with $M^* \leq I-1$ and where $\tilde{z}_{(u)v} = \sqrt{n}\sum_{i,j}p_{ij}a_{iu}b_v(j)$ are asymptotically standard normally distributed random variables. The parentheses around u indicates that the above formulas are concerned with a non-ordered set of row categories. Quantities $\tilde{z}_{(u)v}$ can be written as $\mathbf{Z} = \mathbf{A}^T\mathbf{P}\mathbf{B}_*$ where $\mathbf{A}$ is the $I \times (I-1)$ matrix of left singular vectors. The value of $\tilde{z}_{(u)v}$ means that each principal axis from a simple correspondence analysis can be partitioned into column component values. The Pearson ratio is given by $\alpha_{ij} = \sum_{u=0}^{M^*}\sum_{v=0}^{J-1}a_{iu}(\tilde{z}_{(u)v}/\sqrt{n})b_v(j)$. Eliminating the trivial solution, the matrix Ω can be also rewritten as $\mathbf{D}_I^{-1/2}(\mathbf{P}-\mathbf{D}_I\mathbf{1}_I\mathbf{1}_J^T\mathbf{D}_J)\mathbf{D}_J^{-1/2} = \mathbf{AZB}_*^T$ with $\mathbf{A}^T\mathbf{D}_I\mathbf{A} = \mathbf{I}$ and $\mathbf{B}_*^T\mathbf{D}_J\mathbf{B}_* = \mathbf{I}$. For deeper information refer to [1].

5.2 *Singly Ordered Non Symmetric Correspondence Analysis*

Singly ordinal non-symmetric correspondence analysis [14] allows to combine the summaries obtained from SVD and BMD of the data. The total inertia as well as the partial inertia can be expressed by components that reflect within- and between-variable variation in terms of location, dispersion and higher-order moments. For this approach, the numerator of the Goodman–Kruskal tau index N_τ, is partitioned using generalised correlations. The authors illustrate two distinct approaches: one when the predictor variable consists of ordered categories (SONSCA1) and another when the response variable consists of ordered categories (SONSCA2).

In SONSCA1, the hybrid decomposition uses OPs for the row categories and singular vectors for the nominal column categories such that $\mathbf{D}_I^{-1}(\mathbf{P} - \mathbf{D}_I \mathbf{1}_I \mathbf{1}_J^T \mathbf{D}_J) = \mathbf{A}_* \Lambda \mathbf{B}$ where $\Lambda = \mathbf{A}_*^T \mathbf{D}_I \mathbf{P} \mathbf{B}$ with $\mathbf{A}_*^T \mathbf{D}_I \mathbf{A}_* = \mathbf{I}_{I-1}$ and $\mathbf{B}^T \mathbf{B} = \mathbf{I}$. N_τ is decomposed as $N_\tau = \sum_{u=1}^{I-1} \sum_{m=1}^{M} \lambda_{um}^2$. The significance overall predicability can be tested by $C = \sum_{u=1}^{I-1} \sum_{m=1}^{M} \tilde{\lambda}_{um}^2$ where each term $\tilde{\lambda}_{um}^2 = \lambda_{um}^2 [(n-1)(J-1)/(1 - \sum_{j=1}^{J} p_{.j}^2)]^{1/2}$ is a random variable from an asymptotically standard normal distribution.

Similarly, in SONSCA2 , the hybrid decomposition uses instead singular vectors for the nominal row categories and OPs for the column categories such that $\mathbf{D}_I^{-1}(\mathbf{P} - \mathbf{D}_I \mathbf{1}_I \mathbf{1}_J^T \mathbf{D}_J) = \mathbf{A} \Lambda \mathbf{B}_*$ where $\Lambda = \mathbf{A}^T \mathbf{D}_I \mathbf{P} \mathbf{B}_*$ with $\mathbf{A}^T \mathbf{D}_I \mathbf{A} = \mathbf{I}$ and $\mathbf{B}_*^T \mathbf{B}_* = \mathbf{I}_{J-1}$. N_τ is decomposed as $N_\tau = \sum_{m=1}^{M} \sum_{v=1}^{J-1} \lambda_{mv}^2$. The significance overall predicability is tested by $C = \sum_{m=1}^{M} \sum_{v=1}^{J-1} \tilde{\lambda}_{mv}^2$ where each term $\tilde{\lambda}_{mv}^2 = \lambda_{mv}^2 [(n-1)(J-1)/(1 - \sum_{j=1}^{J} p_{.j}^2)]^{1/2}$ is a random variable from an asymptotically standard normal distribution.

5.3 Hybrid Cumulative Correspondence Analysis

D'Ambra [5] extends the properties of OPs to the Cumulative Correspondence Analysis with the aim to identify the linear, quadratic and higher order components of rows with respect to the aggregate columns categories. The joint effect of these methods leads to the decomposition of the T in power components. In hybrid cumulative correspondence analysis (HCCA) both variables present an ordinal structure and only for the response one (column) it has been considered the cumulative sum. In HCCA, the hybrid decomposition uses OPs for the ordinal row categories and singular vectors for the cumulative column categories such that $\mathbf{D}_I^{-1}(\mathbf{P} - \mathbf{D}_I \mathbf{1}_I \mathbf{1}_J^T \mathbf{D}_J)\mathbf{M}^T \mathbf{W}^{\frac{1}{2}} = \mathbf{A}_* \Lambda \mathbf{B}$ where $\Lambda = \mathbf{A}_*^T \mathbf{D}_I \mathbf{P} \mathbf{B}$ with $\mathbf{A}_*^T \mathbf{D}_I \mathbf{A}_* = \mathbf{I}_{I-1}$ and $\mathbf{B}^T \mathbf{B} = \mathbf{I}$. D'Ambra [5] shows that Taguchi's index T is decomposed as $T = n \sum_{u=1}^{I-1} \sum_{m=1}^{M} \lambda_{um}^2$. Moreover, let $\mathbf{P}_j$ be the $I \times 2$ contingency table obtained by aggregating the first j column categories and the remaining categories $(j+1)$ to J of table $\mathbf{P}$. The linear, quadratic and higher order components of T can be written as sum of the correspondence components of rows computed for each matrix $\mathbf{P}_j$, respectively. For example, the linear components of T_L can be decomposed according to the sum of the linear components ${}^j\chi_L^2$ of matrices $\mathbf{P}_j$ with $j = 1, \ldots, J-1$: $T_L = n \sum_{v=1}^{J-1} \tilde{z}_{1v}^2 = {}^1\chi_L^2 + {}^2\chi_L^2 + \cdots + {}^{J-1}\chi_L^2$. Similarly, the quadratic and higher order components of T can be obtained. The T statistics can be also written as $T = \sum_{j=1}^{J-1} \sum_{i=1}^{I-1} {}^j\tilde{\lambda}_i^2$ where each element ${}^j\tilde{\lambda}_i$ is a statistical variable that follows an asymptotically standard normally distribution. Moreover, each ${}^j\tilde{\lambda}_i^2$ follows a chi-square distribution with 1 degree of freedom.

6 A Unified Approach

Let's consider the following factorization of a matrix into a product of matrices that we name "Generalized Factorization of a Matrix" (GFM).

Let Γ and Φ be given positive definite symmetric matrices of order $(n \times n)$ and $(p \times p)$, respectively. The GFM of matrix $\mathbf{X}$ is defined as $\mathbf{X} = \mathbf{U}\Lambda\mathbf{V}^T$ where the columns of $\mathbf{U}$ and $\mathbf{V}$ are orthonormalized with respect to Γ and Φ (that is $\mathbf{U}^T\Gamma\mathbf{U} = \mathbf{I}$ and $\mathbf{V}^T\Phi\mathbf{V} = \mathbf{I}$), respectively, and Λ is a positive definite matrix. It is noted as $\mathrm{GFM}(\mathbf{X})_{\Omega,\Phi}$. GFM can take several forms. For instance, if the used GFM is the GSVD then matrices $\mathbf{U}$ and $\mathbf{V}$ are given by $\mathbf{A}$ and $\mathbf{B}$ that are orthonormalized with respect to Γ and Φ (that is $\mathbf{A}^T\Gamma\mathbf{A} = \mathbf{I}$ and $\mathbf{B}^T\Phi\mathbf{B} = \mathbf{I}$), respectively, with $\Lambda = \mathbf{D}_\lambda$ is a diagonal and positive definite matrix containing the generalized singular values, ordered from largest to smallest. They can be obtained by means the ordinary SVD. Similarly, if the used GFM is instead the BMD then matrices $\mathbf{U}$ and $\mathbf{V}$ are given by the the vectors of $\mathbf{A}_*$ and $\mathbf{B}_*$ orthonormalized with respect to Γ and Φ (that is $\mathbf{A}_*^T\Gamma\mathbf{A}_* = \mathbf{I}$ and $\mathbf{B}_*^T\Phi\mathbf{B}_* = \mathbf{I}$), respectively, with $\Lambda = \mathbf{A}_*^T\mathbf{P}\mathbf{B}_*$ (Λ in this case is not a diagonal matrix). In order to represent the rows and columns of $\mathbf{N}$ we can consider the following unifying expression

$$\mathrm{GFM}[\mathbf{D}_{(R)}^{-}\hat{\mathbf{R}}(\mathbf{P} - \mathbf{P}_I\mathbf{1}_I\mathbf{1}_J^T\mathbf{P}_J)\hat{\mathbf{C}}^T\mathbf{W}^{\frac{1}{2}}\mathbf{D}_{(C)}^{-}]_{\mathbf{D}_{(R)},\mathbf{D}_{(C)}} \tag{1}$$

that is equivalent to write $\mathbf{D}_{(R)}^{-1/2}\hat{\mathbf{R}}(\mathbf{P} - \mathbf{P}_I\mathbf{1}_I\mathbf{1}_J^T\mathbf{P}_J)\hat{\mathbf{C}}^T\mathbf{W}^{\frac{1}{2}}\mathbf{D}_{(C)}^{-1/2} = \mathbf{U}\Lambda\mathbf{V}^T$. According to type of used factorization (GSVD/SVD, BMD and HMD), this GFM let us to subsume all the previous approaches (see Table 1). Methods for which the ordinal structure of a categorical variable is directly taken into account in their formulations are listed in bold character in Table 1.

For instance, if $\mathbf{D}_{(R)} = \mathbf{D}_I$, $\hat{\mathbf{R}} = \mathbf{I}$, $\hat{\mathbf{C}} = \mathbf{I}$, $\mathbf{W} = \mathbf{I}$ and $\mathbf{D}_{(C)} = \mathbf{D}_J$, with $\mathbf{U} = \mathbf{A}$, $\mathbf{V} = \mathbf{B}$ and $\Lambda = \mathbf{D}_\lambda$, then $\mathrm{GFM}[\mathbf{D}_{(R)}^{-}\hat{\mathbf{R}}(\mathbf{P} - \mathbf{P}_I\mathbf{1}_I\mathbf{1}_J^T\mathbf{P}_J)\hat{\mathbf{C}}^T\mathbf{W}^{\frac{1}{2}}\mathbf{D}_{(C)}^{-}]_{\mathbf{D}_{(R)},\mathbf{D}_{(C)}}$ amounts to ordinary Correspondence Analysis (CA). Double Ordered Correspondence Analysis (DOCA) is instead obtained by using in (1) $\mathbf{D}_{(R)} = \mathbf{D}_I$, $\hat{\mathbf{R}} = \mathbf{I}$, $\hat{\mathbf{C}} = \mathbf{I}$, $\mathbf{W} = \mathbf{I}$ and $\mathbf{D}_{(C)} = \mathbf{D}_J$, with $\mathbf{U} = \mathbf{A}_*$, $\mathbf{V} = \mathbf{B}$ and $\Lambda = \mathbf{A}_*^T\mathbf{P}\mathbf{B}_*$.

Table 1 A unified framework of CA with ordinal categorical data

	GFM$[\mathbf{D}^{-}_{(R)}\hat{\mathbf{R}}(\mathbf{P}-\mathbf{D}_I\mathbf{1}_I\mathbf{1}_J^T\mathbf{D}_J)\hat{\mathbf{C}}^T\mathbf{W}^{\frac{1}{2}}\mathbf{D}^{-}_{(C)}]_{\mathbf{D}_{(R)},\mathbf{D}_{(C)}}$											
			$\mathbf{D}^{-1/2}_{(R)}\hat{\mathbf{R}}(\mathbf{P}-\mathbf{D}_I\mathbf{1}_I\mathbf{1}_J^T\mathbf{D}_J)\hat{\mathbf{C}}^T\mathbf{W}^{\frac{1}{2}}\mathbf{D}^{-1/2}_{(C)} = \mathbf{U}\Lambda\mathbf{V}^T$									
Method	Row	Column	$\mathbf{D}_{(R)}$	$\hat{\mathbf{R}}$	$\hat{\mathbf{C}}$	$\mathbf{W}$	$\mathbf{D}_{(C)}$	$\mathbf{U}$	Λ	$\mathbf{V}$	GFM	Statistic
CA	Nom-Ord	Nom-Ord	$\mathbf{D}_I$	$\mathbf{I}_I$	$\mathbf{I}_J$	$\mathbf{I}_J$	$\mathbf{D}_J$	$\mathbf{A}$	$\mathbf{D}_\lambda$	$\mathbf{B}$	SVD	ϕ^2
NSCA	Nom-Ord	Nom-Ord	$\mathbf{D}_I$	$\mathbf{I}_I$	$\mathbf{I}_J$	$\mathbf{I}_J$	$\mathbf{I}_J$	$\mathbf{A}$	$\mathbf{D}_\lambda$	$\mathbf{B}$	SVD	N_τ
TA	Nom	Ord	$\mathbf{D}_I$	$\mathbf{I}_I$	$\mathbf{M}$	$\mathbf{W}$	$\mathbf{I}_{J-1}$	$\mathbf{A}$	$\mathbf{D}_\lambda$	$\mathbf{B}$	SVD	T
HDA	Ord	Ord	$\mathbf{D}_R$	$\mathbf{R}$	$\mathbf{C}^T$	$\mathbf{I}_{2\times(J-1)}$	$\mathbf{D}_C$	$\mathbf{A}$	$\mathbf{D}_\lambda$	$\mathbf{B}$	SVD	χ^{**2}
DA	Ord	Ord	$\mathbf{D}_I$	$\mathbf{L}$	$\mathbf{M}$	$\mathbf{W}$	$\mathbf{I}_{J-1}$	$\mathbf{A}$	$\mathbf{D}_\lambda$	$\mathbf{B}$	SVD	–
DOCA	Ord	Ord	$\mathbf{D}_I$	$\mathbf{I}_I$	$\mathbf{I}_I$	$\mathbf{I}_J$	$\mathbf{D}_J$	$\mathbf{A}_*$	$\mathbf{A}_*^T\mathbf{P}\mathbf{B}_*$	$\mathbf{B}_*$	BMD	ϕ^2
DONSCA	Ord	Ord	$\mathbf{D}_I$	$\mathbf{I}_I$	$\mathbf{I}_I$	$\mathbf{I}_J$	$\mathbf{I}_J$	$\mathbf{A}_*$	$\mathbf{A}_*^T\mathbf{D}_I\mathbf{P}\mathbf{B}_*$	$\mathbf{B}_*$	BMD	N_τ
SOCA	Nom	Ord	$\mathbf{D}_I$	$\mathbf{I}_I$	$\mathbf{I}_I$	$\mathbf{I}_J$	$\mathbf{D}_J$	$\mathbf{A}$	$\mathbf{A}^T\mathbf{P}\mathbf{B}_*$	$\mathbf{B}_*$	HMD	ϕ^2
SONSCA1	Ord	Nom	$\mathbf{D}_I$	$\mathbf{I}_I$	$\mathbf{I}_I$	$\mathbf{I}_J$	$\mathbf{I}_J$	$\mathbf{A}_*$	$\mathbf{A}_*^T\mathbf{D}_I\mathbf{P}\mathbf{B}$	$\mathbf{B}$	HMD	N_τ
SONSCA2	Nom	Ord	$\mathbf{D}_I$	$\mathbf{I}_I$	$\mathbf{I}_I$	$\mathbf{I}_J$	$\mathbf{I}_J$	$\mathbf{A}$	$\mathbf{A}^T\mathbf{D}_I\mathbf{P}\mathbf{B}_*$	$\mathbf{B}_*$	HMD	N_τ
HCCA	Ord	Ord	$\mathbf{D}_I$	$\mathbf{I}_I$	$\mathbf{M}$	$\mathbf{W}$	$\mathbf{I}_{J-1}$	$\mathbf{A}_*$	$\mathbf{A}_*^T\mathbf{D}_I\mathbf{P}\mathbf{B}$	$\mathbf{B}$	HMD	T

where "Nom" and "Ord" stand for Nominal and Ordinal variable, respectively

7 Conclusion

In multivariate analysis, it is usual to link several methods in a closed expression, which depends on the choices of the nature and characteristics of several matrices. Consider as example the notation of the statistical study $(\mathbf{X}, \mathbf{Q}, \mathbf{D})$ to describe the data and their use. Indeed, this notation subsumes several methods with suitable choices for the data $\mathbf{X}$, and the columns and rows metrics $\mathbf{Q}$ and $\mathbf{D}$, respectively.

Several CA extensions cope with the possible ordinal structure of the variables using several tools for the dimensionality reduction of the involved spaces. These tools enrich the interpretation of the graphical planes providing additional information. A framework of these approach has been then proposed by a formalization of a generalized factorization of a matrix into a product of matrices. This factorization subsumes several tools like SVD, GSVD, BMD and HMD. This framework can aid the user to better understand the methodological differences between these approaches, providing the rationale for an easier their single software implementation.

References

1. Beh, E.J.: Simple correspondence analysis of ordinal cross-classifications using orthogonal polynomials. Biom. J. **39**, 589–613 (1997)
2. Beh, E.J., D'Ambra, L., Simonetti, B.: Correspondence analysis of cumulative frequencies using a decomposition of Taguchi's statistic. Commun. Stat. Theory Methods **40**, 1620–1632 (2011)
3. Beh, E.J., Lombardo, R.: Correspondence Analysis: Theory, Practice and New Strategies. Wiley (2014)
4. Cuadras, C.M., Cuadras, D.: A unified approach for the multivariate analysis of contingency tables. Open J. Stat. **5**, 223–232 (2015)
5. D'Ambra, A.: Cumulative correspondence analysis using orthogonal polynomials. Commun. Stat. Theory Methods (to appear)
6. D'Ambra, L., Beh, E.J., Amenta, P.: CATANOVA for two-way contingency tables with an ordinal response using orthogonal polynomials. Commun. Stat. Theory Methods **34**, 1755–1769 (2005)
7. D'Ambra, L., Beh, E.J., Camminatiello, I.: Cumulative correspondence analysis of two-way ordinal contingency tables. Commun. Stat. Theory Methods **43**(6), 1099–1113 (2014)
8. D'Ambra, L., Lauro, N.: Non symmetrical analysis of three-way contingency tables. In: Coppi, R., Bolasco, S. (eds.) Multiway Data Analysis, pp. 301–315. Elsevier Science Publishers B.V, Amsterdam (1989)
9. Emerson, P.L.: Numerical construction of orthogonal polynomials from a general recurrence formula. Biometrics **24**, 696–701 (1968)
10. Goodman, L.A.: A single general method for the analysis of cross-classified data: reconciliation and synthesis of some methods of Pearson, Yule, and Fisher, and also some methods of correspondence analysis and association analysis. J. Amer. Stat. Assoc. **91**, 408–428 (1996)
11. Goodman, L.A., Kruskal, W.H.: Measures of association for cross-classifications. J. Amer. Stat. Assoc. **49**, 732–764 (1954)
12. Hirotsu, C.: Cumulative chi-squared statistic as a tool for testing goodness of fit. Biometrika **73**, 165–173 (1986)

13. Light, R., Margolin, B.: An analysis of variance for categorical data. J. Amer. Stat. Assoc. **66**(335), 534–544 (1971)
14. Lombardo, R., Beh, E.J., D'Ambra, L.: Non-symmetric correspondence analysis with ordinal variables. Comput. Stat. Data Anal. **52**, 566–577 (2007)
15. Mood, A.M.: On the asymptotic efficiency of certain non-parametric two-sample tests. Ann. Math. Stat. **25**, 514–522 (1954)
16. Nair, V.N.: Chi-squared type tests for ordered alternatives in contingency tables. J. Amer. Stat. Assoc. **82**, 283–291 (1987)
17. Rayner, J.C.W., Best, D.J.: Smooth extensions of Pearson's product moment correlation and Spearman's rho. Stat. Probab. Lett. **30**, 171–177 (1996)
18. Satterthwaite, F.E.: An approximate distribution of estimates of variance components. Biom. Bull. **2**(6), 110–114 (1946)
19. Taguchi, G.: Statistical Analysis. Maruzen, Tokyo (1966)
20. Taguchi, G.: A new statistical analysis for clinical data, the accumulating analysis, in contrast with the chi-square test. Saishin Igaku **29**, 806–813 (1974)

Part IV
Forecasting Time Series

Extended Realized GARCH Models

Richard Gerlach and Giuseppe Storti

Abstract We introduce a new class of models that extends the Realized GARCH models of Hansen et al. (J Appl Econom 27:877–906, 2012, [10]). Our model generalizes the original specification of Hansen et al. (J Appl Econom 27:877–906, 2012, [10]). along three different directions. First, it features a time varying volatility persistence. Namely, the shock response coefficient in the volatility equation adjusts to the time varying accuracy of the associated realized measure. Second, our framework allows to consider, in a parsimonious way, the inclusion of multiple realized measures. Finally, it allows for heteroskedasticity of the noise component in the measurement equation. The appropriateness of the proposed class of models is appraised by means of an application to a set of stock returns data.

Keywords Realized measures · GARCH · Volatility forecasting

1 Motivation and Aim

Aim of this paper is to develop a generalized version of the Realized GARCH model proposed by Hansen et al. [10]. In particular, we propose an alternative parameterization that generalizes the standard realized GARCH in three different directions. First, we allow for time varying persistence in the volatility equation. Namely, we allow the coefficient of the lagged realized volatility (RV) to depend on a measure of its accuracy, the realized quarticity, such in a way that more weight is given to lagged volatilities when they are more accurately measured. This is in line with the recent findings of Bollerslev et al. [6]. Considering a HAR model, they observe that, by letting the volatility persistence to depend on the estimated degree of measurement error, it is possible to remarkably improve the model's predictive performance. Along the same

R. Gerlach
Discipline of Business Analytics, University of Sydney, Sydney, Australia
e-mail: richard.gerlach@usyd.edu.au

G. Storti (✉)
DiSES, University of Salerno, Salerno, Italy
e-mail: storti@unisa.it

C. Perna et al. (eds.), *Studies in Theoretical and Applied Statistics*,
Springer Proceedings in Mathematics & Statistics 227,
https://doi.org/10.1007/978-3-319-73906-9_14

track, Shephard and Xiu [13] find evidence that the magnitude of the response coefficients associated with different RV measures in a GARCHX is related to the quality of the measure itself. Finally, Hansen and Huang [9] observe that the response of the current conditional variance to past unexpected volatility shocks is negatively correlated with the accuracy of the associated realized volatility measure. Second, our framework allows for considering multiple volatility measures in a parsimonious way. Differently from Hansen and Huang [9], this is achieved by considering a single measurement equation where the dependent variable is given by an average of different realized measures whose weights are estimated along with the other model parameters. Third, and last, we allow for heteroskedasticity in the noise component of the measurement equation by introducing an ARCH-style updating equation for the noise variance. Our empirical results show that introducing heteroskedasticity into the model does have a remarkable effect on the estimated parameters and on the out-of-sample forecasting performance of the model.

The paper is organized as follows. In Sect. 2 we recall the standard Realized GARCH model of Hansen et al. [10] while the proposed flexible Realized GARCH specifications are introduced and discussed in Sect. 3. Section 4 presents the results of an application to a time series of stock returns. Section 5 concludes.

2 The Realized GARCH Model

Hansen et al. [10] have proposed the Realized GARCH class of models as an alternative to standard GARCH models where, in the dynamic volatility equation, the squared returns are replaced by a much more efficient proxy such as a realized volatility measure. The standard Realized GARCH model, in its linear formulation, is defined as

$$\begin{aligned} r_t &= h_t z_t \qquad z_t \underset{iid}{\sim} (0,1) \\ h_t^2 &= \omega + \gamma v_{t-1} + \beta h_{t-1}^2 \\ v_t &= \xi + \phi h_t^2 + \tau(z_t) + u_t \qquad u_t \underset{iid}{\sim} (0, \sigma_u^2) \end{aligned}$$

where v_t is the chosen realized volatility measure. The last equation (measurement equation) is justified by the fact that any consistent estimator of the IV can be written as the sum of the conditional variance plus a random innovation. The function $\tau(z)$ can accommodate for leverage effects. A common choice is to set

$$\tau(z) = \tau_1 z + \tau_2(z^2 - 1).$$

It can be easily shown that the model implies an AR(1) representation for h_t^2

$$h_t^2 = \mu + \pi h_{t-1}^2 + w_{t-1}.$$

where

$$\mu = \omega + \gamma\xi$$
$$\pi = \beta + \gamma\phi$$
$$w_t = \gamma\tau(z_t) + \gamma u_t.$$

The condition $|\pi| < 1$ ensures that h_t^2 has a stationary representation.

The model parameters can be easily estimated by standard maximum likelihood. In general, the contribution of the t-th observation to the joint likelihood of the complete data (r_t, v_t), for $t = 2, \ldots, T$, can be decomposed as

$$l_{r,u}(r_t, v_t|I_{t-1}) = l_r(r_t|I_{t-1})l_v(v_t|r_t, I_{t-1})$$

where I_{t-1} is the past information vector including $\{v_1, ..., v_{t-1}, r_1, ..., r_{t-1}\}$. As in Gerlach and Wang [8] we assume that $(r_t|I_{t-1}) \sim t_\nu$ and $(v_t|r_t, I_{t-1}) \sim N(0, \sigma_u^2)$. These choices give

$$l_r(r_t|I_{t-1}) = \left\{ A(\nu) + \frac{1}{2}\log(h_t^2) + \frac{\nu+1}{2\left(1 + \frac{r_t^2}{h_t^2(\nu-2)}\right)} \right\}$$
$$l_v(v_t|r_t, I_{t-1}) = -\frac{1}{2}\left\{ \log(2\pi) + \log(\sigma_u^2) + \frac{u_t^2}{\sigma_u^2} \right\}$$

where

$$A(\nu) = \log\left(\Gamma\left(\frac{\nu+1}{2}\right)\right) - \log\left(\Gamma\left(\frac{\nu}{2}\right)\right) + \log(\pi(\nu-2)).$$

3 Flexible Realized GARCH Models

In this section we present two different generalizations of the standard Realized GARCH model illustrated in Sect. 2. First, in Sect. 3.1 we introduce an adaptive RGARCH model allowing for time varying persistence in the volatility equation. Furthermore, the proposed model allows to account for the presence of heteroskedasticity in the noise component of the measurement equation. Second, in Sect. 3.2 we extend this specification to allow for the inclusion of multiple realized measures. Both specifications nest the standard RGARCH model which can be obtained as a special case under standard parametric restrictions.

3.1 Single Measure Adaptive RGARCH Models

A single measure adaptive RGARCH (SMA-RGARCH) model differs from the standard RGARCH model of Hansen et al. [10] for the structure of its volatility equation which, for a (1,1) model, is given by

$$\begin{aligned} h_t^2 &= \omega + (\gamma + \delta g(RQ_{t-1}))v_{t-1} + \beta h_{t-1}^2 \\ &= \omega + \gamma_t v_{t-1} + \beta h_{t-1}^2 \end{aligned} \tag{1}$$

where $\gamma_t = (\gamma + \delta g(RQ_{t-1}))$. In the above relationship, $g(.)$ is an appropriately defined mathematical function and RQ_t denotes the realized quarticity at time t that, up to a scale factor, is given by

$$RQ_t = \sum_{i=1}^{M} r_{t,i}^4$$

where $r_{t,i}$ indicates the i-th intradaily return on day t, for $i = 1, \ldots, M$, which is the number of intra-daily observations available at the chosen frequency. Although several choices of $g(.)$ are in principle feasible, in this paper we have set

$$g(RQ_t) = log(RQ_t^{1/2}).$$

The transformation to a logarithmic-scale has the advantage of significantly reducing the relative variability of realized quarticity and, on empirical ground, we found this to have a beneficial effect on the variability of the estimates. In addition, in order to further smooth the dynamics of the state variable $g(RQ_t)$, one could also consider pre-smoothing its argument by some linear filter such as an Exponentially Weighted Moving Average (EWMA)

$$\tilde{RQ}_t = (1 - \lambda)\tilde{RQ}_{t-1} + \lambda RQ_{t-1} \tag{2}$$

where the smoothing coefficient $0 < \lambda < 1$ can be estimated along with the other parameters leaving to the data the choice of the optimal amount of smoothing. Values of λ close to 1 will result in a very smooth curve while, on the other side, values close to 0 will return a filtered series much closer to the raw data.

The choice of the realized quarticity as state variable determining the instantaneous volatility persistence is motivated by asymptotic arguments. Following Barndorff-Nielsen and Shephard [4] let us assume

$$dlog(Pt) = \mu_t dt + \sigma_t dWt, \tag{3}$$

where P_t is the price process, μ_t and σ_t are the drift and the instantaneous volatility process and W_t is a standard Brownian motion. The integrated variance at time t is given by $IV_t = \int_{t-1}^{t} \sigma_s^2 ds$. Barndorff-Nielsen and Shephard [4] find that, under standard regularity conditions, IV_t can be consistently estimated by the realized variance

$$RV_t = \sum_{i=1}^{M} r_{t,i}^2$$

where $r_{t,i} = log(P_{t-1+i\Delta}) - log(P_{t-1+(i-1)\Delta})$ is the i-th Δ-period intraday return and $M = 1/\Delta$. In particular they show that, as $\Delta \to 0$, the following holds

$$RV_t = IV_t + \eta_t, \qquad \eta_t \sim N(0, 2\Delta IQ_t) \tag{4}$$

where $IQ_t = \int_{t-1}^{t} \sigma_s^4 ds$ is the integrated quarticity which can be consistently estimated by the realized quarticity $\frac{M}{3}RQ_t$.

Similar results can be shown to hold for other realized measures such as, among others, the medRV estimator and the Bipower and Tripower Variation (see [1]), Subsampled RV [14] and Realized Kernels [5]. For all these realized measures the variance of the estimation error η_t would depend on the value of the integrated quarticity which is inherently time varying. Given that RQ_t takes positive values very close to zero, by far smaller than 1, its log-transform will always be negative. It follows that, if the δ parameter is negative, the model in (1) is giving more weight to the lagged realized volatility measure in periods in which this is more accurately estimated. On the other hand, as the realized quarticity increases and, hence, the accuracy of the volatility measure worsens, $log(RQ_t^{1/2})$ increases leading the model to downweight the lagged volatility measure. We expect this to have a positive effect on the forecasting accuracy of the volatility model.

3.2 Multiple Measure Adaptive RGARCH Models

This section presents an extension of the SMA-RGARCH model which is able to incorporate information from multiple realized volatility measures. In order to save parameters and escape from the incumbent curse of dimensionality, instead of defining and estimating a different measurement equation for each of the realized measures included, we consider a single measurement equation where the dependent variable is a weighted average of the realized measures appearing in the volatility equation. The weights are not predetermined but they can be estimated by maximum likelihood with the other parameters. Although, for ease of exposition we will focus on a model with two realized measures, the generalization to k (≥ 2) measures is immediate.

The volatility equation of a 2 component multiple measure adaptive RGARCH(1,1), MMA-RGARCH(2,1,1), model is given by

$$h_t^2 = \omega + (\gamma + \delta g(RQ_{t-1}))\bar{v}_{t-1} + \beta h_{t-1}^2 \tag{5}$$

where $\bar{v}_t = w_1 v_{1,t} + w_2 v_{2,t}$, with $w_2 = 1 - w_1$, is a weighted average of the two realized measures $v_{1,t}$ and $v_{2,t}$. So, as a by-product of the estimation procedure, the model also returns an optimized volatility measure which is computed as the average of the realized measures appearing in the volatility equation. The model specification is completed by the measurement equation for $\bar{v}_t$

$$\bar{v}_t = \xi + \phi h_t^2 + \tau(z_t) + u_t. \tag{6}$$

It is important to note that, since the measurement equation is unique, the number of additional parameters is fixed and does not increase with k, the number of realized measures included in the model. Furthermore, by simple algebra, Eq. (5) can be rewritten as

$$\begin{aligned} h_t^2 &= \omega + (\gamma w_1 + \delta w_1 g(RQ_{t-1}))v_{1,t-1} \\ &+ (\gamma(1 - w_1) + \delta(1 - w_1)g(RQ_{t-1}))v_{2,t-1} + \beta h_{t-1}^2 \\ &= \omega + \gamma_{1,t} v_{1,t-1} + \gamma_{2,t} v_{2,t-1} + \beta h_{t-1}^2 \end{aligned}$$

where $\gamma_{1,t} = \gamma w_1 + \delta w_1 g(RQ_{t-1})$ and $\gamma_{2,t} = \gamma(1 - w_1) + \delta(1 - w_1)g(RQ_{t-1})$.

4 An Application to Stock Market Data

In this section we illustrate the results of an application to a time series of log-returns on the Allianz stock, traded on the Xetra Market in the German Stock Exchange. The original dataset included tick-by-tick data on transactions taking place over the period from 2/1/2002 to 27/12/2012. The raw data have been cleaned using the procedure described in Brownlees and Gallo [7] and then converted to an equally spaced series of five-minute log-returns. These have been aggregated on a daily basis to compute a time series of 2792 daily open-to-close log-returns and two different realized measures: the standard realized variances (RV_t) and the jump-robust medRV ($medRV_t$) estimator proposed by Andersen et al. [1]. Following Bandi and Russell [2], in order to limit the impact of jumps, the realized quarticity has been computed using a coarser grid based on 20 min sampling.

The data have then been modelled by a SMA-RGARCH(1,1), fitted using both 5 min RV and the medRV estimator, and a MMA-RGARCH(2,1,1) model, blending both volatility measures into a single model. In addition we have considered the heteroskedastic versions of these models where we have assumed that the measurement noise variance is time varying according to the dynamic equation

$$var(u_t) = \sigma_t^2 = e^{\psi_0 + \psi_1 u_{t-1}^2} \qquad t = 2, \dots, T.$$

In the reminder we will indicate these variants as the HSMA-RGARCH and HMMA-RGARCH models, respectively. The results have been compared with those obtained by fitting standard RGARCH models. The full sample estimates of model parameters and associated p-values have been reported in Table 1. The δ parameter is always negative and significant in three cases out of six ($SMA - RG_{mrv}$, $MMA - RG$ and $HMMA - RG$) providing evidence of a state dependent response of h_t^2 to the lagged realized measure. It is particularly interesting to analyze the value of the estimated ϕ in the MMA-RGARCH model. In the homoskedastic model ϕ is $\approx$1.45, coupled with $w_1 \approx 0.80$, suggesting that the optimized measure $\bar{v}_t$ is biased. The situation changes when a heteroskedastic component is included. In this case we have $\delta \approx 1.04$ with w_1, the weight of RV_t, halved. Similar considerations hold for the value of ξ. Hence we expect the proxy $\bar{v}_t$, obtained under the heteroskedasticity assumption, to have nicer statistical properties and, in particular, to be approximately unbiased.

For completeness, for each model we also report the values of $\ell_{r,u}$, and the associated BIC, along with the partial returns log-likelihood ℓ_r. However, we must remark that the overall log-likelihoods $\ell_{r,u}$ are in general not comparable across different models while it makes sense to compare the returns partial likelihoods ℓ_r. Our in-sample results show that the values of ℓ_r are fairly stable across different models failing to indicate a clear winner.

The out-of-sample predictive ability of the fitted models has been assessed by means of a rolling window forecasting exercise using a window of 1000 days. The out-of-sample period goes from June 7, 2005 to the end of the sample and includes 1792 daily observations covering the credit crisis and the turbulence period running from November 2011 to the beginning of 2012. Forecast accuracy is measured by means of the QLIKE loss function. Our choice is motivated by the consideration that, compared to other robust alternatives, such as the Mean Squared Error (MSE), this loss function has revealed to be more powerful in rejecting poorly performing predictors (see [12]). Looking at average values of the QLIKE loss function for each of the models considered, reported in the last row of Table 1, it turns out that the lowest value is obtained for the HMMA-RGARCH model. Furthermore, assessing the significance of differences across different models by the Model Confidence Set (MCS) of Hansen et. al [11], it turns out that the HMMA-RGARCH model is also the only model entering the MCS at the 0.80 confidence level.

Figure 1 compares the in-sample estimated time varying volatility response coefficient (γ_t) for the (H)SMA-RGARCH and (H)MMA-RGARCH models. The SMA-RGARCH models fitted to RV and medRV give very close value of γ_t (to avoid overlapping of the two curves, the γ_t series obtained for medRV has been shifted upwards adding 0.3 to the fitted values). It is also evident that the introduction of the heteroskedastic component has the effect of lowering the variability and, except for $HSMA_{mrv}$, even the level of γ_t. The comparison with the RQ_t plot shows that in all cases more or less pronounced drops of γ_t correspond to the peaks in the RQ_t series.

Table 1 MLE, BIC, log-likelihood ($\ell_{r,u}$) and partial log-likelihood (ℓ_r) values of fitted Realized GARCH models: RGARCH (RG), SMA-RGARCH (SMA-RG), MMA-RGARCH (MMA-RG) and their heteroskedastic counterparts, denoted by the the prefix H (values in brackets are p-values; *: $\times 10^6$; † $\times 10^{-6}$). The last line reports the average value of the QLIKE loss over the chosen out-of-sample period (⋆: model included in the 80% MCS.)

	RG_{rv}	RG_{mrv}	$SMA-RG_{rv}$	$SMA-RG_{mrv}$	$MMA-RG$	HRG_{rv}	HRG_{mrv}	$HSMA-RG_{rv}$	$HSMA-RG_{mrv}$	$HMMA-RG$
ω *	6.4272 (0.2998)	19.588 (0.0833)	3.6715 (0.7597)	0.0001 (1.0000)	0.0010 (0.9999)	0.0470 (0.9858)	3.7866 (0.3396)	0.0000 (1.0000)	0.0028 (0.9997)	2.355 (0.5039)
γ	0.4286 (0.0168)	0.1668 (0.0000)	0.3009 (0.5865)	0.0000 (0.9999)	0.0000 (1.0000)	0.2194 (0.0000)	0.1877 (0.0000)	0.1715 (0.441)	0.6045 (0.1194)	0.1503 (0.0121)
β	0.4408 (0.0525)	0.6733 (0.0000)	0.4257 (0.0672)	0.5088 (0.0011)	0.3999 (0.0036)	0.7307 (0.0000)	0.7870 (0.0000)	0.7207 (0.0000)	0.5443 (0.021)	0.6157 (0.0000)
w_1	–	–	–	–	0.8074 (0.0000)	–	–	–	–	0.4077 (0.0002)
δ	–	–	−0.0310 (0.6553)	−0.0643 (0.0013)	−0.0656 (0.0000)	–	–	−0.0071 (0.8232)	−0.0002 (0.9961)	−0.0238 (0.008)
ξ *	49.5156 (0.0125)	−125.0596 (0.044)	56.1285 (0.0658)	0.0002 (1.0000)	−42.4227 (0.0407)	16.94 (0.2347)	12.5515 (0.461)	0.0000 (1.0000)	96.5967 (0.0000)	4.5973 (0.5822)
ϕ	1.1298 (0.0000)	1.9403 (0.0000)	0.9947 (0.0000)	1.1834 (0.0000)	1.4526 (0.0000)	1.1549 (0.0000)	0.9950 (0.0000)	1.2502 (0.0000)	0.6178 (0.0000)	1.0412 (0.0000)
τ_1*	−36.4781 (0.0014)	−24.5184 (0.0454)	−39.5188 (0.0011)	−26.337 (0.0401)	−36.632 (0.0031)	−6.2075 (0.5754)	0.5214 (0.9719)	−7.7144 (0.5047)	−27.8896 (0.0443)	−12.9031 (0.0000)
τ_2*	56.8369 (0.0000)	56.0213 (0.0000)	63.6834 (0.0000)	65.7912 (0.0000)	58.8661 (0.0000)	33.9352 (0.0003)	48.8029 (0.0038)	34.5477 (0.0006)	73.0413 (0.0000)	31.828 (0.0000)
ψ_0	–	–	–	–	–	−16.2545 (0.0000)	−16.0414 (0.0000)	−16.2527 (0.0000)	−14.8416 (0.0000)	−17.7008 (0.0000)
ψ_1†	–	–	–	–	–	1.1435 (0.0022)	0.7679 (0.0032)	1.14 (0.001)	0.0001 (0.8749)	1.8951 (0.003)
σ_u^2 *	0.2926 (0.0000)	0.3355 (0.0000)	0.2927 (0.0000)	0.3151 (0.0000)	0.2761 (0.0000)	–	–	–	–	–
ν	10.1888 (0.0000)	11.6346 (0.0028)	5.5142 (0.0000)	9.1118 (0.0000)	9.4285 (0.0000)	9.3451 (0.0000)	9.0142 (0.0000)	9.4223 (0.0098)	4.0912 (0.0000)	9.8065 (0.0000)
λ	–	–	0.0064 (0.672)	0.516 (0.0004)	0.6057 (0.0001)	–	–	0.4175 (0.924)	0.0148 (0.4325)	0.5059 (0.0000)
BIC	−49552.0428	−49012.1803	−51988.3113	−51615.9992	−49525.3453	−49296.0386	−51980.7939	−49124.4902	−49661.9466	−55300.1264
$\ell_{r,u}$	24811.7267	24702.3210	26033.8282	25941.2734	24806.3125	24691.6591	26038.0040	24609.8522	24878.5804	27701.6376
ℓ_u	7812.8602	7752.6323	7814.5232	7786.1338	7802.2717	7793.4500	7813.1296	7782.6709	7800.3011	7806.4375
Out-of-sample forecast evaluation										
QLIKE	0.3260	0.3456	0.3442	0.3466	0.4622	0.3456	0.3637	0.3357	0.3296	0.2953 ⋆

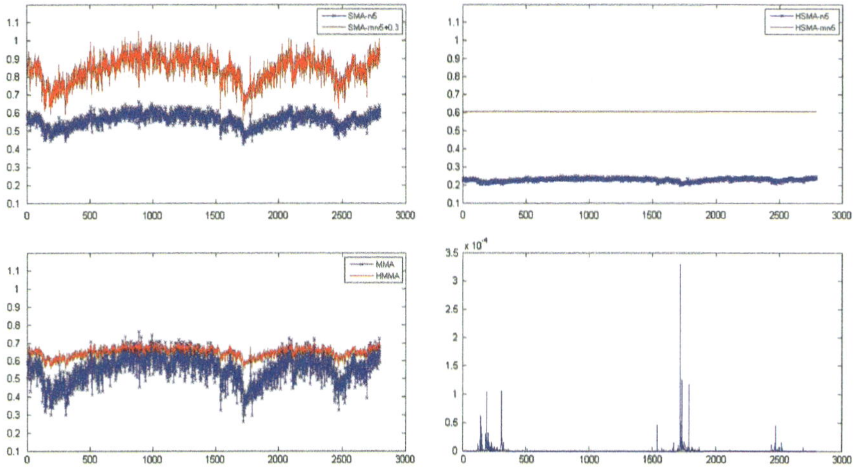

Fig. 1 Estimated time varying volatility response coefficient (γ_t) for (from left to right and top to bottom) SMA-RGARCH, HSMA-RGARCH and (H)MMA-GARCH. The bottom right panel reports the 20 min realized quarticity series. Note: in the top left panel, in order to avoid overlapping with SMA-RV, SMA-MRV has been shifted by adding a constant equal to 0.3

5 Concluding Remarks

Our results suggest that incorporating time varying persistence and multiple realized measures into a RGARCH model can lead to an improvement in its predictive ability. An interesting special case arises when the model incorporates realized variances computed at different frequencies. It can be easily shown that the volatility equation of such a model can be represented as a GARCHX model including a realized measure given by a linear combination, with time-varying weights, of realized variances computed at different frequencies. Such a model structure is close in spirit to the two time scales estimator of Zhang et al. [14] with the difference that the weights assigned to the two realized variances estimators are data driven and time varying. Also, Bandi et al. [3] show that the optimal frequency for computing RV measures is by its own nature time varying and correctly modeling its dynamics can potentially deliver superior forecasts than those obtained by assuming time invariance of the optimal frequency. Investigation of this issue is currently left for future research.

References

1. Andersen, T.G., Dobrev, D., Schaumburg, E.: Jump-robust volatility estimation using nearest neighbor truncation. J. Econom. **169**(1), 75–93 (2012)
2. Bandi, F.M., Russell, J.R.: Microstructure noise, realized variance, and optimal sampling. Rev. Econom. Stud. **75**(2), 339–369 (2008)

3. Bandi, F.M., Russell, J.R., Yang, C.: Realized volatility forecasting in the presence of time varying noise. J. Bus. Econom. Stat. **31**(3), 331–345 (2013)
4. Barndorff-Nielsen, O.E., Shephard, N.: Econometric analysis of realized volatility and its use in estimating stochastic volatility models. J. R. Stat. Soc. Ser. B **64**(2), 253–280 (2002)
5. Barndorff-Nielsen, O.E., Hansen, P.H., Lunde, A., Shephard, N.: Designing realized kernels to measure the expost variation of equity prices in the presence of noise. Econom. Econom. Soc. **76**(6), 1481–1536 (2008)
6. Bollerslev, T., Patton, A., Quaedvlieg, R.: Exploiting the errors: a simple approach for improved volatility forecasting. J. Econom. **192**(1), 1–18 (2016)
7. Brownlees, C.T., Gallo, G.M.: Financial econometric analysis at ultra-high frequency: data handling concerns. Comput. Stat. Data Anal. **51**(4), 2232–2245 (2006)
8. Gerlach, R., Wang, C.: Forecasting risk via realized GARCH, incorporating the realized range. Quant. Financ. **16**(4), 501–511 (2016)
9. Hansen, P.R., Huang, Z.: Exponential GARCH modeling with realized measures of volatility. J. Bus. Econ. Stat. **34**, 269–287 (2016)
10. Hansen, P.R., Huang, Z., Shek, H.H.: Realized GARCH: a joint model for returns and realized measures of volatility. J. Appl. Econom. **27**, 877–906 (2012)
11. Hansen, P.R., Lunde, A., Nason, J.M.: The model confidence set. Econom. Econom. Soc. **79**(2), 453–497 (2011)
12. Liu, L.Y., Patton, A.J., Sheppard, K.: Does anything beat 5-minute RV? A comparison of realized measures across multiple asset classes. J. Econom. **187**(1), 293–311 (2015)
13. Shephard, N., Xiu, D.: Econometric Analysis of Multivariate Realised QML: Efficient Positive Semi-definite Estimators of the Covariation of Equity Prices (2016) (Working Paper)
14. Zhang, L., Mykland, P.A., Ait-Sahalia, Y.: A tale of two time scales: determining integrated volatility with noisy high-frequency data. J. Am. Stat. Assoc. **100**(472), 1394–1411 (2005)

Updating CPI Weights Through Compositional VAR Forecasts: An Application to the Italian Index

Lisa Crosato and Biancamaria Zavanella

Abstract Worldwide, monthly CPIs are mostly calculated as weighted averages of price relatives with fixed base weights. The main source of estimation of CPI weights are National Accounts, whose complexity in terms of data collection, estimation of aggregates and validation procedures leads to several months of delay in the release of the figures. This ends up in a non completely consistent Laspeyres formula since the weights do not refer to the same period as the base prices do, being older by one year and then corrected by the elapsed inflation. In this paper we propose to forecast CPI weights via a compositional VAR model, to obtain more updated weights and, consequently, a more updated measure of inflation through CPIs.

Keywords CPIs · Laspeyres formula · Compositional data analysis · CVAR

1 Introduction

In most OECD countries, Consumer Price Indexes (CPIs) are calculated according to a Laspeyres formula given by a weighted average of price relatives with fixed base weights summing to a constant and measure the change in the price of a basket of fixed composition. Weights are usually derived from Household Final Monetary Consumption Expenditures estimated in National Accounts and related to goods and services belonging to the fixed basket (for a thorough discussion on theory and practice of CPIs see for example [5]).

In Italy, the national CPI[1] is achieved by subsequent aggregation steps going from the price relatives surveyed in local markets for single products to the index

[1]This methodology holds for the main National index, NIC, as well as for the European Harmonized index HCPI. Both are chained Laspeyres indexes.

L. Crosato (✉) · B. Zavanella
Department of Economics, Management and Statistics (DEMS), University of Milano-Bicocca,Via Bicocca degli Arcimboldi, 8, 20126 Milan, Italy
e-mail: lisa.crosato@unimib.it

B. Zavanella
e-mail: biancamaria.zavanella@unimib.it

C. Perna et al. (eds.), *Studies in Theoretical and Applied Statistics*,
Springer Proceedings in Mathematics & Statistics 227,
https://doi.org/10.1007/978-3-319-73906-9_15

corresponding to aggregate products (simply products henceforth), groups of consumption and divisions of consumption, which is the last breakdown before the global index according to the ECOICOP[2] classification.

In each year t, monthly CPIs are calculated with respect to a base period fixed at December of year $t-1$ which, according to Laspeyres formula, also corresponds to the weight reference period. Different kind of weights are applied to products throughout the aggregation mechanism, but the first Laspeyres formula is involved when averaging the elementary national product indexes to obtain the indexes for the subsequent level in the ECOICOP classification (see the ISTAT archive on price indexes for methodological details).

Unfortunately, the data necessary to calculate the weights for the products, mainly the Household Expenditures retrieved from the National Accounts,[3] are not available for December of year $t-1$. For this reason ISTAT uses the Household Expenditures of year $t-2$ updated to December of year $t-1$ using the change in the index of the corresponding product [7]. Once inflated, the expenditures are converted into shares as a proportion of the total expenditure over all considered products.

Therefore in practice, in Italy as well as in other countries (such as Sweden, Ireland, UK, France, Belgium and Austria), the CPI is not a true Laspeyres index because, although for year t the base price is set to December of year $t-1$, the weights cannot be dated back to the same period.

Thus, the idea of this paper arises from the conflict between theory and practice of CPI calculation. Is there a way to produce more updated weights for CPIs? If so, we could calculate the CPIs within a framework more consistent with the Laspeyres formula and, at the same time, account for the latest trends in expenditure shares leaving aside oldest prices movements.

For instance, if ISTAT had the $(t-1)$'s expenditure data available on time for the release of year t's indexes, a rough estimate of the average inflation rate in 2015 would have been 0% instead of the official 0.1%. Given the impact that even small movements in the inflation rate exert on the whole economy, we think it is worthwhile to explore the possibility of updating the CPIs through forecasting.

Our methodology builds on the compositional nature of CPI weights proposing to replace currently used weights, one year older with respect to those the Laspeyres formula allows, with forecasted weights. With compositional time series, we need to face the unit sum and non-negativity constraints that make the use of classical VAR models unfeasible. Therefore, to make predictions for more updated weights, we resort to C-VAR models [2]. As far as we are concerned, there was no earlier attempt on forecasting CPI weights.

In particular, we use the time series of Household Final Monetary Consumption Expenditures to forecast updated weights. Both currently used weights and the predictions are then compared with a benchmark weight constructed in order to preserve the Laspeyres formula.

[2]European Classification of Individual Consumption according to Purpose.

[3]Other sources of information on Household expenditures, both official and nonofficial, are exploited.

The paper is organized as follows: in Sect. 2 we describe our original approach to CPI weight forecasting and the data construction process, in Sect. 3 main results are reported while Sect. 4 concludes.

2 Methodology and Data Description

Compositional data are data whose elements are non-negative and which sum to a constant, usually set to be one (for the basics and more on compositional data analysis see [1]). The sample space of compositional data is called a simplex and is defined as:

$$S^D = \left\{ x = [x_1, \ldots, x_D] | x_i \geq 0 \text{ and } \sum_{j=1}^{D} x_j = k \right\} \tag{1}$$

CPI weights can be well viewed as compositional data that, over time, become compositional time series, since they consist in D non-negative components $x_{1t}, \ldots x_{Dt}$ which at each time t sum to a constant.

Due to the non-negativity and unit-sum constraints, classical time series techniques are unviable because they could lead, for instance, to forecasted weights outside of the simplex. In short, the solution is to exit the simplex using a transformation, apply the traditional techniques and go back to the simplex via the inverse transformation, i.e. apply Compositional multivariate Time Series models (see [2]).

More in detail, the Vectorial Autoregressive model for compositional time series x_t, C-VAR(p), is defined as follows:

$$\left(\mathrm{x}_t \ominus (\xi)\right) \ominus ((\Phi)_{C,1} \odot (\mathrm{x}_{t-1} \ominus (\xi))) \ominus \cdots \ominus \left((\Phi)_{C,p} \odot (\mathrm{x}_{t-p} \ominus (\xi))\right) = \mathrm{w}_t \tag{2}$$

where $(\Phi)_{C,1}, \ldots, (\Phi)_{C,p}$ are A_{DXD} matrices, ξ is the C-mean vector, $w_t \sim WN_C(1_c, \Sigma_C)$ is a white noise and $\ominus$ and $\odot$ are the usual difference perturbation and product operations.

The concise form for this model is described by Eq. 3 as follows:

$$(\Phi)_C(L_C)(\mathrm{x}_t \ominus (\xi)) = \mathrm{w}_t \tag{3}$$

with $(\Phi)_C(L_C) = \mathrm{G}_D \ominus ((\Phi)_{C,1} \odot L_C) \ominus \cdots \ominus ((\Phi)_{C,p} \odot L_C^p)$, where G_D is the centering matrix and L_C is the backshift operator.

The time series of NIC weights at the COICOP consumption class level are available on the ISTAT website for a time span from 1998 to 2016. However, we cannot base our prediction on the original ISTAT weights because of the time structure the forecasted weights would inherit. For instance, keeping the 2015 weights for the sake of comparison, the predicted one-step-ahead weights would share the structure of ISTAT weights for 2015,[4]

[4] All weights indicate by capital W are before normalization just to highlight their temporal structure.

$$\hat{W}^h_{2015} = S^{2013}_h \cdot \frac{I^h_{12,2014}}{I^h_{2013}} \tag{4}$$

where h indicates the generic item, S the related expenditure in the indicated year and I the price index. Our goal, instead, is to obtain a forecast for the 2015 weights with a time structure closer to the one imposed by the Laspeyres formula which would ideally be, considering the impossibility of attributing the whole expenditures to the single month of December 2014,

$$\hat{W}^h_{2015} = S^{2014}_h \cdot \frac{I^h_{12,2014}}{I^h_{2014}} \tag{5}$$

Therefore, to properly forecast the ideal weights, we need a time series of weights with the same ideal structure. In order to forecast the 2015 weights using the ideal weight time series, we just need the data available to ISTAT in order to produce the weights for the same date (National Accounts until 2013).

Applying the C-Var model to the ideal weight series preserves the Laspeyres structure for the one-step-ahead forecast but using only information available at time $t-2$.

Now, since the expenditures necessary to construct the weights are not available at the product level, we have recalculated the time series of ideal weights according to the formula:

$$W^{Ideal,h}_t = S^{t-1}_h \cdot \frac{I^h_{12,t-1}}{I^h_{t-1}} \tag{6}$$

using expenditures at the consumption group level, which are the most disaggregated data on Household final expenditure publicly available.

As can be expected, the time series of official ISTAT weights based on consumption shares derived from the single products, and the recalculated ideal weight time series differ substantially. Both series of weights and a third one, given by weights recalculated according to ISTAT formula but using consumption groups, is represented in Fig. 1 for Division 4 (Housing, Water, Electricity, Gas and other fuels), which serves as a leading example throughout the paper. Two remarks are in order. The first is that this gap is decreasing, most probably because of the deceleration of inflation in the second part of the considered period. Second, the greatest part of the difference seems to be due to the ECOICOP level of expenditures used, rather than to the different formulas, although the ideal weights are somewhat smoother with respect to the other two series (see for example 2012 in the low panels).

Ideal weights and recalculated weights were computed and rescaled to sum to one for eleven out of twelve consumption divisions, in order to work with lowest dimensional compositions. Given the shortness of the time series it would not have been possible to work on the composition of all consumption groups.

Our methodology, therefore, consists in forecasting the ideal weights for the 2015 NIC.

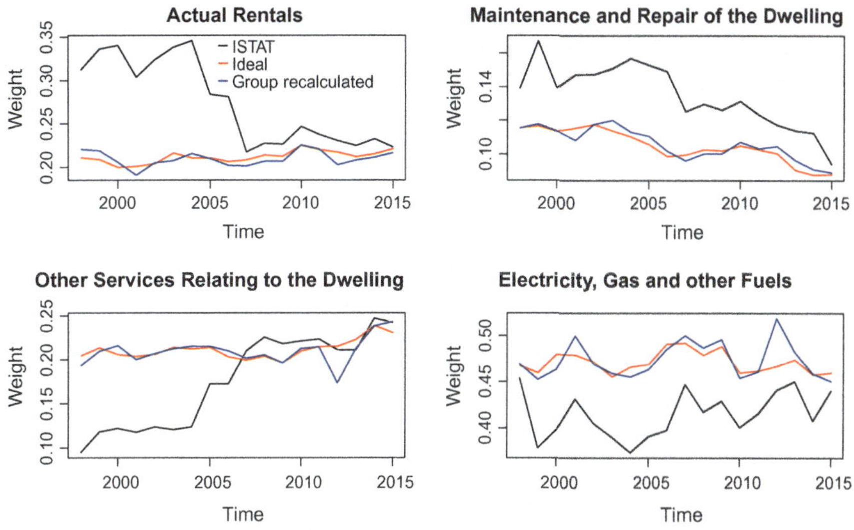

Fig. 1 Time series of the ISTAT, ideal and group recalculated weights, 1998–2015

First, the time series must be transformed to enable us to use traditional time series analysis. We chose an isometric log-ratio (ilr) transformation [4] which preserves distances. The centered log-ratio transformation (clr) was not used due to the singularity of the variance-covariance matrix, which causes problems during the estimation procedure, while the use of additive log-ratio (alr) transformations produced results very similar to the ilr-transformation (for the choice of an appropriate ilr transformation in compositional time series analysis see [6]).

The subsequent steps involve the estimation of a VAR model, which in 4 out of 11 cases collapses to a single AR (two groups only on the simplex). We verify the presence of unit roots using an Augmented Dickey-Fuller test on each of the separate time series. We then use information criteria to establish the lag order of the C-VAR model. Finally, we test for cointegration using the results of the analysis to estimate a Vector Error Correction Model (VECM) which will be used to make our predictions.

Finally, we compare our forecasted weights for the 2015 indexes with both ISTAT's weights and the recalculated weights, in order to eliminate the influence of the different aggregation level on the validity of the results. Our benchmark is the Ideal Weight for 2015, which uses the latest Household Expenditures (year 2014) and is calculated as described by Eq. 7:

$$w_{h,2015}^{ideal} = \frac{S_{h,2014} \cdot \frac{I_{12,2014}^{h}}{I_{2014}^{h}}}{\sum_{h=1}^{n} S_{h,2014} \cdot \frac{I_{12,2014}^{h}}{I_{2014}^{h}}} \tag{7}$$

where h is the group, $S_{h,2014}$ is the expenditure for the year 2014, $I_{12,2014}$ is the NIC index in December of the year 2014 and I_{2014}^{h} is the mean of the NIC index in year

2014. Such data were not available when ISTAT calculated the 2015 weights (which uses 2013 expenditure shares) and this makes the ideal weight more updated with respect to ISTAT's. Finally, to check for the robustness of our fitting and forecasts, we create bootstrap simulated distributions for the forecasted weights.

3 Results

We give a detailed description of all the analyses performed for Division 4 only, due to the similarity of the procedure for the remaining divisions. First of all, we perform an Augmented Dickey-Fuller test on each of the individual ilr-transformed time series to verify the presence of a unit root. The results lead us to conclude that each time series contains one unit root. We proceed by calculating the lag order of the VAR model through information criteria: the AIC and HQ criteria suggest a VAR model with three lags whereas the SC and FPE criteria suggest 1 lag only.

The Johansen cointegration test for the VAR(3) model suggests that the cointegration rank should be equal to one. We decided on the one-lag-one-rank of cointegration combination due to a better fit of the time series and to the shortness of the series at hand, and estimated a VECM model with cointegration rank equal to one. Results were obtained through the `R` statistical software and in particular with the packages `compositions`, `vars` and `tsDyn` and are available on request.

We then use this model to forecast the weights for the year 2015. Figure 2 displays the graphs of the original time series and the fitted values: for all the groups the fitted and original weights seem to get along with one another.

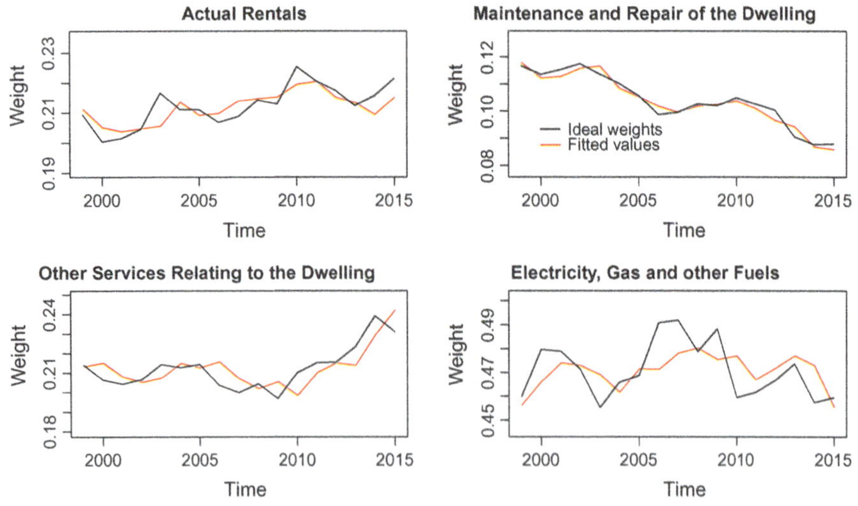

Fig. 2 Ideal weight time series versus fitted data (Division 4). The last value in the fitted data is the forecast for 2015

In order to check for robustness of the forecasts, we bootstrap the model first on the ilr-transformed series parameters and then we go back to the simplex. Thus, we have drawn 1,000 pseudo-random realizations from the CVAR (or VECM) estimated model and re-estimated the model to obtain a set of 1,000 forecasts. Figure 3 displays the densities of the simulated forecasts for each of the groups of Division 4. Along with the densities, we plotted the mean (black vertical line) of the simulated forecasts, the 2015 ISTAT weights (blue solid line), the Group recalculated weights (blue dashed line), our prediction (red line) and finally the ideal weight (green). The proximity of the various weights to the ideal one will be discussed further on for all divisions. Rather, note that the ideal weights seem well placed around the center of the distributions bootstrapped from our C-VAR estimates at least in three out of four cases, confirming that our predictions are not one-shot validity forecasts.

In order to have a complete basis for assessment of the prediction power of our forecasted weights, we compare the predictions, the official ISTAT weights and the recalculated weights with the ideal weights.

All weights will be used for the CPI of the year 2015, although they were obtained using Expenditures of different years, 2013 (ISTAT and recalculated) and 2014 (ideal and forecasted). It is worthwhile to recall that the difference between ISTAT's weights and the recalculated weights is the level of aggregation (products versus groups), while the forecasted weights were obtained as a one-step-ahead prediction after modeling the time series of ideal weights up to 2014.

The absolute difference between forecasted weights and the ideal weights is smaller with respect to the same difference applied to recalculated weights in 67% of

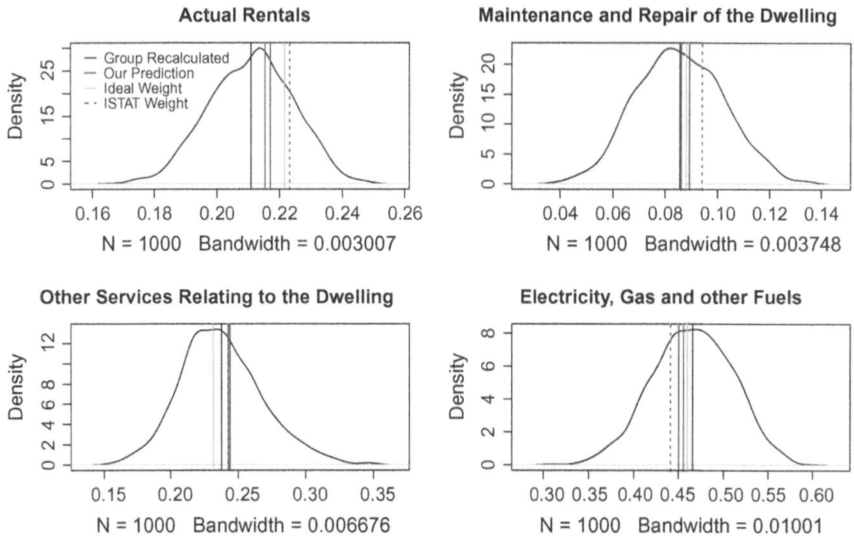

Fig. 3 Simulated weight prediction densities with superimposed ISTAT weights, group recalculated weights, ideal weights and our forecasted weights (all weights to be used for 2015 CPI)

groups (differences multiplied by 1,000,000 are reported in Table 1). The same percentage rises to 79% when we compare the results with the official ISTAT weights, although as we have already specified, in the last case the difference could be explained by the different level of aggregation. The sum of absolute differences confirms the global advantage achieved through forecasted weights: 201,711 ver-

Table 1 Differences between indicated weights and ideal weights by Consumption Groups for 2015s NIC according to ECOICOP classification. All differences have been multiplied by 1,000,000. Gray shades highlight divisions. *Source* the official weights are retrieved directly from ISTAT database on price indexes, the other figures are calculated by the authors

Consumption Groups	Forecasted	Group-Recalculated	Official
1.1 Food	119.35	126.80	1043.93
1.2 Non-alcoholic beverages	119.35	126.80	1043.93
2.1 Alcoholic beverages	5178.50	7915.70	112543.63
2.2 Tobacco	5178.50	7915.70	112543.63
3.1 Clothing	899.30	423.60	10043.31
3.2 Footwear	899.30	423.60	10043.31
4.1 Actual rentals for housing	6091.27	4539.24	1570.95
4.2 Maintenance and repair of the dwelling	1787.91	1165.23	5988.34
4.3 Water supply and misc.	11487.07	12646.67	11232.38
4.4 Electricity, gas and other fuels	3607.79	9272.67	18791.66
5.1 Furniture and furnishings, carpets and other	2600.88	1623.12	5491.38
5.2 Household textiles	2003.18	2907.15	928.57
5.3 Household appliances	475.28	593.71	3432.04
5.4 Glassware, tableware and household utensils	2347.59	777.50	3660.67
5.5 Tools and equipment for house and garden	630.49	638.76	1296.88
5.6 Goods and services for routine household maint.	2790.09	3797.82	14809.53
6.1 Medical products, appliances and equip.	144.70	9016.80	102695.76
6.2 Outpatient services	1345.60	9871.90	56613.27
6.3 Hospital services	1201.00	855.10	46082.50
7.1 Purchase of vehicles	15247.20	3420.50	27956.58
7.2 Operation of personal transport equipment	19883.10	4094.70	18548.13
7.3 Transport services	4636.00	674.30	9408.34
8.1 Postal services	2852.20	3388.29	410.91
8.2 Telephone and telefax equipment	28068.73	36280.87	64491.84
8.3 Telephone and telefax services	25216.52	32892.57	64902.75
9.1 Audio-visual, photographic and inf proc. equip.	3289.62	6977.01	16434.27
9.2 Other major durables for recr. and culture	846.07	1207.71	2384.32
9.3 Other recr. items and equip., gardens and pets	2804.30	5566.99	204.01
9.4 Recreational and cultural services	4948.05	6257.17	2270.94
9.5 Newspapers, books and stationery	3901.51	9283.97	13855.86
9.6 Package holidays	7518.17	2824.48	2280.87
11.1 Catering services	1973.10	3050.60	3771.93
11.2 Accommodation services	1973.10	3050.60	3771.93
12.1 Personal care	4792.75	6726.67	69064.47
12.2 Personal effects n.e.c.	2629.90	1089.02	26543.76
12.3 Social protection	10411.67	3368.01	13485.63
12.4 Insurance	1781.44	12109.70	63530.48
12.5 Financial services n.e.c.	2425.12	5877.72	55290.86
12.6 Other services n.e.c.	7605.15	8981.77	9727.48
Sum	**201,711**	**231,761**	**988,191**

sus 231,761 considering recalculated weights and 988,191 with the official ones (Table 1, last row).

4 Conclusions

This paper builds on two characteristics of consumer price indexes: first, the calculation of CPI weights referring to the single goods and services composing the basket is not as established by the Laspeyres formula which is officially used (for instance, but not only, in Italy), but are one year older than the corresponding base prices; second, the CPI weights constructed as shares of Household Consumption are proper compositional vectors.

These two facts together delineate our aim and methodology. In order to obtain more updated weights for the CPI, we proposed to replace them by forecasted weights. We did so by exploiting both the full time path of Household expenditures and the relative information contained in the vector of weights themselves, using compositional VAR models. We applied the proposed methodology to all the Consumption divisions in the COICOP classification (except for the tenth), forecasting the weights and CPIs for 2015.

For each division, we recalculated the ISTAT weights for the period 1998–2014 using National Accounts data on the consumption groups, then applied the ilr transformation and proceeded to the model estimation and forecasts. Bootstrap robustness graphs were obtained to indicate the solidity of the approach. Forecasted weights were then compared to the 2015 ISTAT and ideal weights, as well as to the Group recalculated weights. Our predictions are closer to the ideal weights in the majority of groups.

This exercise does however have a few limitations. In the first place, our method might lead to more reliable results if expenditure data were available for the aggregate products. If they were, it would be possible to obtain ideal and forecasted weights at all levels of aggregation. Furthermore, the shortness of the time series of weights does not allow us to forecast the weights for all the groups together and, due to the high number of parameters needed to be estimated, forecasts for the whole basket of products cannot be obtained. However, future research could apply some dimension reduction approach, for example a reduced rank regression [8] combined with the compositional VAR estimation. Indeed, this methodology should yield better results with longer time series. In addition, the estimation of several C-VAR starting from the aggregate products could lead to the final estimation of NIC based on forecasted weights.

Above all, we think that the main merit of this work is to show that forecasting CPI weights could lead to a more updated estimation of consumer price indexes with respect to the current practice of indexing the consumption figures from National Accounts of two years earlier. In fact, the forecasted weights based on the recent pattern of expenditure shares could capture upcoming trends in the composition of expenditures.

Finally, the gap between weights and base prices in CPI calculation is not a peculiarity of Italian indices but characterizes many OECD countries (see for example the UK technical manual on CPIs, 2014) so that our method could be applied to CPI and HICP data from several countries. Another even more interesting application might be to Italian Producer Price indexes, whose weights are calculated using data which is two years older with respect to the base of the index.

References

1. Aitchison, J.: The Statistical Analysis of Compositional Data. Monographs on Statistics and Applied Probability, 416 p. Chapman & Hall Ltd., London (1986) (Reprinted in 2003 with additional material by The Blackburn Press)
2. Barcelo-Vidal, C., Aguilar, L., Martín-Fernández, J.A.: Compositional VARIMA Time Series, pp. 91–107. Wiley (2011)
3. Consumer Price Indices Technical Manual, 2014 ed. Office for National Statistics, UK
4. Egozcue, J.J., Pawlowsky-Glahn, V., Mateu-Figueras, G., Barcelo-Vidal, C.: Isometric logratio transformations for compositional data analysis. Math. Geol. **35**(3), 279–300 (2003)
5. ILO Manual on CPI: Consumer Price Index Manual: Theory and Practice. Published for ILO. IMF, OECD, UN, Eurostat, The World Bank by ILO, Geneva (2004)
6. Kynčlová, P., Filzmoser, P., Hron, K.: Modeling compositional time series with vector autoregressive models. J. Forecast. **34**(4), 303–314 (2015)
7. Mostacci, F.: Aspetti teorico-pratici per la costruzione di indici dei prezzi al consumo, Documenti ISTAT (2004)
8. Velu, R., Reinsel, G.C.: Multivariate Reduced-rank Regression: Theory and Applications, vol. 136. Springer Science & Business Media (2013)

Prediction Intervals for Heteroscedastic Series by Holt-Winters Methods

Paolo Chirico

Abstract The paper illustrates a procedure to calculate prediction intervals in case of heteroscedasticity using Holt-Winters methods. The procedure has been applied to the Italian daily electricity prices (PUN) of the year 2014; then the prediction intervals have compared to those provided by an ARIMA-GARCH model. The intervals obtained with HW methods have been very similar to the others, but easier to calculate. Moreover, the HW procedure is more flexible in dealing with periodic volatility as proved in the case study.

Keywords Holt-Winters methods · Heteroscedasticity · Prediction intervals

1 Introduction

The Holt-Winters (HW) methods are forecast methods grounding on recursive formulas for updating the structural components of time series: the local mean level, L_t; the local trend, T_t; the local seasonal index, I_t, if the series is affect by periodic effects. According to the composition model of the structural components, the HW methods can be either additive or multiplicative. In the first case, when a new observation, y_t becomes available, the forecast of y_{t+k}, denoted by $\hat{y}_t(k)$, is:

$$\hat{y}_t(k) = L_t + kT_t + I_{t+k-hS} \quad h = int(k/S) + 1 \tag{1}$$

after having updated the structural components:

$$L_t = \alpha(Y_t - I_{t-s}) + (1 - \alpha)(L_{t-1} + T_{t-1}) \tag{2}$$

$$T_t = \beta(L_t - L_{t-1}) + (1 - \beta)T_{t-1} \tag{3}$$

$$I_t = \gamma(Y_t - L_t) + (1 - \gamma)I_{t-S} \tag{4}$$

P. Chirico (✉)
Department of Economics & Statistics, University of Turin, Turin, Italy
e-mail: paolo.chirico@unito.it

C. Perna et al. (eds.), *Studies in Theoretical and Applied Statistics*, Springer Proceedings in Mathematics & Statistics 227,
https://doi.org/10.1007/978-3-319-73906-9_16

where $0 \leq \alpha, \beta, \gamma \leq 1$ are smoothing coefficients, and S is the length of the seasonal cycle. The additive HW provides optimal forecast if the series derives from a SARIMA process of order $(0, 1, S + 1)(0, 1, 0)_S$ [8].[1]

Originally, the HW methods were not thought for prediction intervals (PI*s*), but over time, some authors have proposed ways to get PI*s*. The proposal of [8] for the additive method allows to get the same PI*s* of the equivalent SARIMA model. They define the k-steps-ahead prediction error as:

$$e_t(k) = \sum_{i=0}^{k-1} v_i e_{t+k-i} \tag{5}$$

where $e_t = y_t - \hat{y}_{t-1}$ (1) is the one-step-ahead (OSA) prediction error, and:

$$v_i = \begin{cases} 1 & i = 0 \\ \alpha + i\alpha\beta & i \neq 0, S, 2S, ... \\ \alpha + i\alpha\beta + (1 - \alpha)\gamma & i = S, 2S, ... \end{cases} \tag{6}$$

If the OSA prediction errors are uncorrelated and Gaussian $N(0, \sigma^2)$, the k-steps-ahead prediction intervals, at $1 - \theta$ confidence level, is:

$$\hat{y}_t(k) \pm z_{(\theta/2)} \sqrt{PMSE_t(k)} \tag{7}$$

$$PMSE_t(k) = \sigma^2 \sum_{i=0}^{k-1} v_i^2 \tag{8}$$

Nevertheless this proposal assumes the OSA prediction errors are homoscedastic, that is not realistic in some cases like electricity prices (see Sect. 3).

2 An HW Method for Heteroscedastic Series

Let's suppose the OSA prediction errors are heteroscedastic, and the volatility level of them, denoted by H_t, is only locally approximative constant. Following the criterion of the Exponential Smoothing, the update of the volatility level may be formulated as:

$$H_t = \lambda e_t^2 + (1 - \lambda)H_{t-1} \tag{9}$$

where $0 \leq \lambda \leq 1$ is a smoothing coefficient.

[1] Instead, the multiplicative HW is not optimal for any linear process. In case of multiplicative composition of the structural components, it is worthwhile to apply the additive HW to the log-serie.

According to the criterion: $\hat{\sigma}^2_{t+1} = H_t$, the volatility forecasts are:

$$\hat{\sigma}^2_{t+1} = \lambda e_t^2 + (1-\lambda)\hat{\sigma}^2_t \tag{10}$$

$$\hat{\sigma}^2_{t+k} = \hat{\sigma}^2_{t+1} \quad k = 2, 3, \ldots \tag{11}$$

The Eq. (10) is known in Statistical Process Control as *Exponentially Weighted Moving Variance*, EWMV [6]. It is formally similar to the IGARCH model, but there are conceptual differences: the IGARCH model, as well as all models of GARCH family, shapes the conditional heteroscedasticity of homoscedastic processes [3]; the method (10) treats a general heteroscedasticity where the volatility can be viewed as a stochastic process. Then the method (10) is conceptually more similar to the methods for stochastic volatility [7].

Taking into account the (11), under the assumption of uncorrelated OSA prediction errors, the $PMSE_t(k)$ becomes:

$$PMSE_t(k) = \sum_{i=0}^{k-1} v_i^2 \sigma^2_{t+k-i} \tag{12}$$

$$= \sigma^2_{t+1} \sum_{i=0}^{k-1} v_i^2 \tag{13}$$

where the coefficients v_is have been defined in (6).

This formula is very similar to (8), but presents a fundamental difference: now the $PMSE_t(k)$ depends on the last predicted volatility; that means more accurate PI*s* in case of heteroscedastic e_t^2.

Periodic Volatility In case of seasonal data, heteroscedasticity can follow a seasonal pattern (*seasonal volatility*): e.g. daily electricity prices present seasonal patterns in the price levels and in the price volatilities. A way to deal with such a case in GARCH framework was proposed by Koopman et al. [4] who suggested to scale volatility by means of seasonal factors. According to that idea, but remaining in HW framework, the method (10) should be changed in the multiplicative following HW method[2]:

$$\begin{aligned} \hat{\sigma}^2_{t+k} &= H_t \cdot X_{t+k-hS} \quad h = int(k/S) + 1 \\ H_t &= \lambda(e_t^2/X_{t-S}) + (1-\lambda)H_{t-1} \\ X_t &= \lambda_x(e_t^2/H_t) + (1-\lambda_x)X_{t-S} \end{aligned} \tag{14}$$

where X_t represents the seasonal effect on volatility, and $0 \leq \lambda_x \leq 1$ is the corresponding smoothing coefficient.

Note that now the formula (12) cannot be reduced into the formula (13) because $\hat{\sigma}^2_{t+k}$ is only periodically constant, then $PMSE_t(k)$ presents cyclical fluctuations.

[2]Here the multiplicative HW should be preferred to the additive one because it avoids negative values for volatility forecasts.

The smoothing coefficients The value $(1-\lambda)$ can be viewed as the persistence degree of volatility: when $(1-\lambda)=1$, volatility (or its level) is constant (max persistence); when $(1-\lambda)=0$, each variance/level does not depend on any previous variances/levels (null persistence). Similarly, $(1-\lambda_x)=1$ means that the seasonal pattern on volatility is constant, and $(1-\lambda_x)=0$ means that there is no seasonal pattern on volatility.

The smoothing coefficients, α, β, γ, λ and λ_x, could be chosen minimising the following sum:

$$\sum[\ln\sigma_t^2+u_t^2] \tag{15}$$

where $u_t=e_t/\sigma_t$ are the standardised OSA prediction errors.

If $u_t\sim N(0,1)$, the criterion (15) is equivalent to the maximum likelihood, but should not be viewed as an estimation method. Indeed, according to the theory of HW methods, the smoothing coefficients are not process parameters, but instruments to perform time series forecasting. Nevertheless, in case of financial prices, the prediction standardised errors are generally leptokurtic/heavy tailed like standardised Student's t-distributions: $t_\nu\sqrt{(\nu-2)/\nu}$. In that case, it is more appropriate to minimise the following sum:

$$\sum[\ln(\sigma_t^2)+\ln(1+\frac{u_t^2}{\nu-2})^{\nu+1}] \tag{16}$$

where ν are the degrees of freedom of the Student's distribution.[3]

3 Prediction Intervals for the Italian PUN

We have considered the series of the daily electricity price (PUN), cleared by the Italian wholesale electricity market in 2014. As all daily electricity market prices, Italian PUN is characterized by strong seasonality affecting the price level and price volatility too (Fig. 1). On the series we have calculated one step ahead prediction intervals using the method described in the previous sections, and using a suitable ARIMA-GARCH model. Then we have compared the results obtained with both approaches.

3.1 The HW Approach

Since the presence of spikes and jumps makes the standardised prediction errors of electricity prices strongly leptokurtic, the smoothing coefficients of the HW method

[3] Minimizing the sum (16) is nearly equivalent to the maximum likelihood in case of Student't distribution [1].

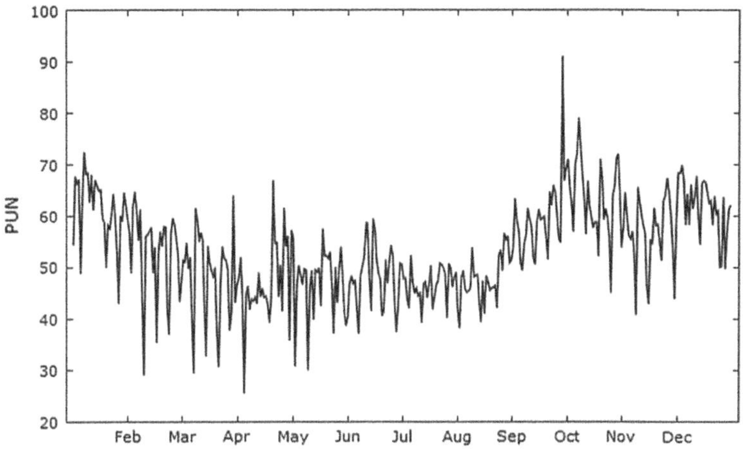

Fig. 1 The Italian PUN in 2014

Table 1 Smoothing coefficients

α	β	γ	λ_1	λ_x	ν	*RMSE*
0.272	0.000	0.000	0.037	0.000	7.40	5.066

have been chosen minimising the sum (16). Table 1 reports the optimal values of the smoothing coefficients, the degree of freedom and the standard deviation of the OSA prediction errors (RMSE).[4]

We can note that the smoothing coefficients present some interesting features: $\beta = 0$ means that the price level does not present any locally persistent drift; $\gamma = 0$ means that the seasonal effects on the price level are periodically constant (deterministic seasonality); $\lambda_x = 0$ means that the seasonality on the volatility is deterministic too.[5] Finally, the low value of the degree of freedom (7.4) confirms the assumption of leptokurtic standardised prediction errors.

We have compared the standardized prediction errors, u_t, to the standardised Student' t-distribution with 7.4 degrees of freedom by means of the quantile-quantile plot (Fig. 2).

We can note that the u_ts correspond almost well to the theoretical distribution except some errors at the distribution extremities. That is due to the presence of spikes and jumps which cannot be consistent with the other data.

[4]The optimal values are reported without inferential information (standard errors, p-values, etc.) because the coefficients are not interpreted here as process parameters.

[5]The daily effects on PUN have become quite constant since 2013, but it was not in the previous years [2].

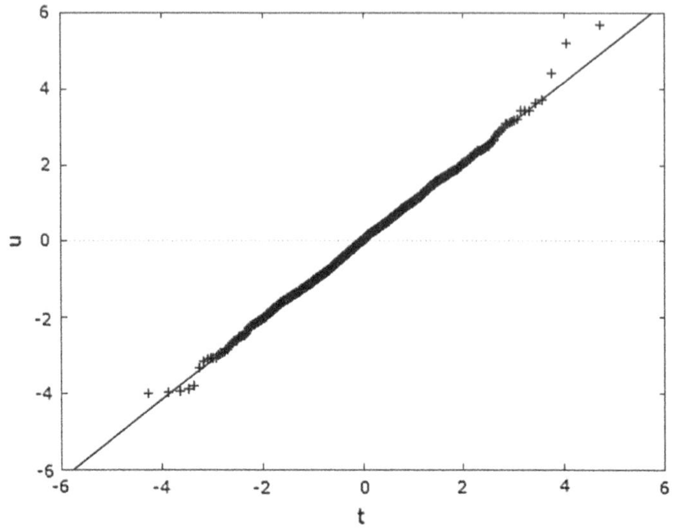

Fig. 2 Q-Q plot of the standardised errors versus standardised Student's t

Therefore the one-step ahead prediction intervals (at $1-\theta$ confidence level) have been calculated with the following formulas:

$$LB: \ \hat{PUN}_t(1) - t^*_{\theta/2}\sqrt{PMSE_t(1)} \tag{17}$$

$$UB: \ \hat{PUN}_t(1) + t^*_{\theta/2}\sqrt{PMSE_t(1)} \tag{18}$$

where $t^*_{\theta/2}$ is the $\theta/2$-th percentile of the standardised Student's t distribution with 7.4 degrees of freedom.

3.2 The ARIMA-GARCH Approach

We have repeated the study in Sect. 3.1 using an ARIMA-GARCH model. Similar to the method illustrated in the previous subsection, we have fitted the PUN series with an ARIMA-EGARCH model[6] with daily regressors (d_j) and Student's standardised errors (u_t):

[6]We have chosen an EGARCH model to avoid the risk of negative variance because that risk is high including seasonal regressors in classic GARCH model.

$$
\begin{aligned}
z_t &= \Delta^d PUN_t = \mu_t + \varepsilon_t \\
\varepsilon_t &= \sigma_t u_t \quad u_t \sim t_v \sqrt{(v-2)/v} \\
\mu_t &= \delta + \sum_{j=1}^{p} \phi_j z_{t-j} + \sum_{j=1}^{q} \theta_j \varepsilon_{t-j} + \sum_{j=1}^{6} s_{1,j} d_j \\
\ln \sigma_t^2 &= \omega + \sum_{j=1}^{P} \beta_j \ln \sigma_{t-j}^2 + \sum_{j=1}^{Q} (\alpha_j |\varepsilon_{t-j}| + \gamma_j \varepsilon_{t-j}) + \sum_{j=1}^{6} s_{2,j} d_j
\end{aligned}
\tag{19}
$$

More specifically awe have identified an ARIMA(6,1,0)-EGARCH(1,1) model with daily regressors, that we have estimated using the *GIG* package of *Gretl* [5]. For space reasons, we don't report here all the estimation output of the model, but only the standard deviation of the OSA prediction errors, $RMSE = 5.062$. This statistic is practically equal to one of the HW approach, that means the HM method fits the series as good as the ARMA-EGARCH model.

Finally, we have calculated the PIs of PUN series in the period September, 2014–December, 2014 using both methods (Fig. 3). Although both intervals capture the 90% of the series in the period, the intervals may be quite different some days. Actually the interval by ARMA-EGARCH model (thin line) seem more sensitive to spike and jumps than one by HW methods does (thick line); this characteristic is typical of EGARCH models because outliers (jumps and spike) produce exponential effects on volatility. On the other hand, the HW procedure provides an interval whose width is quite regular: the seasonality on volatility is periodically constant ($\lambda_x = 0$), and the non-seasonal volatility, H_t, is quite smoothed ($\lambda = 0.037$).

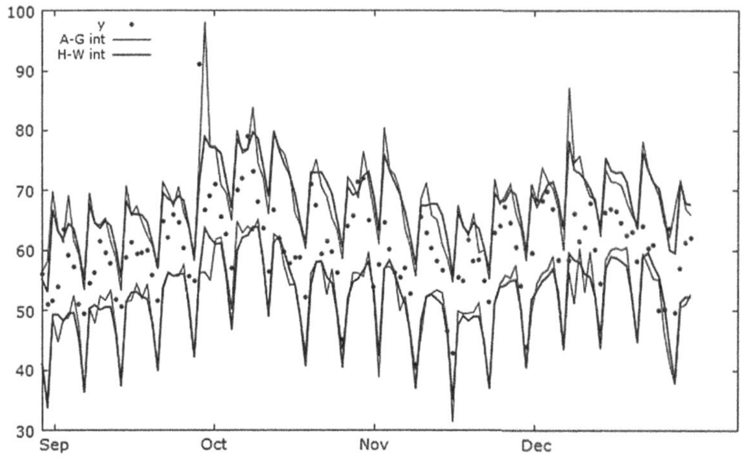

Fig. 3 Prediction intervals

4 Final Considerations

The presence of heteroscedasticity is one of the reasons for preferring ARIMA-GARCH models to HW methods. Nevertheless heteroscedasticity can be treated by HW methods, if some adjustments are taken; the paper illustrates a method to do it. The method has been applied to a case study, and the PI*s* obtained with this method have been similar to the those provided by an ARIMA-EGARCH model; some days the HW procedure seems to work better than ARIMA-EGARCH model too. The most considerable strengths of using HW methods are the ease-of-use and the ability of dealing with several type of heteroscedasticity: stochastic volatility, periodic stochastic volatility, periodically constant volatility. On the other hand, the weakness is that the HW methods are optimal only for specific ARIMA processes although works quite well with a lot of real processes; moreover the HW procedure does not deal with the issue of asymmetry. Nevertheless other versions of the method above-illustrated are possible, and it is our intention to check them.

References

1. Bollerslev, T.: A conditionally heteroskedastic time series model for speculative prices and rates of return. Rev. Econ. Stat. **69**(3), 542–547 (1987)
2. Chirico, P.: Which seasonality in Italian daily electricity prices? A study with state space models. In: Alleva, A., Giommi, G. (eds.) Topics in Theoretical and Applied Statistics, pp. 275–284. Springer (2016)
3. Engle, R.F., Bollerslev, T.: Modelling the persistence of conditional variances. Economet. Rev. **5**(1), 1–50 (1986)
4. Koopman, S.J., Ooms, M., Carnero, M.A.: Periodic seasonal Reg-ARFIMA-GARCH models for daily electricity spot prices. J. Am. Stat. Assoc. **102**(477), 16–27 (2007)
5. Lucchetti, R.J., Balietti, S.: The gig package. Package Guide of Gretl (2016)
6. MacGregor, J.F., Harris, T.J.: The exponentially weighted moving variance. J. Qual. Technol. **25**(2), 106–118 (1993)
7. Taylor, S.J.: Modeling stochastic volatility: a review and comparative study. Math. Financ. **4**(2), 183–204 (1994)
8. Yar, M., Chatfield, C.: Prediction interval for the holt-winters forecasting procedure. Int. J. Forecast. (North-Holland) **6**, 127–137 (1990)

Part V
Spatial Analysis and Issues on Ecological and Environmental Statistics

Measuring Residential Segregation of Selected Foreign Groups with Aspatial and Spatial Evenness Indices. A Case Study

Federico Benassi, Frank Heins, Fabio Lipizzi and Evelina Paluzzi

Abstract Over the last decades there have been important methodological advances in measuring residential segregation, especially concerning spatial indices. After a discussion of the fundamental concepts and approaches some of the numerous indices are introduced. We focus in particular on the most known aspatial and spatial indices in the dimension of evenness namely segregation and dissimilarity indices. The contribution is based on data of the geographic distribution of selected foreign groups resident in the census enumeration areas that form the Local Labour Market Area (LLMA) of Rome. Data refer to the population censuses 2001 and 2011. Applying the indices to the LLMA of Rome serves as a test of the practical and potential usefulness of the proposed measures and their possible interpretation.

Keywords Residential segregation · Foreign population · Evenness indices
Enumeration areas · Rome Local Labour Market Area

F. Benassi (✉) · F. Lipizzi · E. Paluzzi
Italian National Institute of Statistics, Rome, Italy
e-mail: benassi@istat.it

F. Lipizzi
e-mail: lipizzi@istat.it

E. Paluzzi
e-mail: paluzzi@istat.it

F. Heins
Institute of Research on Population and Social Policies,
National Research Council, Rome, Italy
e-mail: f.heins@irpps.cnr.it

C. Perna et al. (eds.), *Studies in Theoretical and Applied Statistics*,
Springer Proceedings in Mathematics & Statistics 227,
https://doi.org/10.1007/978-3-319-73906-9_17

1 Introduction

The presence of foreigners in Italy is not a new phenomenon. The first arrivals of immigrants were in fact registered back in the 1970s and they intensified during the following two decades [5]. However, the recent increase of the foreign population resident in Italy has been unexpectedly high: it grew in fact from approximately 1.3 million people by the end of 2001 to over 4 million by 2011.

As shown by Strozza et al. [24], the growth in the number of foreign residents has followed clear geographical patterns: the foreign population in Italy grew in general, however large cities and their surrounding areas have recorded the greatest increase. One of the key traits related to the foreign population is its spatial distribution and, in particular, the level of its residential segregation. In general residential segregation is in fact seen as detrimental to the development of an open and inclusive society [25, 26].

The residential segregation of foreigners is a phenomena widely studied by scholars of many disciplines like geography, demography and sociology. Numerous studies have been undertaken by scholars during recent years—especially regarding the situation in the U.S. and the U.K.—on this topic. A wide range of measures and indices have been proposed concerning different geographical settings and several geographical scales of analysis: see for instance Massey and Denton [14], Reardon and O'Sullivan [19] and Feitosa et al. [11].

In Italy the academic interest in the study of the settlement patterns of the foreign population and the segregation of foreign groups grew as well since the 1980s. In more recent years several analysis of the residential segregation of the foreign population in Italian metropolitan municipalities were undertaken [4, 6, 12] especially for the case of Rome [1, 7, 8, 17, 18].

The present study intends to contribute to the national and international literature by studying the diachronic evolution (2001–2011) of the residential segregation of the foreign population and selected foreign groups in Rome's Local Labour Market Area.[1]

The contribution concentrates on the most known aspatial and spatial indices that measure the evenness of the distribution of the foreign population. Despite the fact that residential segregation of foreigners is a multidimensional concept, evenness is regarded by most authors as the foremost component of the segregation of a population group and this is why this study concentrates on this later dimension.

The contribution is structured as follows: in the next section a brief description of existing concepts, approaches and indices for measuring segregation is provided.

[1]Defined by Istat on the basis of the commuting patterns observed in 2011 Population and Housing Census, for more details see http://www.istat.it/en/archive/142790.

In the following section the data used and the indices adopted are described. Next the results of the estimation of the aspatial and spatial indices are shown. In the final section of the contribution the results are discussed and then some conclusions are provided.

2 Residential Segregation, Concepts and Measures

The segregation of a given population is a concept used to indicate the separation between different population groups in a given environment. Most often segregation is viewed as a multidimensional process whose depiction requires different indices for each dimension. Massey and Denton [14] were the first scholars to define segregation as a multidimensional concept, identifying the different dimensions that can be measured by different indices: evenness, exposure, concentration, centralization and clustering. In the conceptual model proposed by Massey and Denton evenness and exposure are aspatial dimensions of segregation, while, clustering, centralization and concentration are per se spatial, since they need information about location, shape, and/or size of territorial units. The model of Massey and Denton has been criticized over the years by scholars. The idea behind these criticisms was that the single territorial units are not independent from each other and that the process of segregation cannot be confined to a single unit. Reardon and O'Sullivan [19], for example, argued that segregation should be seen as a complete spatial process and they propose a spatial version of the Massey and Denton model which is composed by just two dimensions: evenness-clustering and exposure-isolation.

Over the years, several measures for each dimension of segregation have been proposed by scholars as well as different approaches to classify them. Even restricting the discussion to the evenness indices the number of existing measures still remains high. Generally speaking indices of evenness measure a group's over or under representation in the spatial units of a given environment. The more unevenly a population group is distributed across these spatial units, the more segregated it is [13].

Before moving to a brief description of some of the most common evenness indices, it is useful to explain the existing criteria normally used to classify the segregation indices. A first fundamental distinction is between one, two or multigroup segregation indices. The one-group indices measure the distribution of a population group compared to the total population. The two-group indices compare the distribution of two different population groups. Finally the multigroup indices analyse the distribution of several population groups simultaneously. Another important criterion is the one that concerns the aspatial or spatial nature of the segregation indices. Aspatial indices don't need information about location, shape, and/or size of territorial units while, on the opposite, spatial indices are based on the spatial setting of spatial units. Finally, it is useful to distinguish among global and local indices. Whereas the global measures provide a summary value for the entire study area the local measures provide one value for each of the territorial units of the study area.

Moving to the description of the most known segregation measures, the first generation were one group and two-group aspatial indices. In the dimension of evenness we recall the segregation index (IS) and the dissimilarity index (ID), both in the version proposed by Duncan and Duncan [9, 10]. To measure segregation among multiple population groups in the same dimension, were introduced, among others, the multigroup dissimilarity index (D) [15, 22]. Nevertheless, all of the aforementioned indices were insensitive to the spatial arrangement of the areal units [16, 27]. To overcome this shortcoming spatial versions of the indices have been advanced. We recall for the one group indices the segregation index adjusted for tract contiguity (IS_{adj}) [16] and the two adjusted versions of the same index proposed by Wong [28]: the segregation index adjusted for tract contiguous boundary lengths (IS_w) and the segregation index adjusted for contiguous tract boundary lengths and perimeter/area ratio (IS_s). The same authors proposed also spatial versions of the two-group indices, namely the index of dissimilarity adjusted for tract contiguity (ID_{adj}) [16], the index of dissimilarity adjusted for contiguous tract boundary lengths (ID_w) and the index of dissimilarity adjusted for contiguous tract boundary lengths and perimeter/area ratio (ID_s) both proposed by Wong [28]. In the same way, spatial versions of the multigroup dissimilarity index have been proposed by Wong [29], Reardon and O'Sullivan [19] and Feitosa et al. [11]: the spatial version of multigroup dissimilarity index (SD), the multigroup spatial dissimilarity index ($\tilde{D}$) and the local version of the multigroup dissimilarity index ($d_{i/m}$) respectively.

The last two indices belong to a class of spatial indices in which it is possible to specify functions that define contiguity. In fact, in the spatial segregation measures the distance between basic territorial units can be treated quite differently: the criteria range from tract contiguity to specific distance functions between centres of the basic units. Obviously, in the absence of a detailed geolocalization of the population all measures are only estimates of the real distance between individuals or families. It must be kept in mind, however, that all measures of segregation are sensitive to definitions of the boundaries of the basic spatial units due to an aggregation or scale effect and to the zoning effect [19]. In this perspective the latest approach for computing spatial measures of segregation—global and local—allows specifying functions that define how population groups interact across boundaries of the basic units, see for instance Reardon and O'Sullivan [19] and, in particular, the contributions of Roberto [20] and of Roberto and Hwang [21].

However, the methods and indices proposed are not always easily applied, because they require information regarding the exact location of individuals and their proximities to one another in the residential space, data usually not available. In order to partially solve these problems, Reardon and O'Sullivan [19] have proposed methods for estimating population densities from aggregated data (for example by the use of the density kernel estimation), but it is also true that these methods imply a high degree of subjective evaluation, for example regarding the definition of the kernel function or the bandwidth of the kernel.

3 Data and Methods

The data used in this contribution are collected in the 2001 and 2011 Italian Population and Housing Census. The census is unique in providing a wide range of information with a socio-demographic and territorial detail in order to capture the different social realities in their peculiarities and specificities.

The daily commuting patterns observed by the censuses are used by Istat to aggregate the municipalities into Local Labour Market Areas. The LLMAs represent socio-economic entities and are a good approximation of the places where people live and work and usually they comprise the social and economic relationships of the individuals and families. The Rome LLMA is with a surface of almost 4,000 km^2 and about 3.5 million residents one of the largest ones. The municipality of Rome dominates the entire Local Labour Market Area, that is composed by a total of 89 municipalities.

The 2001 population and housing census data are here presented in the 2011 boundaries to facilitate the comparison between the two years. It should be noted that between the 2001 and 2011 censuses, the number of enumeration areas in the LLMA of Rome increased from 14,404 (13,335 with at least 10 and 10,532 with at least 50 inhabitants) to 15,242 (14,087 with at least 10 and 11,358 with at least 50 inhabitants). These changes did not affect substantially the computation of the proposed indicators. Most of the changes were concentrated in newly built up areas.

With reference to the adopted measures, the contribution is limited to the use of two aspatial indices, namely the segregation index (IS) and the index of dissimilarity (ID) and to their spatial version, namely the segregation index adjusted for tract contiguity (IS_{adj}) and the index of dissimilarity adjusted for tract contiguity (ID_{adj}).

The four adopted indices are obtained as:

$$IS = \frac{1}{2}\sum_{i=1}^{n}\left|\frac{x_i}{X} - \frac{t_i - x_i}{T - X}\right| \quad IS_{adj} = IS - \left(\frac{\sum_{i=1}^{n}\sum_{j=1}^{n}\left|c_{ij}\left(\frac{x_i}{t_i} - \frac{x_j}{t_j}\right)\right|}{\sum_{i=1}^{n}\sum_{j=1}^{n} c_{ij}}\right) \tag{1}$$

$$ID = \frac{1}{2}\sum_{i=1}^{n}\left|\frac{x_i}{X} - \frac{y_i}{Y}\right| \quad ID_{adj} = ID - \left(\frac{\sum_{i=1}^{n}\sum_{j=1}^{n}\left|c_{ij}\left(\frac{x_i}{t_i} - \frac{x_j}{t_j}\right)\right|}{\sum_{i=1}^{n}\sum_{j=1}^{n} c_{ij}}\right) \tag{2}$$

where c_{ij} is the cell value of a binary contiguity matrix (1 where i and j are contiguous and 0 otherwise); n is the number of spatial units in the metropolitan area; T is the total population in the metropolitan area; t_i is the total population in spatial unit i; t_j is the total population in spatial unit j; X is the population of group X in the metropolitan area; x_i is the total population of group X in spatial unit i; x_j is the total population of group X in spatial unit j; Y is the total population of group Y in the metropolitan area; y_i is the total population of group Y in spatial unit i. The values of the indices vary from 0 to 1 where 0 means absence of segregation and 1 means complete segregation. The results reported were obtained using the Geo-Segregation Analyzer [2].

4 Results

In 2001 Italy is characterized by a wide spectrum of nationalities, whereas in 2011 some nationalities are becoming more dominant, but the variety of nationalities present remains very ample. During the inter-censual decade the foreign presence in Italy has more than tripled. Whereas in 2001 22 nationalities represented 75% of the total foreign population, in 2011 this number dropped to 15. At all territorial levels —Italy, Latium region and the Rome LLMA—very substantial gains for the 5 nationalities considered were recorded, especially for the Romanian and the Bengali communities. From 2001 to 2011 the average annual population growth in the Rome LLMA was 0.6% with most of the growth coming from the foreign population that grew at an average of 8.7% on a yearly base. Whereas the ‘old’ immigration communities like the ones from Poland and the Philippines grew less the number of nationals from Romania and Bangladesh grew most. In 2011 8.8% of the population of the Rome LLMA had a foreign nationality. The 2.8% of the total population were from Romania, 0.8% from the Philippines, 0.5% from Bangladesh, 0.4% from China and 0.4% from Poland.

The empirical example focuses on the Italian and the foreign population, as well as 5 selected foreign communities from Romania, the Philippines, China, Poland and Bangladesh. The aim is to show the working of the indices and the potential of the instruments, but not to provide a complete analysis of the segregation of the foreign communities in the Rome LLMA.

Nationality is certainly a better criterion than continent to form distinct groups, but the authors are aware that citizenship does not signify by definition homogenous groups: ethnic and social differences can persist and different life course stages might play an important role in defining groups with their specific pattern of segregation. The selection includes 3 Asian communities with very different settlement patterns in the Rome LLMA and two European communities with Poles a ‘traditional’ immigrant community (well before Poland became member of the EU in 2004) and Romanians arriving in more recent years, especially after Romania joined the EU in 2007.

The one-group segregation indices (Table 1) show the similarity between the aspatial and the spatial versions of the indices. The following comments are limited to the spatial version of the segregation index: IS_{adj}. The values of the indices are reported including all inhabited census enumeration areas, but the indices were also

Table 1 One-group segregation indices by citizenship, Rome Local Labour Market Area, 2001 and 2011[a]

Country of citizenship	2001		2011	
	IS	IS_{adj}	IS	IS_{adj}
Romania	0.668	0.659	0.505	0.456
Philippines	0.720	0.712	0.650	0.633
Bangladesh	0.921	0.920	0.800	0.793
China	0.908	0.907	0.798	0.791
Poland	0.662	0.657	0.578	0.569
Other foreign countries	0.403	0.364	0.382	0.326
Total foreign population	0.390	0.336	0.357	0.248

Source Our own elaboration on ISTAT data—the 2001 and 2011 Population and Housing Census
[a]*Note* the group of comparison is the total population

estimated including only the ones with at least 50 residents. In most cases the exclusion of enumeration areas with only few residents did, somewhat unexpectedly, not change the value of the segregation indices significantly.

The segregation indices are the lowest for the total foreign population and show a clear contrast between the Asian and the European communities. The residual category 'other foreign communities' follows the total and also Romanians and Poles record low values of the segregation index. The Filipinos are in a medium position, whereas the Bangladeshi and Chinese communities show the highest segregation. Between 2001 and 2011 the values of the indices decreased for all categories of the foreign population. This decrease goes hand in hand with the increase in the numbers of foreigners living in the Rome LLMA, but can certainly not be attributed to it.

The Romanian community seems to have changed the most regarding their segregation: whereas in 2001 Romanians and Poles had similar values of the segregation index, in 2011 the Romanians seem to be the most evenly distributed community over the census enumeration areas of the Rome LLMA.

The two-group indices (Table 2) underline the different settlement patterns of the single nationalities compared to the one of the Italian population. The dissimilarity remains relative low in the case of the Romanian and the Polish population, whereas the highest one exists between the settlement patterns of the Italian population and the nationals of China and Bangladesh. The changes of the segregation indices that occurred between 2001 and 2011 point to important changes in the dissimilarity of the geographic distribution of the Romanian nationals compared to all other nationalities, whereas in the case of the Philippine nationals the pattern seemed to change relatively little. These differences led to the positioning of the Philippine community between the European communities and the other Asian communities, whereas in 2001 the indices tended to be closer to the values of the Romanian and the Polish community.

Table 2 Two-group dissimilarity indices by citizenship, Rome Local Labour Market Area, 2001 and 2011[a]

Country of citizenship	2001		2011	
	ID	ID_{adj}	ID	ID_{adj}
Romania	0.671	0.617	0.509	0.400
Philippines	0.723	0.669	0.650	0.541
Bangladesh	0.923	0.869	0.806	0.697
China	0.910	0.856	0.806	0.697
Poland	0.666	0.612	0.584	0.475
Other foreign countries	0.406	0.352	0.392	0.283
Total foreign population	0.390	0.336	0.357	0.248

Source Our own elaboration on ISTAT data—the 2001 and 2011 Population and Housing Census
[a]*Note* the group of comparison is the Italian population

The values of the indices for the different pairs of foreign communities reveal specific affinities or similarities in the settlement patterns, as well clear dissimilarities. These results indicate that the strongest segregation seems to form not with the Italian population, but between the different foreign communities. Also in the case of the two-group indices, the indices were recalculated for the enumeration areas with at least 50 residents and the resulting values differ only slightly from the ones reported in Table 2. Since for all five communities in the inter-censual decade a decreasing trend in the degree of dissimilarity can be observed, it can be concluded that their spatial distribution is progressively less different from that of the Italian population although still rather high values of segregation do persist.

The significant decrease noted for the Romanians may be influenced by the fact that the Romanian community is the most numerous one. The intermediate position regarding segregation of the Filipino community is probably due to the fact that this is an 'old' immigration community that very often cohabitates with Italians living and working in an Italian household. Probably this was even more true in 2001 than in the last census, perhaps because the Filipino community changed its demographic structure and is today more often characterised by families than single persons that would live with an Italian family. Also the Polish community is an old immigration community in the Rome LLMA, and like the Filipino one, their segregation indices did not decrease as much as those the other nationalities. In summary, we observe a clear polarization between the European communities and the Asian communities, although a general decrease of the dissimilarity, both in respect to the Italian population and in respect to the other nationalities can be observed. Our results confirm existing analysis [1, 3, 7, 17, 18].

The cultural differences between the different immigration communities and the Italian population are certainly an important factor regarding the causes of segregation. It seems central to distinguish the situation where segregation is a process imposed by the majority population (Italians) or where it is a strategy for the advancement of the own community.

The socio-demographic characteristics of the immigrant communities are another important factor influencing their segregation: the gender balance, as well as the age structure of a community, define the degree of its family orientation. Also the time spent in Italy is an important characteristic: about 15% of foreigners belonging to the total of the five communities in question was born in Italy, in particular, 93% of children less than 5 years, three-quarters between 5 and 9 years, more than 40% of youngsters between 10 and 14 years and a share of about 22% of the ones 15–19 years old. The highest values, especially in the later age group, are registered by the Polish community, the Chinese and the Filipinos, indicating their earlier immigration and their stronger rootedness. Also the economic situation of a community is important and of special interest is the owner/tenant status of the individual or the family. The 61% of the members of the five communities are living in rented housing and 20% own their own homes. Almost half of the Chinese nationals are home owners, followed by the Poles with 29.2%, whereas 56.4% of the Filipinos live in rented houses and 28.2% are occupiers in another capacity. The Romanian and Bangladeshi communities are mostly tenants, respectively, 64% and 73%. These numbers indicate that the degree of segregation is not an expression of economic difficulties. The community with the highest segregation is also the community with the highest share of home owners. This might be an indication that the relative high segregation of the Chinese community is a chosen one.

5 Concluding Remarks

One of the most important conclusion is probably the usefulness of the analysis of the distribution of the segregation of specific nationalities, since any policy to further the integration of foreign residents should be focused on specific nationalities. It should be recognized that segregation could be a strategy by some communities to further the advancement of the community [23], so actively sought by the community and not endured passively. There is no question that spatial indices ought to be given priority in analyzing socio-demographic characteristics of the population of small areas. However the advantages seem to remain in the realm of academic studies since they become explicit when comparing different areas. The spatial segregation indices provide an instrument to improve our understanding of the processes behind the forming and persistence of local concentrations of ethnic groups. However they cannot provide us with direct evidence regarding the socio-demographic and socio-economic factors involved in the process of segregation.

References

1. Amico, A., D'Alessandro, G., Di Benedetto, A., Nerli Ballati, E.: Lo sviluppo dei modelli insediativi. Rumeni, filippini e cinesi residenti a Roma. Cambio 123–146 (2013)
2. Apparicio, P., Fournier, É., Apparicio, D.: Geo-Segregation Analyzer: A Multiplatform Application (Version 1.2). Montreal, Spatial Analysis and Regional Economics Laboratory (SAREL), INRS Urbanisation Culture Société (2016)
3. Barbagli, M., Pisati, M.: Dentro e fuori le mura. Città e gruppi sociali dal 1400 a oggi. Il Mulino, Bologna (2012)
4. Benassi, F., Ferrara, R., Gallo, G., Strozza, S.: La presenza straniera nei principali agglomerati urbani italiani: implicazioni demografiche e modelli insediativi. In: Donadio, P., Gabrielli, G., Massari, M. (eds.) Uno come te. Europei e nuovi europei nei percorsi di integrazione, pp. 186–198. Franco Angeli, Milano (2014)
5. Bonifazi, C.: L'immigrazione straniera in Italia. Il Mulino, Bologna (1998)
6. Busetta, A., Mazza, A., Stranges, M.: Residential segregation of foreigners: an analysis of the Italian city of Palermo. Genus 177–198 (2015)
7. Casacchia, O., Natale, L.: L'insediamento degli extracomunitari a Roma: un'analisi sul rione Esquilino. In: Travaglini, C.M. (ed.) I territori di Roma. Storie, popolazioni, geografie, pp. 609–639. Università di Roma La Sapienza, Roma (2002)
8. Cristaldi, C.: Multiethnic Rome: toward residential segregation? GeoJournal 81–90 (2002)
9. Duncan, O.D., Duncan, B.: A methodological analysis of segregation indexes (1995). https://doi.org/10.2307/2088328
10. Duncan, O.D., Duncan, B.: Residential distribution and occupational stratification. Am. J. Sociol. (1955). https://doi.org/10.1086/221609
11. Feitosa, F.F., Câmara, G., Montiero, A.M.V., Koschitki, T., Silva, M.P.S.: Global and local spatial indices of urban segregation. Int. J. Geogr. Inf. Sci. 299–323 (2007)
12. Ferruzza, A., Dardanelli, S., Heins, F., Verrascina, M.: La geografia insediativa degli stranieri residenti: Verona, Firenze e Palermo a confronto. Studi Emigrazione/Migr. Stud. 602–628 (2008)
13. Martori, J.C., Apparicio, P.: Changes in spatial patterns of the immigrant population of a southern European metropolis: the case of the Barcelona metropolitan area (2001–2008). Tijdschrift voor economische en sociale geografie **102**(5), 562–581 (2011)
14. Massey, D.S., Denton, N.A.: The dimension of residential segregation. Soc. Forces (1988). https://doi.org/10.1093/sf/67.2.281
15. Morgan, B.S.: The segregation of socioeconomic groups in urban areas: a comparative analysis. Urban Stud. (1975). https://doi.org/10.1080/00420987520080041
16. Morrill, R.L.: On the measure of geographic segregation. Geogr. Res. Forum 25–36 (1991)
17. Mudu, P.: The people's food: the ingredients of "ethnic" hierarchies and the development of Chinese restaurant in Rome. GeoJournal 195–210 (2007)
18. Natale, L., Casacchia, O., Verdugo, V.: Minority Segregation Processes in an Urban Context: A Comparison Between Paris and Rome, Statistics and Demography: The Legacy of Corrado Gini, Conference of the Italian Statistical Society, Treviso, September 2015
19. Reardon, S.F., O'Sullivan, D.: Measures of spatial segregation. Sociol. Methodol. (2004). https://doi.org/10.1111/j.0081-1750.2004.00150.x
20. Roberto, E.: The Spatial Context of Residential Segregation. arXiv:1509.03678 [physics.soc-ph] (2016)
21. Roberto, E., Hwang, J.: Barriers to Integration: Institutionalized Boundaries and the Spatial Structures of Residential Segregation. arXiv:1509.02574 [physics.soc-ph] (2016)
22. Sakoda, J.M.: A generalized index of dissimilarity. Demography (1981). https://doi.org/10.2307/2061096
23. Simpson, L., Finney, N.: Parallel lives and ghettos in Britain: facts or myths? Geography 124–131 (2010)

24. Strozza, S., Benassi, F., Ferrara, R., Gallo, G.: Recent demographic trends in the major Italian urban agglomerations: the role of foreigners. Spat. Demogr. (2015). https://doi.org/10.1007/s40980-015-0012-2
25. Wacquant, L.: Urban outcasts: stigma and division in the black American ghetto and the French periphery. Int. J. Urban Reg. Res. 366–383 (1993)
26. Waldinger, R.: Black/Immigrant competition re-assessed: new evidence from Los Angeles. Sociol. Perspect. 365–386 (1997)
27. White, M.J.: The measurement of spatial segregation. Am. J. Sociol. 1008–1018 (1983)
28. Wong, D.W.S.: Spatial indices of segregation. Urban Stud. (1993). https://doi.org/10.1080/00420989320080551
29. Wong, D.W.S.: Geostatistics as measures of spatial segregation. Urban Geogr. (1999). https://doi.org/10.2747/0272-3638.20.7.635

Space-Time FPCA Clustering of Multidimensional Curves

Giada Adelfio, Francesca Di Salvo and Marcello Chiodi

Abstract In this paper we focus on finding clusters of multidimensional curves with spatio-temporal structure, applying a variant of a k-means algorithm based on the principal component rotation of data. The main advantage of this approach is to combine the clustering functional analysis of the multidimensional data, with smoothing methods based on generalized additive models, that cope with both the spatial and the temporal variability, and with functional principal components that takes into account the dependency between the curves.

Keywords FPCA · Clustering of multidimensional curves · GAM
Spatio-temporal pattern

1 Introduction

Variable reduction and partitioning of objects in homogeneous groups are the most used analyses for exploring complex structures in the data and obtaining the main information in a limited number of dimensions. These methodologies have been applied to functional data in recent years. Usually, we refer to data varying over a continuum (such as seismic waveforms, financial time series, temperatures recorded by a central source, etc.) as functional data and the continuum is often time. Since in many statistical applications realizations of continuous time series are available as observations of a process recorded in discrete time intervals, one crucial point is to convert discrete data to continuous functions, that is, from vectors to curves or more

G. Adelfio (✉) · F. Di Salvo · M. Chiodi
Dipartimento Scienze Economiche Aziendali e Statistiche,
viale delle Scienze ed 13, 90128 Palermo, Italy
e-mail: giada.adelfio@unipa.it

F. Di Salvo
e-mail: francesca.disalvo@unipa.it

M. Chiodi
e-mail: marcello.chiodi@unipa.it

C. Perna et al. (eds.), *Studies in Theoretical and Applied Statistics*,
Springer Proceedings in Mathematics & Statistics 227,
https://doi.org/10.1007/978-3-319-73906-9_18

generally functions $\mathbf{x}$ in $\mathbb{R}^d, d \geq 1$. When we talk about functional data, we refer to J pairs $(t_j, \mathbf{y}_j)$, $j = 1, ..., J$, where $\mathbf{y}_j$ is the value of an observable d-dimensional function $\mathbf{x}(\cdot)$ at time t_j. The functional predictor is observed with error:

$$\mathbf{y}_{ij} = \mathbf{x}_i(t_j) + \varepsilon_{ij};\quad i = 1, 2, \ldots, N;\ j = 1, 2, ..., J;\ 0 \leq t_j \leq T$$

Therefore, assuming that a functional datum for replication j arrives as a set of discrete measured values the first task is to convert them to functions computable for any t, called functional objects [17]. Several clustering methods for spatial functional data have been proposed in the literature (see [12]). In this paper we combine the aim of finding clusters from a set of individual curves (waveforms) with the functional nature of data in order to highlight some of the common characteristics of data; indeed, functional data analysis can quantify differences throughout waveforms that would not be evident when applying standard statistical methods to discrete variables. Many functional clustering methods have been developed, together with heuristic approaches in this context, such as: variants of the k-means method [23], clustering after transformation and smoothing [22], some model-based procedures, such as clustering sparsely sampled functional data [11], a model-based approach for clustering time-course gene expression data using B-splines in the framework of a mixture model [14], a wavelet-based methodology for clustering time series of smooth curves [3]. Chiodi [4] proposed a method for clustering multivariate short time series, based on the similarities of shapes.

In this paper we provide a new perspective of the Curve Clustering approach proposed in [1], considering multidimensional curves. The first version of the new algorithm, denoted as FPCAC, is based on the trimmed K-means Robust Curve Clustering (by RCC) proposed by Garca-Escudero and Gordaliza [8], and introduces a functional principal component rotation of data (FPCA, [17]), that overcomes the limitation of the cross-correlation among variables.

It also represents an alternative to the spatial clustering methods applied straight to geo-referenced functional data, among which, some ignore the spatial correlation of the curves (see [12]) and others take the spatial dependence into account through proximity measures between curves [9, 18, 21].

2 Clustering of Waveforms by FPCA

Statistical techniques handling huge amounts of information play a basic role in this context and synthesis of objects and variables aims to detect the most relevant information which allows an appropriate interpretation of the data. Various alternative methods, combining cluster analysis and the search for a lower-dimension representation, have been also proposed in the finite dimensional setting [24]. More recently, the use of clustering is considered as a preliminary step for exploring data represented by curves, with a further difficulty associated to the infinite space dimension of data [5, 12, 19]. In this work, we get a procedure dealing with the simultaneous optimal clustering and managing with multidimensional curves.

Waveform clustering may be considered as an issue of clustering of functional data; more generally it could be defined within the wider framework of partition type cluster analysis. Let $\{\mathbf{x}_1, \dots, \mathbf{x}_N\}$ be N multivariate curves, where $\mathbf{x}_i$ are functions in $\mathbb{R}^d, d \geq 1$ and depending on a real parameter t. We seek a partition $\mathcal{P} = \{G_1, G_2, \dots, G_K\}$ of the N curves in K exhaustive clusters with strong internal homogeneity; as usual, we need to define a measure of internal homogeneity, or, alternatively, a measure of distance from a reference curve $\mathbf{C}$ defined in each group. For this purpose a measure of distance between two multivariate curves $\mathbf{x}_i$ and $\mathbf{x}_h$ is defined as:

$$\delta_{ih} = \sum_{r=1}^{d} \ell_2^2(x_{ir} - x_{hr}), \ \forall i \neq h \tag{1}$$

where $\mathbf{x}_i = \{x_{i1}, x_{i2}, \dots, x_{id}\}$, with x_{ir} a single curve, and $\ell(\cdot)$ is defined as $\ell_2(\mathbf{x}) = \left(\int_0^T \mathbf{x}(t)^2\, dt\right)^{1/2}$. In the presence of discrete data the above integration will be substituted by a suitable quadrature formula or summation.

It is easy to see that if the curves are standardized, in order to be centered and scaled to unity, then $\delta_{ih} = 2 - 2\rho(\mathbf{x}_i, \mathbf{x}_h)$, where $\rho(\mathbf{x}_i, \mathbf{x}_h)$ is the correlation function between curves $\mathbf{x}_i$ and $\mathbf{x}_h, \forall i \neq h$. Then, minimizing the sum of distances defined in (1) for standardized curves is equivalent to maximizing the sum of correlations.

Some sort of curve registration procedure can also improve the overall similarity of this kind of data. The distance of a generic curve $\mathbf{x}_i$ from a group G_m is defined as the distance of the curve $\mathbf{x}_i$ from a reference curve $\mathbf{C}_m$ of G_m (in our implementation $\mathbf{C}_m$ is the average of the curves $\mathbf{x}_i$ belonging to G_m, although other choices are plausible, like median curves, medoids, etc. (see [20]). Then, given a partition $\mathcal{P}$, we denote by $g(\cdot)$ an associated labelling function, such that $g(i) = m : \mathbf{x}_i \in G_m \ \forall i$, define an overall measure of distances as the average of the N distances of the curves from their group template, that is:

$$\Delta_{\mathcal{P}} = \frac{\sum_{i=1}^{N} \delta\left(\mathbf{x}_i, \mathbf{C}_{g(i)}\right)}{N} \tag{2}$$

Theoretically, given the number of groups K, a curves clustering procedure looks for a partition $\mathcal{P}^*$ and m reference d-dimensional curves $\mathbf{C}_m^*$, which jointly minimize the value of (2), which minimize the value of (1).

In [2] the authors proposed an algorithm, structured as the K-means method, that deals with the simultaneous optimal clustering and warping of curves by means of a Procrustes fitting criterion, implemented by alternating expectation and maximization steps, as in an EM-type algorithm [16, 20]) since data are replaced by estimated curves taking into account the space-time nature of the original data, here we do not account for warping of curves.

2.1 The Used Robust Clustering of Curve Algorithm

Briefly, the Robust Clustering of Curve algorithm (RCC) is a kind of robust version of the K-means methodology through a trimming procedure. In few words, given a d-dimensional data sample $\boldsymbol{X}_1, \boldsymbol{X}_2, ..., \boldsymbol{X}_n$ with $\boldsymbol{X}_i = X_{i1}, ..., X_{iq}$, and fixed the number of clusters K, the trimmed K-means clustering algorithm looks for the K centers $C_1, ..., C_K$ that are solution of the minimization problem:

$$O_k(\alpha) = \min_{Y} \min_{C_1,\ldots,C_K} \frac{1}{[n(1-\alpha)]} \sum_{X_i \in Y} \inf_{1\leq j\leq k} ||X_i - C_j||^2 \tag{3}$$

with α the trimming size and Y is the set of subsets of $\boldsymbol{X}_1, \ldots, \boldsymbol{X}_n$ containing $[n(1-\alpha)]$ data points, where $[x]$ is the integer part of x. This method allocates each non-trimmed observation to the cluster identified by its closest center C_j, dealing with possible outliers by the given proportion of observations to be discarded α. This curve clustering procedure is based on a least-squares fit to cubic B-spline functions bases [10], applying the trimmed K-means clustering as Eq. 3 on the resulting coefficients

The proposed functional PCA-based clustering approach (FPCAC) is a variation of the RCC algorithm, looking for clusters of functions according to the direction of largest variance. The starting clusters K are determined by a random procedure. In this paper we use a modified version of the trimmed K-means algorithm, that considers a preprocessing of data accounting for their multivariate nature and applies the clustering algorithm to the results of the eigen-analysis performed on the curves, instead of using straightforward the coefficients of the linear fitting to B-spline bases. The proposed approach has the advantage of an immediate use of PCA for functional data [17] avoiding some objective choices as in RCC. Simulations and applications suggested also the well behavior of the FPCAC algorithm, both in terms of stable and easily interpretable results. The proposed multivariate clustering algorithm was performed by using the R software package [15].

3 Functional Principal Components for multivariate spatio-temporal data

The idea behind most dimension reduction methods is to reduce the original set of variables to a few of new transformed variables incorporating most of the information contained in the original ones and, as for the Principal Component Analysis, with the advantage to be uncorrelated. Functional Principal Component Analysis (FPCA) determines the uncorrelated linear combinations of the original dimensions, that account for most of the variability expressed by the covariance functions.

In [6] the authors propose an extension of the FPCA proposed by Ramsay and Silverman [17] to study possible spatial modes of variability and their changes in more than one dimension. This approach, based on the definition of functional data as functions of space and time, incorporates the spatial and time main effects through Generalized Additive Model estimation (GAM, [10, 25]); then, by mean of an eigen-analysis of the Variance function, the approach provides a reduction of the dimensions.

Let consider multivariate curves observed at N sites with coordinates $\mathbf{s} = (s_1, s_2)$. The D dimensional vector $\mathbf{y}_{st}$ observed in a site $\mathbf{s}$ and time t, is considered as realization of a function of both space and time and affected by noise:

$$y_{st} = x_{st} + \varepsilon_{st}$$

Let x_{st} $\left\{s \in S \subseteq R^2;\ t \in T \subseteq R\right\}$ be a non-stationary functional random field and the set ε_{st} be a stationary Gaussian process with a zero first moment and isotropic spherical covariance. The underlying functional predictor is assumed to be smooth and square integrable in its domain and is represented in terms of a linear combination of orthonormal polynomial basis and coefficients:

$$x_{st}^d = \Phi(s,t)\theta^d \tag{4}$$

or in matrix terms: $X = \Phi\Theta$, where Φ, with $(N \times T)$ rows and M columns, is the space-time smoothing basis matrix defined as the Kronecker product of the Basis matrices Φ_s, referred to space, and Φ_t referred to time; Θ is the matrix of coefficients $(M \times D)$, and M is the number of parameters. The Variance functions $COV(X^d(s,t), X^{d'}(s^*, t^*))$ quantify the dynamics of the space-time structures of the smoothed data in the dimensions d and d'. It is estimated on the basis of the linear expansion (4) as an M order penalized Variance in terms of the bases Φ and coefficients Θ and a Penalty matrix $\mathbf{P}$, with anisotropic smoothing structures (see [6, 13]):

$$\mathbf{V} = \{\Phi\left(\Phi'\Phi + \mathbf{P}\right)^{-1}\}'\Phi\Theta\Theta'\Phi\{\Phi\left(\Phi'\Phi + \mathbf{P}\right)^{-1}\}, \tag{5}$$

Under suitable regularity conditions, there exists an eigen-decomposition of the estimator (5) of the Variance function; the eigenfunctions, as solutions of the eigen-analysis, synthesize over the D dimensions and retain the most relevant information on the variations in space and time. This space-time structures is exploited in the cluster algorithm, that is applied on a given number of the eigenfunctions, selected according to the percentage of explained variance.

4 Application to Data and Conclusion

Multivariate spatio-temporal structures are actually very common in literature: some of the most common examples are environmental data, signals observed in multiple dimension (e.g. the components of a seismic wave: UpDown, NorthSouth and East-

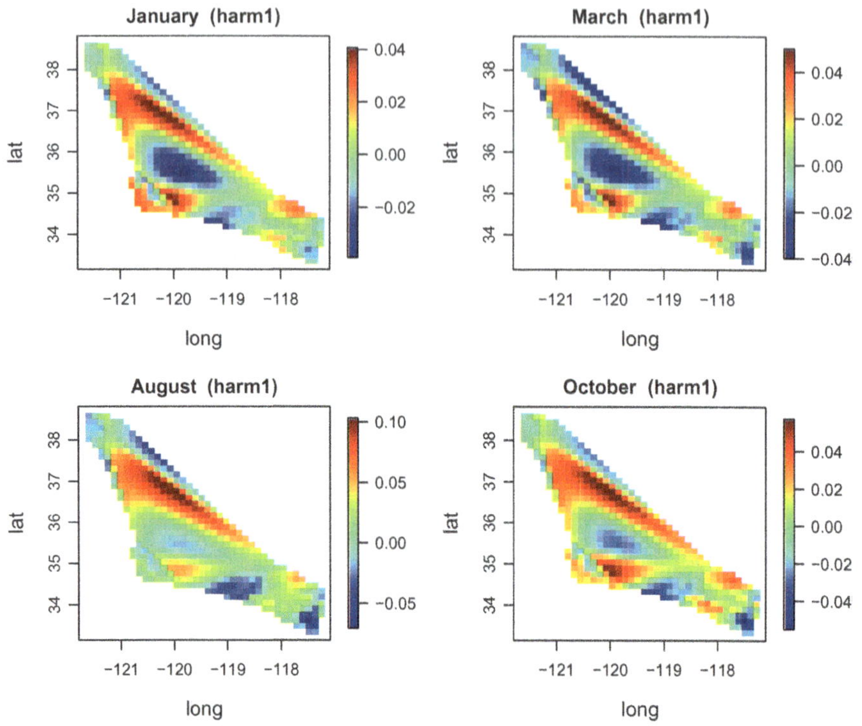

Fig. 1 Map of the first eigenfunction for four months: synthesis by variable: the time (month)-specific eigenfunction allows to identify months with a higher variability among pollutants in space

West). A spatio-temporal multivariate dataset related to air quality is here considered in order to prove the effectiveness of the proposal. Data concern concentrations of five main pollutants (CO, NO_2, O_3, PM10, SO_2) recorded during a year (2011) at different monitoring stations dislocated along the California state. Two different point of views can be considered: we can consider a synthesis by time, and the variable (pollutant in our example)-specific eigenfunctions allow to verify higher (or lower) variability along time (months). Spatial plots of the eigenfunctions, conditioning on pollutants, highlight the areas characterized by higher variability along time. On the other hand, when a synthesis by variable is considered, the time (month)-specific eigenfunctions allow to identify months with a higher variability among pollutants: as an example, in Fig. 1 we consider a synthesis by variable, such that the map of the time (months)-specific eigenfunctions allow to identify areas with a higher variability among pollutants.

The present version of the algorithm, reference curves have been estimated simply by an average curve for each cluster. Since the curve of arithmetic mean minimizes the sum of squares along a generic curve, this is in agreement to the minimization of squared distance used in the clustering criterion. Although a reasonable starting partition could be obtained cutting the tree diagram resulting from a hierarchical agglomerative clustering procedure [7], we know that this choice is usually considerably better than a random starting partition and is computationally acceptable when dealing with hundreds of waves. In our case we prefer to start with a random partition, in order to get also a random approach for the choice of K. Indeed, for finding a 'good' number of clusters, we studied the correlation inside the groups, for $K = 4, 5, 6, 7, 8, 9$, since values of $K > 9$ would produce also empty groups and $K < 4$ forced and spurious clusters, and therefore, useless results. For each value of K, we sampled 100 random partitions of the 59 curves in K groups, computing on each of them the average correlation r_K, computed for each pair of curves inside each cluster. For each K, we computed also the average r_{m_K} and the standard deviation r_{s_K} of the 100 simulated values r_K, and used them to standardize the sequence of r_K; than, we look for the number of clusters such that the median of the standardized correlation $\dot{r}_K = \frac{r_k - r_{m_k}}{r_{s_K}}$ inside them is maximized. The values $\dot{r}_K$ are plotted against K in Fig. 2. The choice of the value K = 7, corresponding to local minimum values, is suggested. Once we have chosen the number of clusters, we can apply the proposed algorithm, moving from FDA results of the GAM-PCA procedure proposed by Di Salvo et al. [6]. The FPCAC is performed and a FPCs subset is retained and subjected to varimax rotation, considering three harmonics, that account for the 85%of the variance; the Fig. 3, on the left, shows the seven clusters on the space of the first

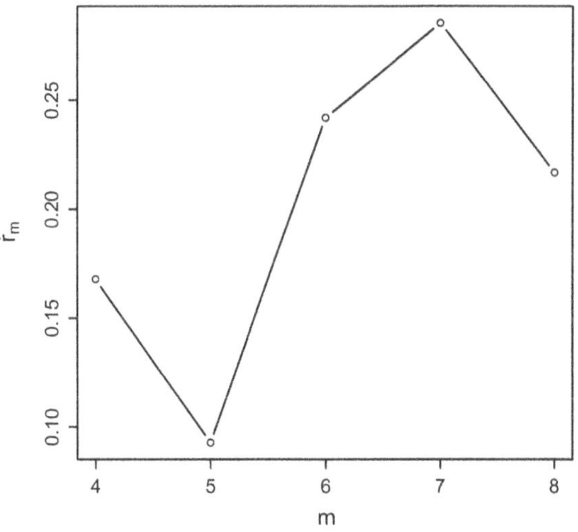

Fig. 2 Values of $\dot{r}_K$ against K, K = 4, ..., 8, for the choice of the number of groups

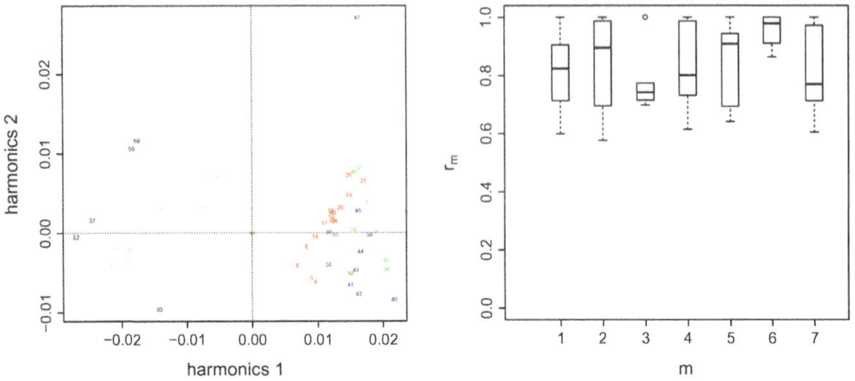

Fig. 3 Plot of the first two eigenfunction aggregate by time—clusters are identified by different colors (on the left). Boxplot of correlations among curves for the seven clusters (on the right)

Table 1 Mean correlation among the curves of each cluster and dimensions of the seven identified clusters

m	1	2	3	4	5	6	7
$\bar{r}_m$	0.867	0.859	0.893	0.839	0.883	0.951	0.802
n_m	13	6	5	7	5	4	7

two eigenfunctions. These seven clusters contain signals with internal mean correlation ($\bar{r}_m =$, $m = 1, 2, .., 7$, greater than 0.87. The boxplot of correlation among curves (r_m) for the seven clusters is reported in Fig. 3 on the right. The $\bar{r}_m$ values, together with cluster dimensions n_m, are reported in Table 1. The lines, conditioned to each pollutant, correspondent to the more numerous clusters (m $= 1, 2, 4, 7$ with 13, 6, 7 and 7 clustered curves respectively) of 59 curves, together with the average curve and the three estimated harmonics are reported in Fig. 4.

The application of the proposed method provides interesting results in finding reasonable clusters of curves combined with an immediate use of FPCA. The approach uses a P-spline smoothing as a regularization step and the simultaneous reduction of objects and dimensions has the advantage of defining the clusters along the directions in which the curves have the highest variability; overcoming the limitation of the cross-correlation approaches, it turns out to be more suitable in presence of a considerable number of dimensions.

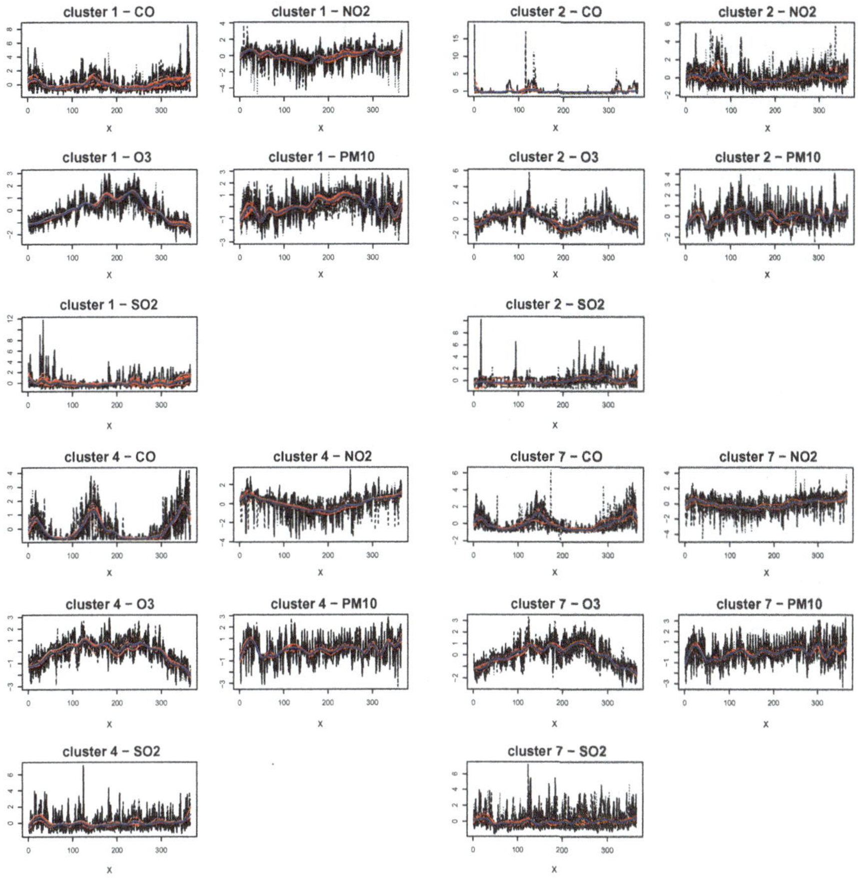

Fig. 4 Main clusters (m = 1, 2, 4, 7) considering as input of the FPCAC approach the result of the GAM-PCA procedure, for the five pollutants. The blu lines correspond to average curves and the red lines refer to the three estimated harmonics

References

1. Adelfio, G., Chiodi, M., D'Alessandro, A., Luzio, D.: FPCA algorithm for waveform clustering. J. Commun. Comput. **8**, 494–502 (2011)
2. Adelfio, G., Chiodi, M., D'Alessandro, A., Luzio, D., D'Anna, G., Mangano, G.: Simultaneous seismic wave clustering and registration. Comput. Geosci. **8**(44), 6069 (2012)
3. Antoniadis, A., Paparoditis, E., Sapatinas, T.: A functional waveletkernel approach for time series prediction. J. R. Stat. Soci. Ser. B Stat. Methodol. **68**(5), 837 (2006)
4. Chiodi, M.: The clustering of longitudinal data when time series are short, Multivar. Data Anal. 445–453 (1989)
5. Delicado, P., Giraldo, R., Comas, C., Mateu, J.: Statistics for Spatial Functional Data: Some Recent Contributions, Environmetrics. Published Online in Wiley InterScience (2009). https://doi.org/10.1002/env.1003, https://www.interscience.wiley.com

6. Di Salvo, F., Ruggieri, M., Plaia, A.: Functional principal component analysis for multivariate multidimensional environmental data. Env. Ecol. Stat. **22**, 739–757 (2015)
7. Everitt, B.: Cluster Analysis. Wiley, New York (1993)
8. Garca-Escudero, L.A., Gordaliza, A.: A proposal for robust curve clustering. J. Classif. **22**, 185–201 (2005)
9. Giraldo, R., Delicado, P., Comas, C., Mateu, J.: Hierarchical clustering of spatially correlated functional data. Stat. Neerl. **66**, 403–421 (2011)
10. Hastie, T.J., Tibshirani, R.J.: Generalized Additive Models. Chapman and Hall, London (1990)
11. James, G.M., Sugar, C.A.: Clustering for sparsely sampled functional data. J. Am. Stat. Assoc. **98**(462), 397–408 (2003)
12. Jacques, J., Preda, C.: Functional data clustering: a survey. Advances in Data Analysis and Classification, vol. 8(3), pp. 231–255. Springer (2014)
13. Lee, D.J., Durban, M.: P-spline ANOVA-type interaction models for spatio-temporal smoothing. Stat. Model. **11**(1), 49–69 (2011)
14. Luan, Y., Li, H.: Clustering of time-course gene expression data using a mixed-effects model with B-splines. Bioinformatics **19**(4), 474–82 (2003)
15. R Core Team: R: A language and environment for statistical computing. R Foundation for Statistical Computing, Vienna, Austria (2015). https://www.R-project.org/
16. Ramsay, J.O., Li, X.: Curve registration. J. R. Stat. Soc. Sect. B **60**, 351–363 (1998)
17. Ramsay, J.O., Silverman, B.W.: Functional Data Analysis. Springer, New York (2005)
18. Romano, E., Balzanella, A., Verde, R.: Spatial variability clustering for spatially dependent functional data. Stat. Comput. (2016)
19. Romano, E., Mateu, J., Giraldo, R.: On the performance of two clustering methods for spatial functional data. Advances in Data Analysis and Classification, pp. 467–492. Springer (2015)
20. Sangalli, L.M., Secchi, P., Vantini, S., Veneziani, A.: A case study in explorative functional data analysis: geometrical features of the Internal Carotid Artery. J. Am. Stat. Assoc. **104**(485), 37–48 (2009)
21. Secchi, C., Vantini, S., Vitelli, V.: Spatial Clustering of Functional Data (2011) Recent Advances in Functional Data Analysis and Related Topics, pp 283–289 (2011)
22. Serban, N., Wasserman, L.: CATS: cluster analysis by transformation and smoothing. J. Am. Stat. Assoc. **100**(471), 990–999 (2005)
23. Tarpey, T., Kinateder, K.K.J.: Clustering functional data. J. Classif. **20**, 3–114 (2003)
24. Vichi, M., Saporta, G.: Clustering and disjoint principal component analysis. Comput. Stat. Data Anal. (2008). https://doi.org/10.1016/j.csda.2008.05.028
25. Wood, S.N.: Generalized Additive Models: an Introduction with R. Chapman & Hall (2006)

The Power of Generalized Entropy for Biodiversity Assessment by Remote Sensing: An Open Source Approach

Duccio Rocchini, Luca Delucchi and Giovanni Bacaro

Abstract The assessment of species diversity in relatively large areas has always been a challenging task for ecologists, mainly because of the intrinsic difficulty to judge the completeness of species lists and to undertake sufficient and appropriate sampling. Since the variability of remotely sensed signal is expected to be related to landscape diversity, it could be used as a good proxy of diversity at species level. It has been demonstrated that the relation between species and landscape diversity measured from remotely sensed data or land use maps varies with scale. While traditional metrics supply point descriptions of diversity, generalized entropy's framework offers a continuum of possible diversity measures, which differ in their sensitivity to rare and abundant reflectance values. In this paper, we aim at: (i) discussing the ecological background beyond the importance of measuring diversity based on generalized entropy and (ii) providing a test on an Open Source tool with its source code for calculating it. We expect that the subject of this paper will stimulate

D. Rocchini (✉)
Center Agriculture Food Environment (C3A), University of Trento, Via E. Mach 1, 38010 S. Michele all'Adige, Trentino, Italy
e-mail: duccio.rocchini@unitn.it

D. Rocchini
Centre for Integrative Biology, University of Trento, Via Sommarive, 14, 38123 Povo, Trentino, Italy

D. Rocchini · L. Delucchi
Fondazione Edmund Mach, Department of Biodiversity and Molecular Ecology, Research and Innovation Centre, Via E. Mach 1, 38010 S. Michele all'Adige, Trentino, Italy
e-mail: luca.delucchi@fmach.it

G. Bacaro
Department of Life Sciences, University of Trieste, Via L. Giorgieri 10, 34127 Trieste, Italy
e-mail: giovanni.bacaro@units.it

C. Perna et al. (eds.), *Studies in Theoretical and Applied Statistics*, Springer Proceedings in Mathematics & Statistics 227, https://doi.org/10.1007/978-3-319-73906-9_19

discussions on the opportunities offered by Free and Open Source Software to calculate landscape diversity indices.

Keywords Biodiversity · Generalized entropy · Rényi entropy · Remote sensing

1 Introduction

Species-based measures of diversity like species richness (alpha-diversity) or species turnover (beta-diversity) are the most commonly used metrics for quantifying the diversity of an area. Nonetheless, the assessment of species richness in relatively large areas has always been a challenging task for ecologists, mainly because of the intrinsic difficulty in judging the completeness of species lists and in undertaking sufficient appropriate sampling [1]. Inventorying species over large regions is hampered by the effort required for field sampling and complications resulting from changes in species composition through time. Therefore, different methods have been proposed to overcome these issues. These methods include the use of habitats as a proxy for estimating species diversity [2], or the examination of remotely-sensed proxies for richness [3].

Given the difficulties of field-based data collection, the use of remote sensing for estimating environmental heterogeneity and (subsequently) species diversity represents a powerful tool since it allows for a synoptic view of an area with a high temporal resolution. As an example, the availability of satellite-derived data like those achieved by the Landsat program makes it feasible to study all parts of the globe up to a resolution of 30 m. This is particularly relevant in view of the availability of recent Open Source systems for the analysis of remotely-sensed imagery [4].

In this paper we will present one of the most straightforward measures of spatial complexity available in the Free and Open Source Software GRASS GIS, namely the Rényi entropy measure to estimate key ecological variability patterns and related processes.

2 Rényi Calculation in GRASS GIS

2.1 Software Used

GRASS GIS (Geographical Resources Analysis Support System, [5]) represents one of the most powerful free and open source tools for geographical mapping, which includes more than 450 modules for managing and analyzing geographical data. GRASS was created in 1982 by the U.S. Army Construction Engineering Research Laboratories, and nowadays it is one of the cutting-edge projects of the Open Source Geospatial Foundation (OSGeo, founded in 2006). The adoption of the free open source software (FOSS) license in 1999 and the introduction of an online source code repository (Concurrent Versioning System at that time) changed the development process of GRASS, thus allowing worldwide contributions from the scientific community.

In this manuscript we will deal with the most powerful functions in GRASS GIS to measure spatial complexity under an ecological perspective.

2.2 *Rényi Entropy as a Local Diversity Measure in a Neighbourhood*

Calculating local diversity is important to detect spots of diversity at a local scale. As an example, in biodiversity research, this is known as α-diversity and it is a widely-used metric in ecology [3].

GRASS GIS is capable of handling common Information-theory based statistics such as Boltzman or Shannon-Weaver entropy H [6], Pielou evenness [7] and Simpson's reversed dominance (1-D, [8]). Such diversity measures are generally used to summarise large multivariate data sets, providing for one potentially meaningful single value. Such an approach inevitably results in information loss, since no single summary statistic can characterize in an unequivocal manner all aspects of diversity [10]. Rocchini and Neteler [11] addressed such problems when measuring diversity from a satellite image relying on the richness and relative abundance of Digital Numbers (DNs), by only using entropy-based metrics. In particular, they observed: (i) the intrinsic impossibility of discriminating among different ecological situations with one single diversity index, and (ii) the impossibility of understanding whether diversity of different sites is more related to differences in richness or in relative abundance of DN values. As an example, they provided a theoretical case in which the same value of the Shannon index would actually be related to very different situations in terms of DNs richness and abundances (see Fig. 2 in [11]). In general, to solve this issue, combining these entropy-based indices with evenness-based metrics might lead to an increase in their information content. In this regard, the most commonly-used metric is the Pielou evenness index $J = \frac{-\sum p \times ln(p)}{ln(N)}$ [7], which can be rewritten as: $J = \frac{H}{Hmax}$ since it contains the maximum possible diversity (ln(N)), for N DNs.

All the previously described metrics based on Information theory only supply point descriptions of diversity. By contrast, Rényi [9] firstly introduced a generalized entropy metric, $H_\alpha = \frac{1}{1-\alpha} \times ln \sum p^\alpha$ which shows a high flexibility and power because a number of popular diversity indices are special cases of H_α. In mathematical terms, if we consider e.g. the variation of α from 0 to 2:

$$H\alpha = \begin{cases} \alpha = 0, H_0 = ln(N) \\ \alpha \to 1, H_1 = -\sum(p \times ln(p)) \\ \alpha = 2, H_2 = ln(1/D) \end{cases} \tag{1}$$

where N = number of Digital Numbers (DNs), p = relative abundance of each DN value, D = Simpson index.

Concerning the results attained when $\alpha = 1$, the Shannon index is derived according to the L'Hôpital's rule of calculus [10]. Rényi generalized entropy represents a continuum of diversity measures, meaning that it is possible to maintain sensitivity to both rare and abundant DNs, and it is more responsive to the commonest DNs while α increases. Varying α can be viewed as a scaling operation, not in a real space but in the data space.

As far as we know, GRASS GIS is the only software capable of calculating generalized measures of diversity in 2-D such as the Rényi formula, based on the following function:

```
r.li.renyi conf=conf3 in=input_image out=output_renyi
alpha=2
```

Changing the parameter α will change the behaviour of the formula, generating different maps of diversity, representing a continuum of diversity values over space instead of single measures. Increasing alpha values in the Rényi diversity index will weight differences in relative abundance more heavily than differences in simple richness. In the aforementioned code a moving window of 3×3 cells passes over

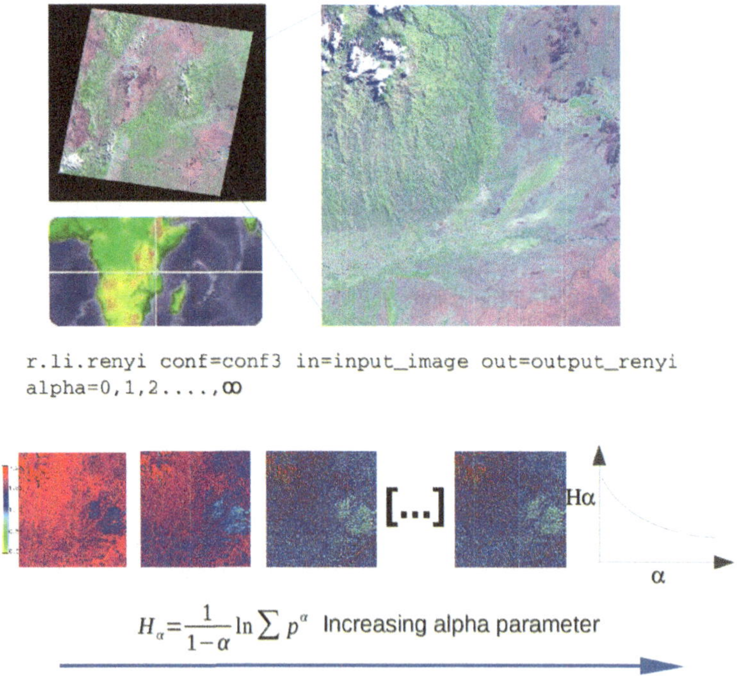

Fig. 1 A Landsat ETM + image (band combination 7:4:2) may represent the spectral behavior of an area composed by very different habitats. In this example a tropical forest over the Tanzania region is shown. The Rényi entropy formula applied in GRASS GIS might lead to a diversity profile, by changing the alpha parameter

the original image and caculateds the Rényi value at fixed space lags. Applying the aforementioned code in GRASS GIS to a MODIS image the Rényi leads to different maps of diversity shown in in Fig. 1.

3 Discussion

In this paper we presented a GRASS GIS-based tool for calculating landscape diversity to be potentially related to biodiversity studies at different spatial scales (moving window sizes). The proposed automatic and freely available tool for calculating landscape diversity has the potential towards, e.g., (i) the generation of species-diversity proxies or indicators [12], (ii) an increase of species-inventory efficiency [13], (iii) quantitative comparison of different areas with different degrees of diversity at multiple scales [14]. Remotely sensed information can function as a driver for developing field sampling design strategies. As previously stated, sampling species in the field has a number of drawbacks such as (i) observer bias [15], (ii) the definition of statistical population when developing sampling designs [16], (iii) reproducibility [17], (iv) spatial errors [18], (v) historical bias about species distribution records [19]. Refer to [20] for a complete dissertation about imperfectness of species field data considering both quality (data labelling) and quantity (sample size).

Hence, although we are unable to directly detect organisms remotely, proxies for community properties provide a valuable data source for the study of species diversity. Therefore, the use of indirect remote sensing techniques for estimating diversity of landscapes shows promise to forecast species diversity over different spatial scales. In this view, r.li.renyi allows measuring landscape diversity and relating it to species diversity at multiple moving window sizes the analysis process, from fine scale field sampling units to the entire study area.

While a number of tailored software tools exist for calculating landscape diversity metrics like FRAGSTATS [21] or Patch Analyst [22] they do not allow users to access and/or review directly the source code, thus hampering the straightforward development of new metrics by several researchers at a time. In this view, the introduction of a concurrent versioning system of GRASS GIS in 1999 enabled different institutions and individuals to contribute to the code base simultaneously from different countries around the world, along with real time peer review of the submitted changes. The modular software design of GRASS facilitates the introduction of new functionalities without affecting the overall performance of the system. Moreover, recent improvements also allow GRASS users and developers to make use of the Python programming language [23] to introduce new features.

Diversity cannot be reduced to single index information, since one can never capture all aspects of diversity in a single statistic [24]. As an example, [25], dealing with the Shannon H index and the Simpson diversity index 1D, reports the case of discordant diversity patterns obtained by considering different indices. Such information may remain hidden once only one index is considered. Hence, a restricted set of non redundant indices could reach significant aspects on the spatial patterns.

For this reason, future work will be devoted to the development of a continuum of diversity measures such as the Rényi entropy presented in this paper. Such measures are particularly important since they are not redundant and they allow to consider several measures at a time, by varying one parameter therein, like the alpha parameter in the generalized Rényi entropy. The very aim of using the Rényi entropy in ecology does not consist in selecting the most appropriate parameter (if any) that best explains the problem under study, but rather in constructing a diversity profile showing how parametric diversity responds to changes in the parameter sensitivity to rare and abundant DNs. As far as we know, this is the first example in which Rényi entropy is provided in an Open Source framework. Hence r.li.renyi's code is available from the GRASS GIS source code repository (https://grass.osgeo.org/grass70/manuals/r.li.renyi.html) for further modifications, improvements, if needed, bug fixing, and the reuse for a development of new indices based on new or still underused mathematical theory.

Understanding the ecological processes which shape diversity over space at different spatial scales may be done by the quantification of surface gradients. Spatial gradients may be quantified by relying on the variability—or diversity—of a surface over space, being such surface for instance a remotely sensed image. In this view, r.li.renyi in GRASS GIS can be a valuable tool, on the strength of its major advantages like: (i) the availability of a continuous variation of multiple indices at a time, (ii) the possibility to create new indices directly reusing the code, (iii) the possibility to calculate landscape diversity at multiple spatial scales in an explicit way based on varying moving windows, and thus (iv) reducing the problems of hidden patterns of the relation between field- and landscape-based diversity due to scale mismatch.

References

1. Palmer, M.W.: Distance decay in an old-growth neotropical forest. J. Veg. Sci. **16**, 161–166 (2005)
2. Goetz, S., Steinberg, D., Dubayah, R., Blair, B.: Laser remote sensing of canopy habitat heterogeneity as a predictor of bird species richness in an eastern temperate forest. USA. Remote Sens. Environ. **108**, 254–263 (2007)
3. Rocchini, D.: Effects of spatial and spectral resolution in estimating ecosystem -diversity by satellite imagery. Remote Sens. Environ. **111**, 423–434 (2007)
4. Rocchini, D., Neteler, M.: Let the four freedoms paradigm apply to ecology. Trends Ecol. Evol. **27**, 310–311 (2012)
5. Neteler, M., Bowman, M.H., Landa, M., Metz, M.: GRASS GIS: a multi-purpose open source GIS. Environ. Model. Softw. **31**, 124–130 (2012)
6. Shannon, C.: A mathematical theory of communication. Bell Syst. Tech. J. **27**, 379–423 (1948)
7. Pielou, E.C.: An Introduction to Mathematical Ecology. Wiley, New York (1969)
8. Simpson, E.H.: Measurement of diversity. Nature **163**, 688 (1949)
9. Rényi, A.: Probability Theory. North Holland Publishing Company, Amsterdam (1970)
10. Ricotta, C.: On possible measures for evaluating the degree of uncertainty of fuzzy thematic maps. Int. J. Remote Sens. **26**, 5573–5583 (2005)
11. Rocchini, D., Neteler, M.: Spectral rank-abundance for measuring landscape diversity. Int. J. Remote Sens. **33**, 4458–4470 (2012)

12. Feilhauer, H., Faude, U., Schmidtlein, S.: Combining Isomap ordination and imaging spectroscopy to map continuous floristic gradients in a heterogeneous landscape. Remote Sens. Environ. **115**, 2513–2524 (2011)
13. Rocchini, D., Andreini Butini, S., Chiarucci, A.: Maximizing plant species inventory efficiency by means of remotely sensed spectral distances. Glob. Ecol. Biogeogr. **14**, 431–437 (2005)
14. Oldeland, J., Wesuls, D., Rocchini, D., Schmidt, M., Jgens, N.: Does using species abundance data improve estimates of species diversity from remotely sensed spectral heterogeneity? Ecol. Indic. **10**, 390–396 (2010)
15. Bacaro, G., Baragatti, E., Chiarucci, A.: Using taxonomic data to assess and monitor biodiversity: are the tribes still fighting? J. Environ. Monit. **11**, 798–801 (2009)
16. Chiarucci, A.: To sample or not to sample? That is the question... for the vegetation scientist. Folia Geobot. **42**, 209–216 (2007)
17. Ferretti, M., Chiarucci, A.: Design concepts adopted in longterm forest monitoring programs in Europe: problems for the future? Sci. Total Environ. **310**, 171–178 (2003)
18. Graham, C.H., Elith, J., Hijmans, R.J., Guisan, A., Peterson, A.T., Loiselle, B.A.: The influence of spatial errors in species occurrence data used in distribution models. J. Appl. Ecol. **45**, 239–247 (2008)
19. Hortal, J., Jimenez-Valverde, A., Gomez, J.F., Lobo, J.M., Baselga, A.: Historical bias in biodiversity inventories affects the observed environmental niche of the species. Oikos **117**, 847–858 (2008)
20. Foody, G.M.: Impacts of imperfect reference data on the apparent accuracy of species presence-absence models and their predictions. Glob. Ecol. Biogeogr. **20**, 498–508 (2011)
21. McGarigal, K., Marks, B.J.: FRAGSTATS: Spatial pattern analysis program for quantifying landscape structure. General Technical Report PNW-GTR-351. U.S. Department of Agriculture, Forest Service, Pacific Northwest Research Station, Portland, OR (1995)
22. Elkie, P., Rempel, R., Carr, A.: Patch analyst user's manual. Ont. Min. Natur. Resour. Northwest Sci. Technol. Thunder Bay Ont. TM-002 (1999)
23. van Rossum, G.: Python Library Reference. CWI Report CS-R9524 (1995)
24. Gorelick, R.: Combining richness and abundance into a single diversity index using matrix analogues of Shannon's and Simpson's indices. Ecography **29**, 525–530 (2006)
25. Nagendra, H.: Opposite trends in response for the Shannon and Simpson indices of landscape diversity. Appl. Geogr. **22**, 175–186 (2002)

An Empirical Approach to Monitoring Ship CO_2 Emissions via Partial Least-Squares Regression

Antonio Lepore, Biagio Palumbo and Christian Capezza

Abstract Kyoto Protocol and competitiveness of the shipping market have been urging shipping companies to pay increasing attention to ship energy efficiency monitoring. At the same time, new monitoring data acquisition systems on modern ships have brought to a navigation data overload that have to be fully utilized via statistical methodologies. For this purpose, an empirical approach based on Partial Least-Squares regression is introduced by means of a real case study in order to give practical indications on CO_2 emission control and for supporting prognosis of faults.

Keywords Partial least-squares · CO_2 emission monitoring · Prediction error Multivariate control chart

1 Introduction

The new Regulation of the European Union [3] in accordance with the International Maritime Organization (IMO) guidelines urges shipping companies to set up a system for Monitoring, Reporting and Verification (MRV) of CO_2 emissions based on ship fuel consumption [11]. However, in nowadays market there is a lack of effective tools that can be adopted in a real environment. With respect to this problem, the marine engineering literature mainly relies on the use of speed-power curves [16], which describe an empirical relation between the engine power and the vessel speed. In fact, the engine power can be put in relationship with the fuel consumption through the specific fuel consumption coefficient [2]. However, these curves are based on experiments obtained from specific tests which usually

A. Lepore · B. Palumbo (✉) · C. Capezza
University of Naples Federico II, Naples, Italy
e-mail: biagio.palumbo@unina.it

A. Lepore
e-mail: antonio.lepore@unina.it

C. Capezza
e-mail: christian.capezza@unina.it

C. Perna et al. (eds.), *Studies in Theoretical and Applied Statistics*,
Springer Proceedings in Mathematics & Statistics 227,
https://doi.org/10.1007/978-3-319-73906-9_20

overlook further factors the vessel may be subject to (e.g. different routes, weather conditions, sea weaves, displacement, trim, engine operation mode). Therefore, they give poor predictions in real cases and result unreliable for supporting verification of fuel consumption and harmful emissions [1]. The purpose of this paper is to introduce a statistical approach that can exploit the massive navigation data collected on-board by modern Data AcQuisition (DAQ) systems to better characterize the *ship operating conditions* via the Partial Least-Squares (PLS) regression. As is known, PLS techniques naturally deal with large data matrices that unavoidably have issues with classical regression assumptions. The new approach is illustrated by means of a relevant case study to directly demonstrate its practical applicability into MRV of CO_2 emissions and engineering prognosis of faults.

2 Ship Navigation Data Analysis with a Real Case Study

PLS techniques have been used in several areas after firstly appeared in the Econometric field. They are a good alternative to the classical multiple linear regression, ridge regression and other well known regression techniques because of the stability of predictors [9] and the robustness of the parameter estimators to new reference samples [19, 25].

The model starts off with a matrix $\mathbf{X}$ of n observations for m predictor variables (Table 1) and a vector $\mathbf{y}$ of n observations for the response variable. Incidentally, variables are standardized (i.e. divided by the sample standard deviation) and mean-centered (i.e. the average value of each variable is subtracted from the data) to give them equal weight in the data analysis. This improves the interpretability of the model [6] and allows regarding inner products as covariances [21].

2.1 Data Description

Navigation data are collected from a cruise ship owned by Grimaldi Group that operates on standard commercial routes in the Adriatic Sea. Ship name, voyage dates and port names are intentionally omitted for confidentiality reasons. All the voyages are identified by a Voyage progressive Number (VN).

According to the international laws, master and chief engineer must fill the voyage report at the end of each voyage and deliver it to the Energy Saving Department (ESD) on land. Unfortunately, those reports result mostly unuseful for managerial decision-making.

Therefore, the shipping company installed on-board a DAQ system (patented by CETENA S.p.A) that provides, at each voyage, summary statistics of measurements collected by a large-scale sensor network, thus also removing the human intervention. Table 1 reports the complete set of $m = 19$ predictor variables of engineering interest that describe the *ship operating conditions*.

Table 1 Predictor variables considered to build the model

	Variable	Description
1	V	Speed over ground (SOG) cubed [kn]
2	σ_V	SOG standard deviation [kn]
3	W_H	Head wind [kn]
4	W_F	Following wind [kn]
5	W_S	Side wind [kn]
6	SG_P	Shaft generator power (port) [kW]
7	SG_S	Shaft generator power (starboard) [kW]
8	ΔP	Power difference between port and starboard propeller shafts [kW]
9	T_{FD}	Departure draught (fore perpendicular) [m]
10	T_{AD}	Departure draught (aft perpendicular) [m]
11	T_{PD}	Departure draught (midship section—port) [m]
12	T_{SD}	Departure draught (midship section—starboard) [m]
13	T_{FA}	Arrival draught (fore perpendicular) [m]
14	T_{AA}	Arrival draught (aft perpendicular) [m]
15	T_{PA}	Arrival draught (midship section—port) [m]
16	T_{SA}	Arrival draught (midship section—starboard) [m]
17	$Trim_D$	Departure trim [m]
18	$Trim_A$	Arrival trim [m]
19	Δ	Displacement [Mt]

Predictor variables Physical variables reported in Table 1 are used as predictors to build the PLS model. Strictly speaking, those variables are function of (and not the pure measurements collected by) the sensor signals. Variable definition and some information on data acquisition mode are given below. Further details can be found in [1]. Explicitly note that statistics recorded by the DAQ system refer to the actual voyage navigation time, which is defined as the time between the finished with engine order (when the ship leaves the departure port) and the stand by engine order (when the ships enters the arrival port) [10].

Speed Over Ground (SOG) is obtained as the ratio between the sailed distance over ground (calculated by DAQ system from latitude and longitude by using the Haversine formula [23]) and the actual navigation time.

SOG standard deviation is the sample standard deviation of SOG observations recorded every 5 min on the actual voyage navigation time. It takes into account SOG variation (acceleration).

Head, following and side wind are defined based upon comprehensive engineering considerations on the wind component influence. Let V_{WT} denote the true wind speed and Ψ_{WT} the difference between the true wind angle (in the earth system) and the course over ground [12]. Then W_H, W_F and W_S are obtained as the mean value of

$$\tilde{W}_H = \begin{cases} 0 & \text{if } 90^\circ \leq \Psi_{WT} \leq 270^\circ \\ V_{WT}\cos(\Psi_{WT}) & \text{otherwise} \end{cases}, \tag{1}$$

$$\tilde{W}_F = \begin{cases} -V_{WT}\cos(\Psi_{WT}) & \text{if } 90^\circ \leq \Psi_{WT} \leq 270^\circ \\ 0 & \text{otherwise} \end{cases}, \tag{2}$$

and $\tilde{W}_S = |V_{WT}sin(\Psi_{WT})|$, respectively.

Port and starboard shaft generator power allow taking into account the different modes of navigation (constant and combinator mode).

Power difference between two propeller shafts allows discovering anomalies or malfunctioning in the main engines.

Departure and arrival trim are obtained through the inclinometer measurements and the geometric features of the ship.

Draughts are measured both at departure and arrival ports by four submersible transmitters located at fore and aft perpendiculars, and at port and starboard midship sections [1].

Displacement is derived from the hydrostatic data on the basis of the mean draught at midship and trim.

Response variable Hourly CO_2 emission at each voyage is chosen as response variable. It is calculated based on fuel consumption through a fuel mass to CO_2 conversion factor for heavy fuel oil [11], where fuel consumption is obtained as described in [7].

2.2 *Model Building*

The idea in PLS algorithms is projecting the m original variables into a smaller number a of orthogonal components (i.e. latent variables), whose observations are collected into score vectors $\mathbf{t}_i$ $(i = 1, \ldots, a)$. The criteria for the determination of a is illustrated later in this section.

The proposed approach is based on the Nonlinear Iterative PArtial Least-Squares (NIPALS) algorithm [8] for a single response variable that is summarized in the following steps. Initialize the algorithm by setting $\mathbf{X}_0 = \mathbf{X}$ and $\mathbf{y}_0 = \mathbf{y}$. Then, at each iteration $i = 1, \ldots, a$:

1. calculate the weight vector $\mathbf{w}_i = \mathbf{X}_{i-1}^T \mathbf{y}_{i-1}/(\mathbf{y}_{i-1}^T \mathbf{y}_{i-1})$;
2. scale $\mathbf{w}_i$ to be unit length;
3. calculate the score vector as linear combination of $\mathbf{X}$-columns $\mathbf{t}_i = \mathbf{X}_{i-1}\mathbf{w}_i$;
4. regress $\mathbf{y}_{i-1}$ on $\mathbf{t}_i$ obtaining the coefficient $b_i = \mathbf{y}_{i-1}^T \mathbf{t}_i/(\mathbf{t}_i^T \mathbf{t}_i)$;
5. calculate the loading vector $\mathbf{p}_i = \mathbf{X}_{i-1}^T \mathbf{t}_i/(\mathbf{t}_i^T \mathbf{t}_i)$;
6. deflate the matrices $\mathbf{X}_i = \mathbf{X}_{i-1} - \mathbf{t}_i\mathbf{p}_i^T$ and $\mathbf{y}_i = \mathbf{y}_{i-1} - b_i\mathbf{t}_i$ and start from step 1 to find a new component.

In such a way, at each iteration i the objective function of PLS is such that $\mathbf{t}_i$ does not only explain as much variance in $\mathbf{X}$ as possible, but is also highly correlated with $\mathbf{y}$ [5]. Then, the following matrices are obtained: $\mathbf{T} = (\mathbf{t}_1, \ldots, \mathbf{t}_a)$, $\mathbf{P} = (\mathbf{p}_1, \ldots, \mathbf{p}_a)$, $\mathbf{W} = (\mathbf{w}_1, \ldots, \mathbf{w}_a)$ and $\mathbf{b} = (b_1, \ldots, b_a)$. Note that the vectors $\mathbf{t}_i$'s are orthogonal, as well as $\mathbf{w}_i$'s [9] and usually a few components can express most of the variability in the data when the variables are highly correlated. The error terms $\mathbf{E} = \mathbf{X}_a$ and $\mathbf{f} = \mathbf{y}_a$ are assumed to be independent and normally distributed random variables.

Other ways of computing PLS latent vectors are discussed by Hoskuldsson [9]. However, Dayal and MacGregor [4] presented a faster kernel algorithm based on the proof that only one of either $\mathbf{X}$ or $\mathbf{y}$ needs to be deflated in PLS algorithms during the sequential process of computing latent vectors.

In order to determine, as anticipated, the number of components a to retain in the final model, a cross-validation procedure is used as follows. For each (tentative) number of components $j = 1, \ldots, m$, the data set is divided into g parts $(1 \leq g \leq n)$. Then, the prediction residual is calculated for each part through the model built on the remaining $g - 1$ ones, providing the residual vector $\mathbf{f}_{(g)}$, and the Prediction REsiduals Sum of Squares (PRESS) is calculated as $\mathbf{f}_{(g)}^T \mathbf{f}_{(g)}$. The number of components a is the one that corresponds to the lowest PRESS. In the proposed approach, the cross-validation procedure is done by setting $g = n$ (leave-one-out cross-validation). In this case study, $a = 10$ components are included into the model based on one year's worth of data (525 voyages). Note that reference voyages are obtained by excluding outliers that could be confirmed as exceptional by the ESD engineers.

Figure 1a shows the score values corresponding to the first two components, whereas the loading plot (Fig. 1b) shows the orientation of the obtained plane in relation to the original variables. Variables that are grouped together (e.g. draughts and displacement) contribute similar information because they are correlated. Influential, variables are found on the periphery of the loading plot, whereas uninfluential ones are around the origin. As expected by speed-power curves, the kinematic variables V and σ_V are confirmed to have the most influence.

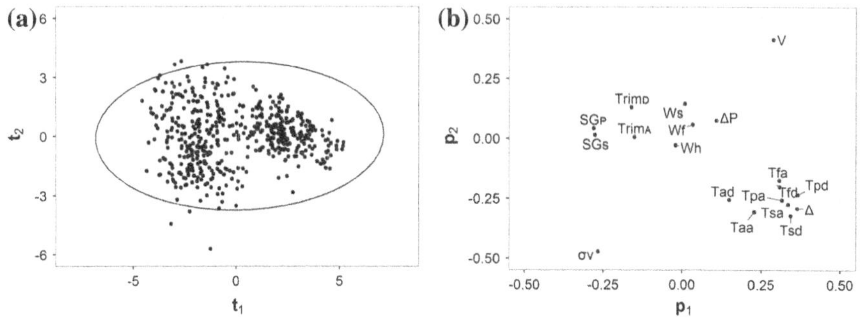

Fig. 1 Score plot $\mathbf{t}_1$ versus $\mathbf{t}_2$ (**a**) and loading plot $\mathbf{p}_1$ versus $\mathbf{p}_2$ (**b**) for reference model

2.3 Monitoring Phase

At the end of a new voyage, predictor variable observations can be arranged into a row vector $\mathbf{x}_{new}^T$ and then the corresponding score vector $\mathbf{t}_{new}^T$ can be calculated as $\mathbf{t}_{new}^T = \mathbf{x}_{new}^T\mathbf{W}(\mathbf{P}^T\mathbf{W})^{-1}$ [20]. Therefore, $\mathbf{t}_{new}^T$ can be used to predict the new response value $\hat{y}$. If $\mathbf{t}_{new}^T$ is close to the origin, then the voyage is plausibly similar to those used as reference to build the model.

The proposed approach encourages to monitor new observations through the following control charts.

Prediction error chart The error in predicting the response variable, defined as the difference between the observation y and its prediction $\hat{y}$, can be usefully monitored in order to visualize trends and/or patterns over time (i.e. consecutive voyages). A control chart can be then obtained by plotting, at each VN, $y - \hat{y}$ and the approximate $100(1-\alpha)$ percent prediction intervals derived by [18] as $\pm t_{n-a-1,\alpha/2}\hat{\sigma}\sqrt{(1+\mathbf{t}_{new}^T(\mathbf{T}^T\mathbf{T})^{-1}\mathbf{t}_{new})}$, where $t_{n-a-1,\alpha/2}$ is the $100\alpha/2$ percentile of a Student's distribution with $n-a-1$ degrees of freedom. The estimate $\hat{\sigma}$ is calculated as $\mathbf{f}^T\mathbf{f}/(n-a-1)$ and can be regarded as the mean squared error of the residuals in the reference model. However, before evaluating prediction error at new voyages, predictor variables have to show no unusual variation.

This can be monitored via the two following classical multivariate control charts.

Hotelling T^2 chart The Hotelling control chart on scores monitors the variation inside the latent variable model. The Hotelling T^2 statistic for a new observation is calculated as $T^2 = \mathbf{t}_{new}^T\mathbf{S}^{-1}\mathbf{t}_{new}n(n-a)/(a(n^2-1))$ [17, 22], where $\mathbf{S}$ is the estimated covariance matrix of the scores in reference model. T^2 follows a F-distribution with a and $n-a$ degrees of freedom because scores are linear combination of measurement variables (see step 3 of NIPALS algorithm) and can be assumed as normally distributed. Therefore, T^2 can be used to build a multivariate control chart on the scores, by defining an upper control limit able to test whether a new observation remains within the normal operating region in the projection space or not [15]. Figure 2a shows the T^2 control chart with 99% control limit for new voyages 2042 ÷ 2050 that do not show unusual inside variation. This can also be verified in the $\mathbf{t}_1$ versus $\mathbf{t}_2$ plot depicted in Fig. 2b, where no voyage has a significant distance from the origin.

Squared Prediction Error of X chart (SPE_X) The SPE_X control chart detects the occurrence of any new event which causes the *ship operating conditions* to move away from the hyperplane defined by the reference model. A valid specification in the reduced space involves not only monitoring the scores of the properties from new observations, but also requires monitoring the residuals or the distance of the multivariate properties from the PLS model [5]. The Squared Prediction Error (SPE) of the residuals for predictor variables, also called Q-statistic [13], is defined for a new observation as $SPE_X = \sum_{j=1}^{m}(x_{new,j} - \hat{x}_{new,j})^2$, where $\hat{x}_{new,j}$ is the value of the new observation for the j-th variable predicted by the latent

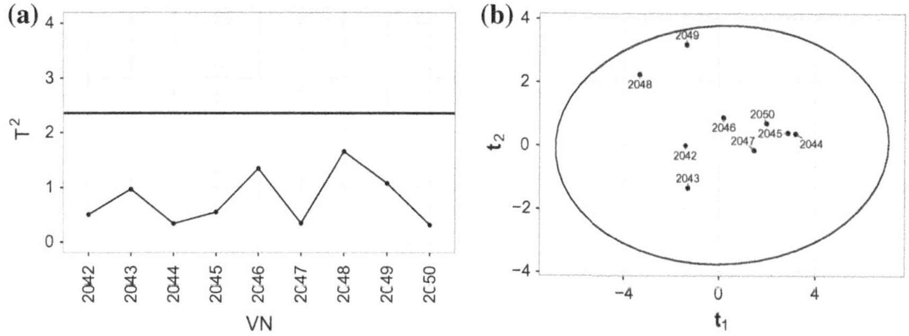

Fig. 2 Hotelling T^2 control chart (**a**) and plot of $\mathbf{t}_1$ versus $\mathbf{t}_2$ (**b**) for voyages 2042÷2050

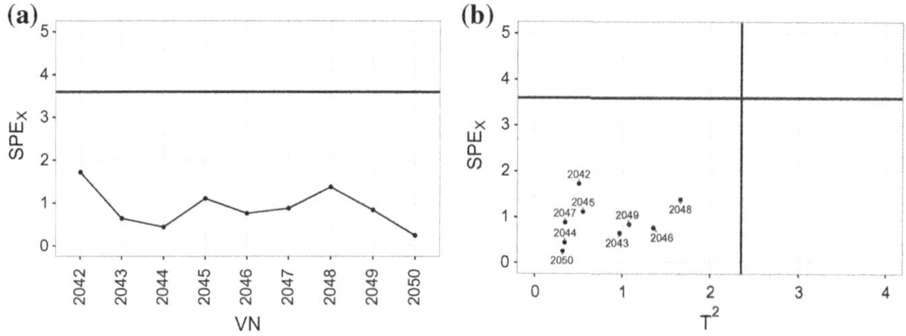

Fig. 3 SPE_X control chart (**a**) and SPE_X versus T^2 chart (**b**) for voyages 2042 ÷ 2050

variable model. In other words, it represents the squared perpendicular distance of an observation from the projection space and gives a measure of how close the observation is to the reduced space [15]. Upper control limit for this statistic is calculated assuming that it is approximately distributed as a weighted chi-squared [13, 17]. The expression of the control limit Q with a significance level equal to α is $Q_\alpha = \theta_1[z_\alpha(2\theta_2 h_0^2)^{1/2}/\theta_1 + \theta_2 h_0(h_0 - 1)/\theta_1^2 + 1]^{1/h_0}$ [14], where θ_i can be estimated from the covariance matrix $\mathbf{V}$ of $\mathbf{E}$ (i.e. the $\mathbf{X}$-residuals matrix), as $\theta_i = trace(\mathbf{V}^i)$, $h_0 = 1 - 2\theta_1\theta_3/3\theta_2^2$, and z_α is the $100(1-\alpha)$ percentile of the standard normal distribution. Figure 3a shows that for the new voyages 2042 ÷ 2050, all the observations do not fall outside the upper 99% limit, i.e. the distance from the projection hyperplane is in control.

In those conditions, i.e. both T^2 and SPE_X are in control (it can also be verified in Fig. 3b), the prediction error chart for the response variable can be used to monitor CO_2 emissions.

Prediction error chart for voyages 2042 ÷ 2050 with $\alpha = 0.05$ is shown in Fig. 4. All the observations for prediction errors fall within the prediction limits, but from VN 2047 to 2049, hourly CO_2 emissions values fall over the upper control limit. Since there is no unusual variation in the predictors (Figs. 2 and 3), further engineer-

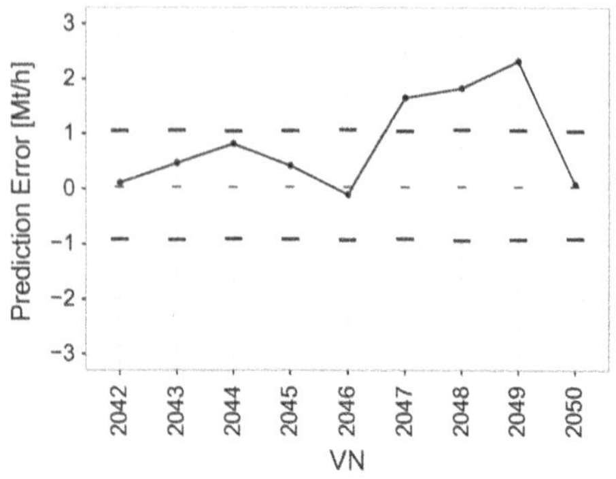

Fig. 4 Prediction error chart for voyages 2042 ÷ 2050

ing investigation that plausibly involves different variables than those considered in Table 1 has been encouraged to detect whether any anomaly occurred. The higher emissions have in fact been due to a malfunctioning in two of the four main engines that does not have any effect on the considered variables. This shows a case not detected by the classical multivariate control charts (e.g. T^2 and SPE_X) in which the response variable may not meet the performance standards.

However, if any observation falls over the upper control limit in either T^2 or SPE_X control chart, contribution plots can be utilized in order to identify variables, among the predictors used in the model, that show unusual variation [24].

On-the-fly maintenance of engines involved in malfunctioning has allowed solving the problem. Figure 4 shows that CO_2 emissions came back into control limits after voyage 2049.

3 Conclusion

The proposed approach based on Partial Least-Squares regression is shown to be capable of monitoring CO_2 emissions through a latent variable model that characterizes *ship operating conditions* better than the classical regression models by exploiting the massive navigation data available from modern DAQ systems.

A real case study has demonstrated the approach to be effective for controlling CO_2 emissions even when no classical multivariate control charts (T^2 and SPE_X) have displayed any anomalies for the predictor variables. In particular, the use of the additional prediction error control chart has supported engineers in technological investigations on variables outside the monitored ones and has allowed for faster diagnosis of technical causes.

Acknowledgements The authors are grateful to the Grimaldi Group Energy Saving Department engineers Dario Bocchetti and Andrea D'Ambra.

References

1. Bocchetti, D., Lepore, A., Palumbo, B., Vitiello, L.: A statistical approach to ship fuel consumption monitoring. J. Ship Res. **59**(3), 162–171 (2015)
2. Corbett, J.J., Koehler, H.W.: Updated emissions from ocean shipping. J. Geophys. Res. **108**(D20) (2003)
3. Council of European Union: Regulation (eu) 2015/757 of the European Parliament and of the Council of 29 april 2015 on the Monitoring, Reporting and Verification of Carbon Dioxide Emissions from Maritime Transport, and Amending Directive 2009/16/ec (2015)
4. Dayal, B., MacGregor, J.F., et al.: Improved PLS algorithms. J. Chemom. **11**(1), 73–85 (1997)
5. Duchesne, C., MacGregor, J.F.: Establishing multivariate specification regions for incoming materials. J. Qual. Technol. **36**(1), 78–94 (2004)
6. Eriksson, L., Kettaneh-Wold, N., Trygg, J., Wikström, C., Wold, S.: Multi-and Megavariate Data Analysis: Part I: Basic Principles and Applications. Umetrics Inc (2006)
7. Erto, P., Lepore, A., Palumbo, B., Vitiello, L.: A procedure for predicting and controlling the ship fuel consumption: its implementation and test. Qual. Reliab. Eng. Int. **31**(7), 1177–1184 (2015)
8. Geladi, P., Kowalski, B.R.: Partial least-squares regression: a tutorial. Anal. Chim. Acta **185**, 1–17 (1986)
9. Höskuldsson, A.: PLS regression methods. J. Chemom. **2**(3), 211–228 (1988)
10. IMO: Imo standard marine communication phrases (SMCP) (2000). http://www.segeln.co.at/media/pdf/smcp.pdf
11. IMO: IMO REF. T5/1.01 MEPC.1/Circ.684 17 Aug 2009 (2009)
12. ITTC: Dictionary of ship hydrodynamics. R. Inst. Nav. Archit. (2008)
13. Jackson, J.E.: A User's Guide to Principal Components, vol. 587. Wiley (2005)
14. Jackson, J.E., Mudholkar, G.S.: Control procedures for residuals associated with principal component analysis. Technometrics **21**(3), 341–349 (1979)
15. Kourti, T., MacGregor, J.F.: Multivariate SPC methods for process and product monitoring. J. Qual. Technol. **28**(4), 409–428 (1996)
16. Lewis, E.V.: Principles of Naval Architecture, Second Revision, vol. II. Resistance Propulsion and Vibration, Society of Naval Architects and Marine Engineers (1988)
17. Nomikos, P., MacGregor, J.F.: Multi-way partial least squares in monitoring batch processes. Chemom. Intell. Lab Syst. **30**(1), 97–108 (1995)
18. Nomikos, P., MacGregor, J.F.: Multivariate SPC charts for monitoring batch processes. Technometrics **37**(1), 41–59 (1995)
19. Otto, M., Wegscheider, W.: Spectrophotometric multicomponent analysis applied to trace metal determinations. Anal. Chem. **57**(1), 63–69 (1985)
20. Rosipal, R., Krämer, N.: Overview and recent advances in partial least squares. In: Subspace, Latent Structure and Feature Selection, pp. 34–51. Springer (2006)
21. Ter Braak, C.J., de Jong, S.: The objective function of partial least squares regression. J. Chemom. **12**(1), 41–54 (1998)
22. Tracy, N.D., Young, J.C., Mason, R.L.: Multivariate control charts for individual observations. J. Qual. Technol. **24**(2), 88–95 (1992)
23. Veness, C.: Calculate distance, bearing and more between two latitude/longitude points. Institute of Geophysics Texas (2007)
24. Wise, B.M., Gallagher, N.B.: The process chemometrics approach to process monitoring and fault detection. J. Process. Control **6**(6), 329–348 (1996)
25. Wold, S., Ruhe, A., Wold, H., Dunn III, W.: The collinearity problem in linear regression. The partial least squares (PLS) approach to generalized inverses. SIAM. J. Sci. Comput. **5**(3), 735–743 (1984)

Part VI
Statistics and the Education System

Promoting Statistical Literacy to University Students: A New Approach Adopted by Istat

Alessandro Valentini, Monica Carbonara and Giulia De Candia

Abstract Istat, Italian National Statistical Institute, has been pursuing the aim of promoting statistical literacy for many years. Recently (2013) the constitution of a territorial network of experts in disseminating activities is a further effort towards this direction. A new project is devoted to university students. The new approach is gradual: (i) to assess statistical literacy of students; (ii) to intercept statistical requirements of professors; (iii) to design standardized educational packages aimed at improving students ability to read data and statistical information; (iv) to guide students towards statistical thinking through laboratories. Implications of the new approach are discussed.

Keywords Statistical literacy · University students · Microdata

1 Background

Modern society is characterized by a large amount of information freely available to all citizens, mainly linked to the enormous development of internet. In 2015 internet users overtook the quota of 3 billion. The amount of information produced is impressive: above others, a particularly significant figure is that of 4.3 billion of Facebook messages daily posted. The number of internet researches is even more relevant (every day more than 6 billion of Google queries). These new and simple methods of data retrieval are only apparently signs of progress: the risk is that this growing dissemination of data becomes incomprehensible, unnecessary, misleading, or even dangerous. The data deluge may generate uncertainty and confusion in users who need to read, interpreter or use data. More and more frequently, for the same phenomenon, different figures are provided by various institutes, newspapers or even citizens, without being properly explained and reconciled. This confusion is often enhanced by politicians and opinion leaders that often argue their reasoning

A. Valentini (✉) · M. Carbonara · G. De Candia
Istat, Italian National Statistical Institute, Rome, Italy
e-mail: alvalent@istat.it

C. Perna et al. (eds.), *Studies in Theoretical and Applied Statistics*,
Springer Proceedings in Mathematics & Statistics 227,
https://doi.org/10.1007/978-3-319-73906-9_21

by citing figures, results of surveys or opinion polls without verifying the reliability and accuracy of sources.

In this cultural framework actions to increase statistical literacy become more and more urgent. As well known, statistical literacy is intended as the ability to understanding and using the basic language and tools of statistics [2, 3, 7]. This is more than numeracy but also the capability to communicate statistics and to use statistics for professional and personal decisions.

With the scope to strengthen statistical literacy, according to statistical agenda Stat2015 [1], Italian National Statistical Institute, Istat, recently (2013) constituted the "Territorial network of experts in promoting statistical literacy" (NPSL) [5]. The main scope of NPSL is to design and implement actions, activities and instruments to support the development of statistical literacy in the various Italian regions. The network is composed by a staff of more than 80 Istat researchers (working for the project for around 20% of their time) located in the various Territorial Offices of the institute.

Recipients of this initiative are: students, teachers of primary and secondary schools, civil servant, stakeholders and the general public. In the first three years of activity (2013–2015) the network produced more than 500 events reaching around 40,000 participants; moreover 34 learning packages for school students were released on the institutional site of Istat.

This article focuses on the promotion of statistical literacy among university students. This target is particularly significant for at least two reasons. Firstly this is the last opportunity to educate young adults to numeracy, to communicate statistics and to use statistics to make professional or personal choices. Secondly, university students will play a key role in society in the near future. The new approach of NPSL for university students is to propose paths for increasing their critical thinking, instruments for understanding statistics about society. Actions take into account that students are digital natives with a long educational experience. So traditional education is accompanied by new practices.

The main idea is that knowledge of basic elements of statistics will allow students to read and interpret graphs, tables, and data, as a prerequisite for the modern citizen. More in detail, according scientific methods, the study and analysis of each discipline requires to collect and/or analyze data and to transform data into information. Unfortunately, this operation is too often realized in a superficial and misleading fashion. Academic studies have, instead, to provide specific data analysis skills. The statistical skills can no longer be niche competencies, but they should be part of the cultural baggage of a modern citizen.

The new multifaceted strategy adopted by Istat to promote statistical literacy among university students is articulated in 4 steps. The first two steps are the assessment of students' statistical literacy (Sect. 2) and the analysis of statistical requirements of professors (Sect. 3). These activities defined the framework of the new strategy and concluded in 2015. The other two steps were designed according to results of the previous activities and produced materials for students in terms of: packages useful to better understand official statistics and statistical methods (Sect. 4); proposals of laboratory activities based on the use of survey microdata

(Sect. 5). The last two activities are only partially concluded: materials have to be extended in terms of contents and further improved in their ability to catch the interest of university students; laboratories need to be tested before the release.

2 Measurement of Statistical Literacy

Istat with the collaboration of the statistical departments of the three universities of Firenze, Pisa and Siena assessed the level of statistical literacy of the university students. An on-line questionnaire, QValStat [4, 6], was administered to more than 10,000 first-year under-graduated students of Tuscany using a funny approach: slogan of the operation was "Play with statistics … and you will discover your profile!" (Fig. 1). Profiles were defined as *Seed*, *Germ*, *Shoot*, *Tree*, *Statistical tree* and each profile was described by a sentence.[1] The profile was assigned to each respondent in proportion to the number of correct answers, according to the positioning of the pointer (an angle between 0 and 180°) on a "statistical dashboard".

More than 3,000 students (31.9%) answered the questionnaire. The highest response rate was recorded in Siena (37.3%), with the peak of 39.1% for students of Economics and Statistics. The response rates of Pisa (31.9%) and Florence (28.7%) were lower, with the remarkable exception of the Science group (42.5%) in Pisa.

The modal profile was Tree (43.7%), followed by the profile of Shoot (33.5%). Percentages were lower for Statistical Tree (11.7%), Germ (10.6%) and Seed (0.5%).

Standardizing results on a scale from 0 to 100, the mean level of statistical literacy was 63.7 points. Differences were accounted among individual covariates (gender, previous scholastic curriculum) and between the various typologies of degree groups. The highest result was that of students of scientific groups (78.0), followed by groups of engineering (69.9) and medicine (67.5). The lowest results were accounted by students of legal (53.1), literacy-humanistic-linguistic (56.1) and socio-political (58.0) groups.

A deeper two-level analysis was realized to assess the effect of the various covariates. The first step consisted in a One-way Anova test within the following groups: individual characteristics (gender; citizenship), university choice (place of university; group of degree course), scholastic curricula (high school specialization; high school final grade).

[1]The slogan associated to each profile is the following: *Seed*: Your statistical culture needs to be improved; *Germ*: You started to cultivate your statistical culture; *Shoot*: Your statistical literacy is growing; *Tree*: Your statistical literacy is putting down its roots!; *Statistical tree*: Congratulations, you are in full flowering!

Fig. 1 QValStat: available profiles

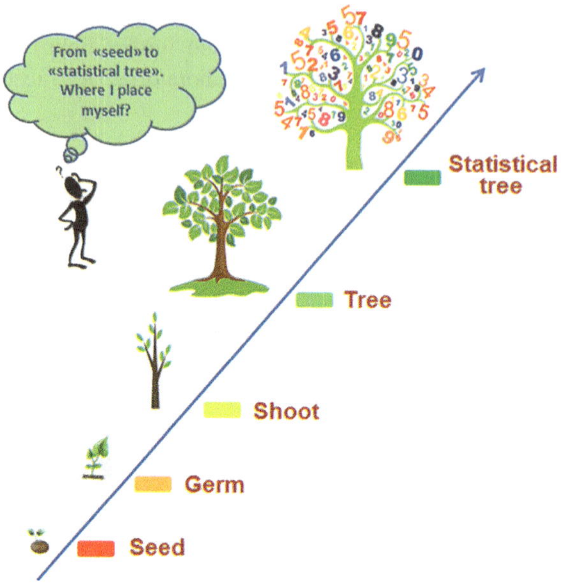

The second step analysis was a logit model with the covariates significant at the first step[2]: gender, high school specialization, high school final grade. To apply the logit model, the reference category for each of the three independent variables was the one with the highest level of statistical literacy in the univariate case: (in the order) males, students that attended Lyceum and students with the highest final evaluation (98 and more).

The most significant effect among different variables concerned the final grade and consisted in the switch between the highest and the lowest levels (98 and more vs. 60–74); this switch halved the odds ratio. A similar effect is that of gender: odds ratio reduces of 42% from male to female. Even the effect on odds ratio of high school specialisation was interesting: limited (−10%) the shift from Lyceum to a Technical school, much more significant the one to Artistic (−23%), Pedagogic (−24%) and Vocational (−34%) school.

According to the survey results, the level of statistical literacy is quite unsatisfactory for the most part of university students, especially for those that attended an Artistic or a Vocational school and for those with a low high school final grade. NPSL tries to solve these criticisms approaching to university with a new strategy specifically designed to catch Academic needs.

[2]Note that even the group of degree course is significant at the first step analysis. This variable was excluded from the second step because of the significant ($P < 0.0001$) connection with the high school specialization.

3 Identification of Statistical Requirements

During 2015 Istat realized an exploratory survey in order to catch statistical requirements of university teachers. The survey was conducted in 10 Universities located throughout the Country to identify potential areas of collaboration. 85 co-ordinators of degree courses answered the survey: 32.9% of economic science and statistics, 36.5% of social science, 20% of base and applied sciences and 10.6% of life sciences.

Results of the inquiry are interesting. The main statistical requirement concerns data sources: which official statistics are available, where data are located and how they can be correctly extracted and used. According to the professors, students have to become familiar with statistical sources, in order to be able to distinguish between good data and fallacious data. Moreover students have to learn to look for metadata and to understand their meaning. This implies a targeted approach to the official sources and the adoption of the proper instruments in order to illustrate data bases and to show examples. The requirement of professors is addressed to focus on statistical procedures adopted by Istat to build socio-economic indicators and to focus on data quality processes.

It also emerges the need of developing and promoting the features and the specificities of micro data treatment. The first approach of students with official micro data is often arduous: some guidelines tailored for students and some laboratory on data analysis conducted by Istat's experts can be useful at this scope.

Finally, the last requirement emerges is that of "story-telling" in order to enhancing the potential of quantitative communication and to build small stories about social, demographic and economic transformations using time-series data. The oral and written presentation of statistical analysis carried out by the students reinforces and clarifies its understanding [3].

4 Educational Packages

A first NPSL response to the information needs expressed by the professors consists of educational materials (packages) useful for students of different areas. The packages are structured in the form of slide in PowerPoint with the following features:

- flexibility: the topics covered and the level of detail achieved must be appropriate for students of all faculties;
- hypertextuality: the slides contain hypertext with links to in-depth documentation;
- self-consistency: the packages contains all information necessary for an understanding of subject matter;
- completeness: in the notes are inserted information that helps the user to understand the train of thought and (or) to gain detailed information.

Table 1 Packages released to university students (2015)

No.	Title	Content
1	Official statistics, Sistan and Istat	Description of Italian NSI
2	Statistical surveys and data quality	Introduction to the various phases of a survey
3.1	Istat dissemination system	Access to statistical information Istat
3.2	I.Stat: the datawarehouse of Istat	Instructions/examples to navigate the database
4	Evolution of the Census in Italy	History and perspectives of census in Italy
5	Socio-demographic profile of Italy	Socio-demographic aspects at 2011 Census
6.1	Inflation: what is it and how it is measured	Description of the method adopted by Istat to produce information on inflation
6.2	The recent inflation in Italy	Recent trends in index of consumer prices

Packages deal with statistical issues from the point of view of an official statistics' producer. At this scope packages are much applied, abundant of examples and referred to Istat's surveys and data analysis methods and results. Methodological aspects are illustrated in terms of description of the statistical processes. Results are referred to the use of official indicators as instruments able to read the socio-economic conditions of our country. At present, six packages were released (Table 1) on the following themes:

- Methodological aspects of official statistics and surveys;
- Data sources of Istat;
- Socio-demographic profile (from Population Census);
- Dynamics of prices.

Packages are available in the platform of Statistical training and promotion, recently released on the web (https://formazione.istat.it/).

Before the release, the packages are tested by some universities in Lombardy, Tuscany, Puglia and Calabria. The test involved 20 professors and 40 undergraduate students from different degree programs, in particular Economics, Political Science and Medicine. Professors and students participating in the test were asked to express their feedback by filling in an online questionnaire. Aspects to evaluate were: effectiveness of communication, language adopted, balance between theory and practice, strengths and weaknesses, integration into the curriculum.

Both professors and students have preferred to deal with the package on statistics and the quality of data, on Istat dissemination system and I.Istat data warehouse, and on inflation. The judgment was positive: on a scale of 1–10 the judgment varies in the range from 7 to 8 both for professors and students. Moreover, to improve statistical literacy, all have expressed the need to have more packages on the data analysis and the planning and implementation of indicators.

5 Directions for Future Work

In the last three years NPSL has been engaged to promote statistical literacy to university students both by producing standardized educational materials and by organizing training sessions and seminars on the territory. At the moment, new packages are under construction; they will be released in the next future in order to completely fulfil the requirements of professors. Prototypal seminars on official statistics have been performed in various universities and more standardized projects are being implemented.

The particular relevance of the target represented by university students spurs NPSL to increase the efficacy and the efficiency of actions to promote statistical literacy.

A particularly promising line of action is that of laboratories for the analysis of micro-data. Students can experience the whole statistical process from micro-data analysis to macro-data production. Data treatment can be performed using statistical packages with free license available on the net. In the laboratories students can develop abilities to manage collections of elementary data, to read record layout and metadata, to analyse data using statistical methods, to read and interpret their findings. Laboratories are thought to be used by professors even without Istat support. Documentation on the meaning of micro-data, the role of macro-data and some hints for data elaboration is available for students. A series of research questions will be requested to students on a dataset freely downloadable. Questions will facilitate students for the reading of data and for their interpretation.

Efficacy of laboratories will be tested in sample courses before the release. Test will consist in replicating the experience of measuring statistical literacy at the start and at the end of courses to two groups of students randomly selected. The laboratories will be followed only by students of the first group. In this way, it should be possible to evaluate the improvement in statistical literacy due to the activities in the laboratories by comparing the two groups.

References

1. Baldacci, E.: Un 2012 ricco di novità: verso il modello Stat2015, *NewsStat* n 4 (2012)
2. Gal, I.: Adults' statistical literacy: meanings, components, responsibilities. Int. Stat. Rev. **70**, 1–51 (2002)
3. Unece: Making Data Meaningful Part 4: How to Improve Statistical Literacy: A Guide for Statistical Organizations. United Nation Economic Commission for Europe, Geneva (2014)
4. Valentini, A.: Come è possibile misurare la cultura statistica? Resoconto di un'esperienza condotta in Toscana, Induzioni **50**(2015), 79–90 (2015)

5. Valentini, A., Cortese, P.F.: Il nuovo approccio alla promozione della cultura statistica da parte della rete territoriale Istat. Induzioni **48**(1), 79–93 (2014)
6. Valentini, A., Pratesi, M., Martelli, B.M.: Promozione e misurazione della cultura statistica negli Atenei della Toscana: alcune evidenze empiriche. Statistica Società **2**, 36–42 (2015)
7. Wallman, K.K.: Enhancing statistical literacy: enriching our society. J. Am. Statist. Assoc. **88**, 1–8 (1993)

From South to North? Mobility of Southern Italian Students at the Transition from the First to the Second Level University Degree

Marco Enea

Abstract In the last decades, the Italian University System has encountered several structural reforms aimed at making it more internationally competitive. Among them, the introduction of the University financial autonomy has triggered an "internal" competition among Universities to attract students from the entire country. Students' enrollment at the first level has decreased significantly especially after the economic crisis of 2008, while the students' migration from the South to the Central and Northern regions of the country has increased. These phenomena have created further inequalities within the country and a cultural and socio-economic loss for the South that does not appear to slow down. While Italian internal mobility at the first level has been previously investigated, second level mobility has received little attention. This work attempts to fill this gap, by analyzing the transition from first to second level university degree courses of the Southern Italian students in terms of macro-regional mobility. The data were provided by the Italian Ministry of Education, University and Research. They are a national level longitudinal administrative micro-data on educational careers of the freshmen enrolled at the first level Italian university degree course in 2008–09 and followed up to 2014. We will use a discrete-time competing risk model with the aim to detect the determinants of the choices of Southern Italian students after their bachelor degree: discontinuing university; enrolling at the second level University degree course in a Southern university, or (moving) to Central or Northern universities. We will analyze the role played by demographic variables, time elapsed to get the first level degree, the performance in the previous schooling career, etc. in order to provide mover or stayer profiles of Southern bachelors.

Keywords Students' mobility · Time to event · Discrete-time competing risk model

M. Enea (✉)
Dipartimento di Scienze Economiche, Aziendali e Statistiche, University of Palermo,
Viale delle Scienze, Edificio 13,
90128 Palermo, Italy
e-mail: marco.enea@unipa.it

C. Perna et al. (eds.), *Studies in Theoretical and Applied Statistics*,
Springer Proceedings in Mathematics & Statistics 227,
https://doi.org/10.1007/978-3-319-73906-9_22

1 Introduction

In the last decades, the Italian University System has encountered several structural reforms that produced many changes, with the aim to make the system more efficient and internationally competitive. Among these reforms, the university financial autonomy (1994) was introduced as a tool to reduce the inefficiencies of many universities, particularly those of the Southern Italy. However, in the years these reforms were accompanied by repeated cuts to public spending. This process has triggered an "internal" competition among universities to receive public funds through a rewarding mechanism, so inserting public higher education in a market perspective. Central and Northern Italy universities, located in most favorable territorial contexts than the Southern ones, have easily won this competition. Therefore, Southern universities are not only deprived of financial resources but also of human capital, accelerating a students' migration process toward the North and abroad. From this view, students' mobility (SM) analysis has become particularly of interest in the last years. The study of SM is aimed at detecting the determinants of the migration at the level of student, of University and region. For example, students might be motivated by their own socio-cultural background, by the type of training offered by an extra-regional University and/or by possible future job opportunities of the more wealthy regions. On the grounds of the data availability, such a study usually follows two major approaches proposed in the literature: the analysis of aggregate (or flow) data (AD) and that of individual student data (ISD). The first one is concerned with flows of "moving" students and usually performed through the so-called spatial interaction model, originated by an adaption of Newtons law of gravitation to the students mobility [1–3]. The second approach analyzes the effect of the individual features, such as the talent of the student [4–7], on the decision of moving. In both the approaches the distance between the place of origin and the chosen University has been demonstrated to have a frictional effect [8]. However, while Italian internal mobility has been investigated for the transition from the high school to the first level degree course (first level mobility), the second level mobility, that is the bachelors' mobility to enroll at a second level degree course has received little attention.

This work deals with the analysis of the Southern Italian SM. There are two novelties in this paper. The first one is filling this lack of contributions on the analysis of the transition from first to second level university degree courses of the Southern Italian Bachelors (SIBs) in terms of macro-regional mobility. The second novelty is proposing the use of an alternative model, the discrete-time competing risk one, to detect the determinants of the SIBs' choices. The competing risk are: discontinuing university; enrolling at the second level University degree course in a Southern university, or (moving) to Central or Northern universities. The data were provided by the Italian Ministry of Education, University and Research. They are a national level longitudinal administrative micro-data on educational careers of the freshmen enrolled at the first level Italian university degree course in 2008–09 and followed up to 2014. We analyze the role played by demographic variables, time elapsed to get the first level degree, the performance in the previous schooling career, etc. in order to provide mover or stayer profiles of Southern bachelors. Another main question is concerned with "regular" students, that is those students who get their degree in at most four years, according to the Italian rules. Are they more inclined than non regular students at continuing university at the master level? Moreover, does their performance affect the choice of where enrolling?

The paper organized as follows. In Sect. 2, we first briefly introduce the discrete-time survival analysis and then we focus on competing risk methodology. Then, Sect. 3 describes the transition from first to second level. Section 4 contains the model estimates and the results of the analysis, while the conclusions are in Sect. 5.

2 Methodological Approach

In discrete-time survival analysis [9] event (or failure) times are measured in discrete intervals, where $t = 1, 2, 3, \dots,$ (for instance months, years). Interest is usually focused on the *hazard function* $h(t)$, and how it depends on covariates. By denoting with T the event time, such function is defined as $h(t) = \Pr(T = t|T \geq t)$, that is the probability that an event occurs during interval t, given that no event occurred before t. Another quantity of interest is the *survivor function* $S(t) = \Pr(T \geq t)$, that is the probability that an event has not occurred before t. The hazard and the survivor functions are linked by the relation $S(t) = [1 - h(1)] \times [1 - h(2)] \times \cdots \times [1 - h(t - 1)] = S(t - 1) \times [1 - h(t - 1)]$. The hazard function is usually made depend on a set of p covariates Z through a proportional-hazard (PH) model $h(t|Z) = h_0(t) \exp(\beta' Z)$, where $h_0(t)$ is the baseline hazard at time t, and β a vector of regression parameters describing the effect of the covariates on the risk of failure. Let K be the follow-up time, a common choice to specify the hazard baseline is $h_0(t) = \exp(\alpha_1 D_1 + \alpha_2 D_2 +, \dots, \alpha_K D_K)$, where α_t, $t = 1, \dots K$, is the coefficient of the dummy variable D_t for time t. The PH model assumes that the effect of β on the hazard baseline is constant over time. It is shown this model can be fitted by using a logistic regression model after transforming data in the *person-period* format [10].

However, when the risk of failure comes from different causes the modelling approach used is the competing risk one. Discrete-time examples are in educational research, where at any time university students are at risk to get a degree or to drop out, but only one event can occur. Suppose there are J competing risks, the *event-specific hazard* (or cause-specific hazard), that is the hazard of failing from jth $(j = 1, \dots, J)$, event is

$$h^{(j)}(t) = \Pr(\text{event of type } j \text{ at time } t|T \geq t), \tag{1}$$

while the hazard that no event occurs at time T, given survival at time t, is $h^{(0)}(t) = 1 - \sum_{j=1}^{J} h^{(j)}(t)$. Accordingly, the probability of survival to any event is $S(t) = h^{(0)}(1) \times h^{(0)}(2) \times \cdots \times h^{(0)}(t - 1)$, and the event-specific hazard model follows a proportional-hazard model

$$h^{(j)}(t|Z^{(j)}) = h_0^{(j)}(t)\exp(\beta^{(j)\prime}Z^{(j)}), \tag{2}$$

where $h_0^{(j)}$, $\beta^{(j)}$ and $Z^{(j)}$ are the baseline hazard, the regression coefficients and the set of covariates for the jth event. In the same way, by specifying $h_0^{(j)}(t) = \exp(\alpha_0^{(j)}(t)) = \exp(\alpha_1^{(j)}D_1^{(j)} + \alpha_2^{(j)}D_2^{(j)}+, \cdots, +\alpha_K^{(j)}D_K^{(j)})$, and by denoting $h^{(j)} = h^{(j)}(t|Z^j)$, model (2) can be fitted within the framework of a multinomial logit model:

$$\log\left(\frac{h^{(j)}}{h^{(0)}}\right) = \alpha_0^{(j)}(t) + \beta^{(j)\prime}Z^{(j)}, \tag{3}$$

where $h^{(0)} = 1 - \sum_{r=1}^{J} h^{(j)}(t|Z^{(j)})$ is the hazard for the reference category. Parameter interpretation is made in terms of *relative risk ratio* i.e. $\exp(\beta^{(j)})$ is the multiplicative effect of a 1-unit increase in z on the risk of event type j versus the risk that no event occurs [11]. It can be also useful to look at the predicted event-specific hazard directly:

$$h^{(j)} = \frac{\exp(\alpha_0^{(j)}(t) + \beta^{(j)\prime}Z^{(j)})}{1 + \sum_{l=1}^{J}\exp(\alpha_0^{(l)}(t) + \beta^{(l)\prime}Z^{(l)})}, \tag{4}$$

$j = 1, 2, \ldots, J$, representing the risk that jth event occurs given the presence of all other types of events.

In order to check for the PH assumption, one could include time-varying coefficients into the linear predictor:

$$h^{(j)}(t|Z^{(j)}) = h_0^{(j)}(t)\exp(\beta^{(j)\prime}Z^{(j)} + \beta^{(j)\prime}(t)X^{(j)}), \tag{5}$$

where $\beta^{(j)}(t)$ is the vector of time-varying coefficients and $X^{(j)}$ indicates the set of interaction terms between t (or a function thereof) and the variables which the PH assumption is checked for. If the time varying coefficients are not significant there is evidence that the PH assumption holds. If they are, these coefficients remain into the model and the quantity (4) is modified accordingly.

An alternative approach to fit model (2) is using a logit model for each event. Although estimated coefficients using this latter method will, in general, be different as the reference category is not the same, the predicted probabilities will be similar.

3 Modelling the Second Level Degree Students' Mobility

The objects needed to apply the model described previously are: the population at risk, the follow-up time, the time and the censored units. *The population at risk* is the sub-cohort of SIBs from the 2008/09 cohort, with a *follow-up time* of 4 years. The *time t* is defined in terms of *academic years passed since getting the first level degree*. For example, an undergraduate who gets his/her bachelor in 3 years, i.e. in the academic year (AY) '10/'11 ($t = 0$), can enroll at a master degree course in one of the following AYs: '11/'12 ($t = 1$); '12/'13 ($t = 2$); '13/'14 ($t = 3$) and '14/'15 ($t = 4$).

At the end of the follow-up time, students who did not enroll anywhere are *censored*. Moreover, there are students who delay one, two or three years in getting their degree, thus we can observe them for at most three, two and one year, respectively, before being *censored*. We assume that all the students are at risk of enrollment during the four-year observation time. This means that students who decide to enroll either abroad or at another first level or five/six-year degree course are, for simplicity, considered at risk as well.

The competing risk structure is depicted in Fig. 1. At any academic year (hereafter year), a SIB can enroll at university to attend a second level degree course in either a Southern or a Central or a Northern Italy university or, otherwise, he/she can remain at risk of enrolling successively. We observed 34164 SIBs who are potentially at risk of enrollment in one of the three Italian macro-regions within four years of observation. At the end of the fourth year of observation, the bachelors who decided to enroll were 21753, that is about 64%. Table 1 reports their cross-classification according to the time of observation and the macro-region.

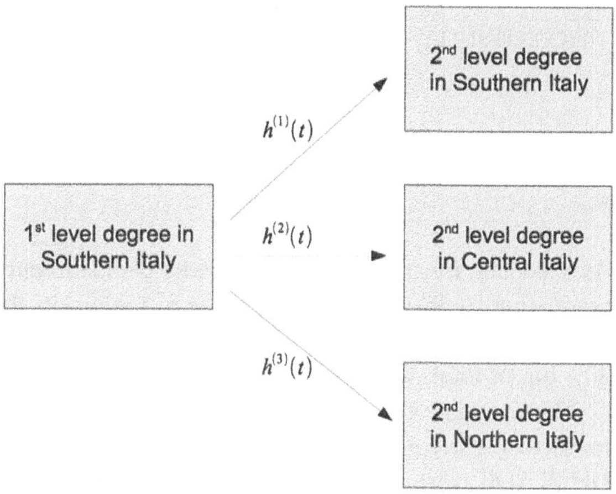

Fig. 1 Structure of the discrete-time competing risk model. At each year a SIB can enroll to attend a second level degree course in either a southern or a central or a northern Italy university or, otherwise, he/she can remain at risk of enrolling successively

Table 1 Cross-classification of the bachelors who decide to enroll at a second level degree course according time of observation and macro-region

Time since graduation	Center	North	South	Total
$t = 1$	629	1256	13419	15304
$t = 2$	441	687	3587	4715
$t = 3$	247	251	645	1143
$t = 4$	100	144	347	591
Total	1417	2338	17998	21753

Table 2 Covariate coding and percentages/medians for the southern Italian bachelor's data

Variable	Level/Range	%/Median	Description
SEX	Female	64.0	Student's sex
	Male	36.0	
AGE	17; 80	19	Student's age at the enrollment
DEGTIME	≤4	64.4	Time to first level graduation in years
	=5	22.0	
	=6	13.6	
HSGRADE	−23.5; 18.5	0	High school final grade scaled at the mean 83.5
LICEO	Yes	56.6	Classical or scientific high schools?
	No	43.4	
HSPRIV	Yes	3.3	Private high school?
	No	96.7	
DFGRADE	−30; 10	0	First level degree final grade scaled at the mean 101
DISCONTINUITY	Yes	5.8	More than one year delay to enroll at university since high school?
	No	94.2	

As expected, most of the bachelors who decide to enroll does it at the fist year since graduation, with a strong preference to remain in the Southern Italy, followed by Northern and Central Italy respectively. However, there are 6449 students, about 30% out of total, who enroll after the first year.

The covariate coding is reported in Table 2, along with the observed percentages for the categorical covariates or the medians for the continuous ones. Both AGE and DFGRADE were correlated to other covariates, so we they were not included into the model. The first one is correlated to variable DISCONTINUITY, while the second one to variables HSGRADE and LICEO. In fact, in many analyses of students' performances [12, 13] both HSGRADE and LICEO result to be good predictors of DFGRADE.

4 Model Estimates

Before model fitting, data were transformed into the person-period format. We used two R packages [14, 15] for data preparation and model fitting, respectively. Starting from a model with interaction between all covariates and DEGTIME, we first performed an AIC-based backward model selection. The selected model (called mod0) contains all main effects except HSPRIV, and the interaction between DEGTIME and LICEO. Then, we used likelihood ratio (LR) tests to check mod0 for proportional hazard assumption. In particular, we considered new interactions between the covariates of mod0 and the time t (considered as continuous). Because some of such interactions were significant, we included them into the new final model called mod1 (Table 3).

As expected, at any time the baseline hazard to enroll in the South is the highest among the macro-regions and shows a sensible decreasing trend. Estimated main effects of DEGTIME are all negative and significant only for the enrollment in the South, or the North when the graduation is at most at the fifth year. Bachelors coming from liceo have relative risks of exp(0.684 − 0.245) = 1.55, 1.79 and 2.19 times higher than the other high schools, to enroll in the first year at a master degree course in the Southern, Central and Northern Italy, respectively. However, this effect significantly decreases every year in both South and North. At the fourth year, these relative risk ratios are 0.74 and 1.07 for South and North, respectively. Moreover, by looking at the positive interactions between DEGTIME and LICEO, it seems that these bachelors don't care about the delay in getting their graduation, with a multiplicative effect on the relative risk ratio of enrolling in the Center or in the North of about 1.5 for both. When the degree is gotten in six years the only significant chances to enroll at the master are in the North. An higher high school final grade

Table 3 Model estimates

	South → South		South → Center		South → North	
	$\hat{\beta}$	p-value	$\hat{\beta}$	p-value	$\hat{\beta}$	p-value
$t = 1$	−0.699	<0.001	−3.787	<0.001	−3.487	<0.001
$t = 2$	−1.425	<0.001	−3.980	<0.001	−3.543	<0.001
$t = 3$	−3.498	<0.001	−5.967	<0.001	−6.246	<0.001
$t = 4$	−4.006	<0.001	−8.288	<0.001	−6.125	<0.001
DEGTIME_5	−0.090	0.017	−0.262	0.071	−0.355	0.003
DEGTIME_6	−0.066	0.021	−0.177	0.407	−0.326	0.074
LICEO	0.684	<0.001	0.581	<0.007	1.023	<0.001
LICEO: t	−0.245	<0.001	−0.181	0.181	−0.240	0.019
DEGTIME_5: LICEO	0.149	0.003	0.419	0.019	0.402	0.004
DEGTIME_6: LICEO	0.126	0.072	0.227	0.392	0.438	0.038
HSGRADE	0.047	<0.001	0.014	0.080	0.053	<0.001
HSGRADE: t	−0.015	<0.001	−0.002	0.706	−0.014	0.001
SEX_M	0.545	<0.001	0.303	0.122	0.832	<0.001
SEX_M: t	−0.228	<0.001	0.008	0.954	−0.182	0.060
DISCONTINUITY	−0.876	<0.001	−1.847	<0.001	−1.864	<0.001

has a significant positive effect on the decision to enroll but only in the Southern and in the Northern Italy. However its effect slightly decreases as the time elapsed since graduation increases. Male bachelors appear to be more inclined to enroll and move than females in the first year, in the Southern or in the Northern Italy with relative risks of 1.37 and 1.91, respectively. However, at the fourth year these risks reduce to 0.69 and 1.11. Finally, enrollment at the first level degree after more than one year since high school graduation has strongly negative effects on the decision to enroll in anywhere.

From these estimates we construct the predicted hazards of enrollment in the macro-regions for different student's profiles. Beyond the baseline profile, we considered a student with the highest risk to enroll and the lowest one. These hazards and their cumulative version are graphically reported in Fig. 2.

As expected, the hazards of enrollment are decreasing for all competing risks, with a change point in $t = 3$ that set them down to almost zero. Even for the high risk profile, the probability that bachelors choose to enroll in the Central Italy is already low at $t = 1$, while Northern Italy has about 0.1 probability to be chosen. However, the hazard of enrollment in the Southern Italy in the first year is about 0.6, for the high risk student, 0.4 for the baseline one and 0.1 for the low risk one.

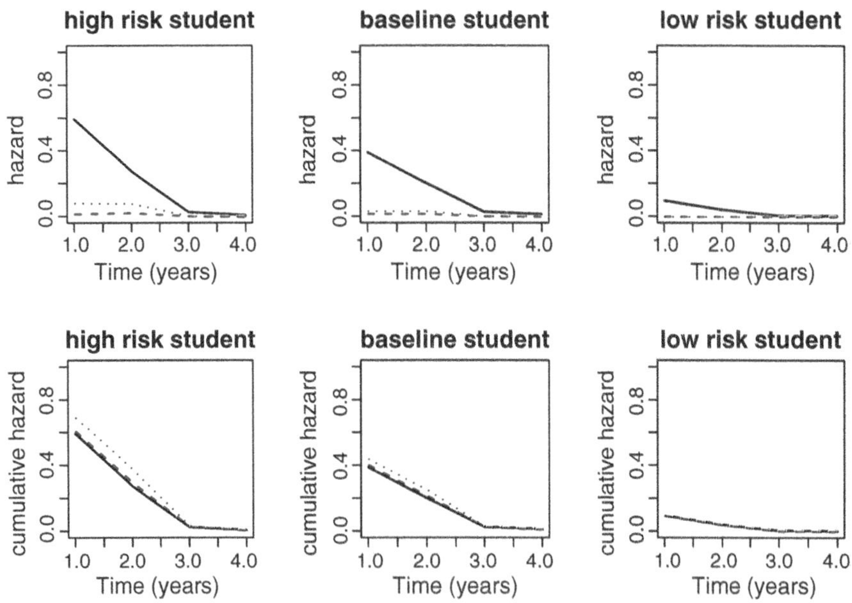

Fig. 2 Predicted hazard (top plots) and and cumulative hazard (bottom) to enroll at the second level degree versus years since first level graduation in either southern (solid line) or central (dashed line) or northern (dotted line) Italy, for the three different risk profiles of southern Italian bachelors

The cumulative hazard of enrollment in the first year, summing up all competing risks, is about 0.7 in the first year and 0.4 in the second one. In the low risk profile only 10 % students enroll in the first year, but in a Southern Italy university.

Figure 3 reports the predicted cumulative hazards conditioned to enrollments, to highlight the hazard of choosing among the three macro-regions, once the bachelor decided to enroll. Observe that this conditional cumulative hazard is no more decreasing, as it contemplates only those bachelors who decided to enroll. The lines appear to be stacked and the vertical distance between them is just the conditional hazard. Thus, we can say that the conditional hazard to move is, in average, approximately 0.2 for the high risk student and 0.15 for the baseline student.

Figure 4 reports the predicted cumulative hazard conditioned to the moving students, to underline difference between the probability to choose Central or Northern Italy, once bachelors decided to move.

By averaging the conditional hazards of choosing the Central Italy over the time, the high risk bachelors have an approximately 0.75 conditional hazard to choose the Northern Italy, and 0.65 and 0.55 for the baseline and the low risk ones, respectively.

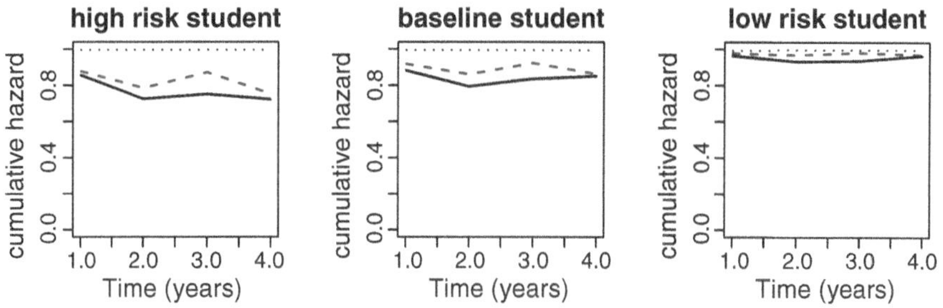

Fig. 3 Predicted cumulative hazard, *conditioned to enrollments*, to choice whether staying in southern Italy (solid line) or moving to central (dashed line) or northern (dotted line) Italy to enroll at the second level degree versus years since first level graduation, for the three different risk profiles of southern Italian bachelors

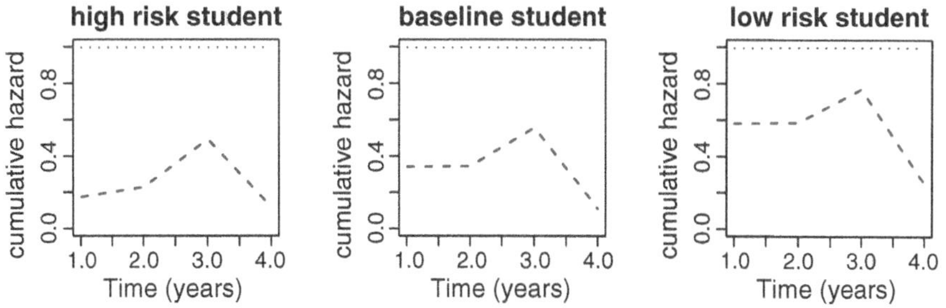

Fig. 4 Predicted cumulative hazard, *conditioned to the moving students*, to choice between central (dashed line) or northern (dotted line) Italy to enroll at the second level degree versus years since first level graduation, for the three different risk profiles of southern Italian bachelors

5 Conclusions

This work focused on the students' mobility analysis in the transition from the first to the second level degree course in Italy. Our target population was the Southern Italian bachelors. We proposed the use of a discrete-time competing risk model as a novel and alternative modelling approach. It was assumed that bachelors can enroll at the master in a Southern university, or (moving) to Central or Northern universities. If not, they stay at risk of enrollment for the successive years. The model was based on a multinomial scheme and included some time-varying effects, as the PH assumption did not hold for some covariates. As main results, we found that the hazard of enrollment decreases and sets almost to zero in the third year since getting the first level degree. Time to get a first level degree resulted to be significant in the choice of whether discontinuing university. The analysis confirmed the role that classical and scientific high schools, as well as high school final grade, play in determining best students, which have the high risk to enroll and move. These have approximately a 70% chance to enroll at a second level degree course in the first after graduation. If they do, then have a 80% chance to stay in the South, while the remaining 20% choose to move. Conditioned to the decision of moving, a Northern University is chosen with a 75% chance. Although the SIBs' mobility rates are not very high, these students are selected anyway. This means the Southern universities loose their best students in favor of the Northern ones. Also considering that moving students choose universities located in regions with a dynamic market labour, the chances that they return after graduation are very low [3], with the result of an high-quality human resource pauperization in the already lagged Southern regions.

The covariates we considered were mostly based on students' registry data. We are working to include more information at the level of university and macro regions.

References

1. Bruno, G., Genovese, A.: A spatial interaction model for the representation of the mobility of University students on the Italian territory. Netw. Spat. Econ. (2010). https://doi.org/10.1007/s11067-010-9142-7
2. Dal Bianco, A., Spairani, A., Ricciari, V.: La mobilitá degli studenti in Italia: un'analisi empirica. Rivista di Economia e Statistica del Territorio **1**(1), 123–143 (2010)
3. Dotti, N.F., Fratesi, U., Lenzi C., Percoco M.: Local labour markets and the interregional mobility of Italian university students, working paper (2010)
4. Sá, C., Florax, R.J.G.M., Rietveld, P.: Does accessibility to higher education matter? Choice behaviour of high school graduates in the Netherlands. Spat. Econ. Anal. **1**(2), 155–174 (2006)
5. Lupi, C., Ordine, P.: Family income and students' mobility. Giornale degli economisti **68**(1), 1–23 (2009)
6. Pitti, M., Pipitone, V., Fulantelli, G., Allegra, M.: La scelta universitaria in Italia: differenze fra Nord e Sud. Rivista economica del Mezzogiorno, a. **XXV**(4), 943–966 (2011)
7. Enea, M., Plaia, A., Capursi, V.: Modelling confidential data via modified hurdle mixed models. In: Proceeding of the 28th International Workshop of Statistical Modeling (IWSM), vol. I. ISBN 978-88-96251-47-8 (2013)

8. Frenette, M.: Too far to go on? Distance to school and University partecipation. Educ. Econ. **13**(4), 31–58 (2006)
9. Singer, J.D., Willett, J.B.: Applied longitudinal data analysis: modelling changes and event occurrence. Oxford University Press, London (2003)
10. Singer, J.D., Willett, J.B.: It's about time: using discrete-time survival analysis to study duration and the timing of events. J. Educ. Stat. **18**(2), 155–195 (1993)
11. Steele, F.: Event history analysis. NCRM Methods Review Papers, NCRM\ 004. http://eprints.ncrm.ac.uk/88/1/MethodsReviewPaperNCRM-004.pdf (2005)
12. Enea, M., Attanasio, A.: An association model for bivariate data with application to the analysis of university students' success. J. Appl. Stat. **43**(1), 46–57 (2016). https://doi.org/10.1080/02664763.2014.998407
13. Adelfio, G., Boscaino, G., Capursi, V.: Erratum to: a new indicator for higher education student performance. Higher Educ. **70**(3), 609–609 (2015); Erratum to Giada, A., Giovanni, B., Vincenza C.: A new indicator for higher education student performance. Higher Educ. **68**(5), 653–668 (2014)
14. Welchowski T., Schmid, M.: discSurv: discrete time survival analysis. R package version 1.1.2. http://CRAN.R-project.org/package=discSurv (2015)
15. Yee, T.W.: The VGAM package for categorical data analysis. J. Stat. Softw. **32**(10), 1–34. http://www.jstatsoft.org/v32/i10/ (2010)

Monitoring School Performance Using Value-Added and Value-Table Models: Lessons from the UK

George Leckie and Harvey Goldstein

Abstract Since 1992, the UK Government has published so-called 'school league tables' summarizing the average attainment and progress made by students in each state-funded secondary school in England. In this article, we statistically critique and compare prominent past, current and forthcoming value-added and value-table measures of school performance. We discuss the advantages and disadvantages of these different measures as well as their underlying statistical models.

Keywords Value-added model · Value-table · Multilevel model · School accountability

1 Introduction

The UK has a long history of publishing 'school league tables' summarising students' examination and test results [6, 12]. Over time, increasingly sophisticated measures have been introduced culminating in 2006 with contextual value-added (CVA), a 'value-added' multilevel modelling based approach. However, in 2011 the Government withdrew CVA replacing it with expected progress (EP), a simpler 'value-table' descriptive statistic approach. In this paper we: statistically critique the Government's reasons for withdrawing CVA; we argue that EP suffers from serious design flaws; and we show that CVA and EP lead to very different rankings and therefore that choice of school performance measure has important ramifications for school accountability.

While we focus on school league tables and the specific performance measures published in England, the issues we discuss are pertinent to other education sys-

G. Leckie (✉) · H. Goldstein
Centre for Multilevel Modelling and School of Education,
University of Bristol, Bristol, UK
e-mail: g.leckie@bristol.ac.uk

H. Goldstein
e-mail: h.goldstein@bristol.ac.uk

C. Perna et al. (eds.), *Studies in Theoretical and Applied Statistics*,
Springer Proceedings in Mathematics & Statistics 227,
https://doi.org/10.1007/978-3-319-73906-9_23

tems which use student test score data to monitor school performance. Prominent examples are the Tennessee Value-Added Assessment System in the US [17], the My School website in Australia [14] and the School Value-added Information System in [16] Hong Kong. Various book-length treatments are available on statistical models for measuring school performance [2] and on the many difficulties surrounding the use of value-added in school and teacher accountability systems [1].

2 Background to National Tests, School Performance Measures and School League Tables in England

The English education system consists of a primary phase of education (ages 4–11) followed by a secondary phase of education (ages 11–16). Effectively all students change schools at the transition between the two phases. At the end of primary schooling, all students sit national Key Stage 2 (KS2) tests in English and maths. These are measured using continuous point scores, but are discretised into ordinal categories or levels for reporting purposes: W (working below level 1), 1, 2, 3, 4, 5. At the end of secondary schooling, all students sit national ‘GCSE’ examinations in English, mathematics as well as in a range of other subjects of their choosing. Attainment in each subject is measured using GCSE grades: U, G, F, E, D, C, B, A, A*. School league tables are then constructed summarising schools’ performances in these KS2 tests and GCSE examinations [6]. Our focus is on secondary school performance measures and therefore schools’ GCSE performances, and especially the average progress made by students in each school during secondary schooling between their KS2 tests and their GCSE examinations.

The Government gives three main justifications for publishing school league tables. First school league tables are published to support parental school choice based on schools’ ability to teach the national curriculum, and to therefore create competition and a free market in education. As a result, these tables are routinely republished by the media and so have a very high national profile [9, 10]. Second, they are published to enable school accountability; publically funded schools should be held publically accountable. Schools whose results do not improve face takeover by neighbouring schools or ultimately closure. Third, they are published to promote school improvement via school self-reflection and the identification of effective practices being employed in successful schools. Indeed, a number of commercial and charitable companies now sell to schools student performance monitoring software and other services based on the same data which underlies the Government’s tables.

An important distinction to be made is between ‘attainment’ and ‘progress’ school performance measures. Attainment measures aim to report the average ‘status’ of students at the end of secondary schooling. The headline attainment measure in England for effectively the last 20 years has been the percentage of students achieving 5 or more A*-C GCSE grades (5 A*-C). Attainment measures such as this may give useful information regarding school inequalities, but it is crucial to realise that in England and other education systems more generally they reflect differences in school intake composition more than school processes. In contrast, progress measures (e.g., CVA and EP) aim to report the average 'growth' or 'improvement' made by students during secondary schooling. Progress measures are generally considered the fairer and more meaningful way to measure school performance for school choice, accountability and improvement purposes.

3 Contextual Value-Added (2006–2010)

The Government's CVA measure is based on the standard approach to modelling value-added in the school-effectiveness literature which is to fit a two-level student-within-schools random-intercept model to students' final attainment adjusting for student prior attainment and other student socioeconomic and demographic characteristics [7]. The CVA model can be written as

$$y_{ij} = \mathbf{x}_{ij}'\boldsymbol{\beta} + u_j + e_{ij}, \quad u_j \sim N\left(0, \sigma_u^2\right), \quad e_{ij} \sim N\left(0, \sigma_e^2\right) \tag{1}$$

The response y_{ij} denotes the GCSE score of student i $(i = 1, ..., n_j)$ in school j $(j = 1, ..., J)$, $\mathbf{x}_{ij}$ denotes the vector of student- and school-level covariates with associated coefficients $\boldsymbol{\beta}$, u_j denotes the school random intercept effect, and e_{ij} denotes the student residual. The vector of covariates includes KS2 score as a flexible polynomial as well as range of additional factors such as student age, gender, ethnicity, special education needs status and residential deprivation score. The GCSE score is summed over students' best eight GCSE results, while their KS2 score is averaged across their separate English and mathematics scores.The reported CVA scores are empirical Bayes predictions of u_j

$$\tilde{u}_j = \hat{R}_j \left\{ \frac{1}{n_j} \sum_{i=1}^{n_j} \left(y_{ij} - \mathbf{x}_{ij}'\hat{\boldsymbol{\beta}} \right) \right\}, \quad \text{where} \quad 0 < \hat{R}_j \equiv \frac{\hat{\sigma}_u^{\,2}}{\hat{\sigma}_u^{\,2} + \frac{\hat{\sigma}_e^2}{n_j}} < 1. \tag{2}$$

Conceptually, CVA scores and value-added scores more generally can therefore be viewed as school-level averages (albeit shrunken towards the overall average as a function of their reliabilities, $\hat{R}_j$) of the differences between students' actual and predicted GCSE scores. These scores are presented with 95% confidence intervals to communicate their statistical uncertainty.

3.1 The Government's Justifications for Withdrawing CVA

The Government withdrew CVA in 2010 citing several key justifications [4]. First they argued that '[CVA] is difficult for the public to understand' [4, p. 68]. Clearly

CVA is more complex than simply reporting school average exam scores. However, the notion of making adjustments for differences in schools' student compositions in terms of their prior attainment and other factors isn't in itself intrinsically difficult to understand. There is no need for the public to understand the statistical details of the model in order to interpret the adjusted school-mean scores. Interestingly, the underlying multilevel model is the simplest possible, a two-level random-intercept model. Many more realistically complex models have been proposed in the literature [1, 2], including, for example, models which take into account neighborhood and school-district effects [15], student mobility [8], the instabilty of school effects over time [18], and which allow schools to have different effects on different student groups. Perhaps the real problem is that the Government did not do enough to explain and communicate CVA? For example, one had to delve deep into the technical documentation to find out what the CVA unit of measurement was. Clearly the notion of 95% confidence intervals is also hard for the public to understand, however perhaps the Government should have explored various graphical approaches for communicating statistical uncertainty rather than simply reporting the confidence intervals in tabular form [11, 13].

The Government's second reason for ending CVA was that 'recent research shows [CVA] to be a less strong predictor of success than raw attainment measures' [4, p. 68]. It is not entirely clear what the Government are trying to say here (they don't cite the research they refer to). One possible interpretation is that a school's average GCSE performance ('success') is more strongly predicted by their students' average KS2 performance ('raw attainment measures') than by their school's CVA score. Why this may well be the case, such a result does not in itself mean that CVA is a poor measure of school effectiveness. Indeed, it would more be a reflection of the relatively small influence that schools have on student progress in England versus the substantial influence of school differences in the composition of student prior attainment [15].

The Government's third reason for ending CVA was that '[CVA] also has the effect of expecting different levels of progress from different groups of students on the basis of their ethnic background, or family circumstances, which we think is wrong in principle' [4, p. 68]. However, CVA did not apriori expect different levels of progress from different student groups, rather it adjusted for such differences if they arose. The reality is that some student groups do make less progress than others and that this must be adjusted for if we are to make fair comparisons between schools. Failure to do so leads to 'comparing apples and oranges'.

Expanding on this theme, the Government argue that 'It is morally wrong to have an attainment measure which entrenches low aspirations for children because of their background' [4, p. 68]. The Government are arguing that by adjusting for student background, CVA led to a system-level acceptance that socially and other disadvantaged student groups will make less progress than their more advantage peers. Although not stated explicitly, the real concern appears to be that some schools started to use the published CVA model to set differential GCSE targets for current students based on their background. This was never the purpose of CVA and reflects the perverse incentives that so often arise with high-stakes school league tables.

4 Expected Progress (2011–2015)

In contrast to CVA, the Government's EP measure is based on value-table methodology [2]. EP is published separately for English and mathematics. EP is calculated simply as the percentage of students making three levels of progress between KS2 and GCSE; it ignores students' socioeconomic and demographic characteristics. The Government's introduction of EP can be seen as an explicit attempt to address the flaws they perceived in CVA. Specifically, EP is designed to be both easy for the public to understand and blind to all differences between schools' intakes other than their prior attainment.

Every student is effectively set a target GCSE grade in English and separately in mathematics as a function of their KS2 levels in those subjects. Table 1 presents this idea in tabular form. Thus, for example, low prior attainers (those who achieved KS2 level 3) are expected to achieve a D GCSE grade or higher, while middle prior attainers (KS2 level 4) are expected to achieve a C or higher, and high prior attainers (KS2 level 5) are expected to achieve a B or higher. Essentially, all students are expected to progress by 3 (or more) levels during the five year duration of secondary schooling.

We can write this value-table model for English (or equally for mathematics) as

$$EP_{ij} = \mathrm{I}(y_{ij} - x_{ij} \geq 3), \quad \overline{EP}_{.j} = \frac{1}{n_j}\sum_{i=1}^{n_j} EP_{ij} \tag{3}$$

where y_{ij} denotes the GCSE level associated with the English grade of student i ($i = 1, ..., n_j$) in school j ($j = 1, ..., J$), x_{ij} denotes their English KS2 level, EP_{ij} denotes whether they made expected progress in English (i.e., 3 or more levels of progress), and $\overline{EP}_{.j}$ denotes the school proportion of students making expected progress in English.

Table 1 Value-table showing how expected progress in English and Mathematics is calculated [5]

KS2 level	GCSE target grade/level								
	U	G	F	E	D	C	B	A	A*
	2	3	4	5	6	7	8	9	10
W	0	1	1	1	1	1	1	1	1
1	0	0	1	1	1	1	1	1	1
2	0	0	0	1	1	1	1	1	1
3	0	0	0	0	1	1	1	1	1
4	0	0	0	0	0	1	1	1	1
5	0	0	0	0	0	0	1	1	1

W = Working towards level 1; 0 = EP not made; 1 = EP made

4.1 Statistical Concerns with Expected Progress

Our first statistical concern with EP is that it will bring about perverse incentives whereby schools concentrate their efforts on those students who are borderline in terms of potentially making EP (i.e., those students operating just below the No/Yes boundary in Table 1). This perverse incentive is largely driven by the fact that the transition values of the value table are binary (EP is a threshold measure). There are no partial rewards for just missing target grades; no additional rewards for surpassing target grades.

Our second statistical concern with EP is that, nationally, there is a strong dependency on prior attainment. Figure 1 plots a scatterplot of the percentage of students making EP against the underlying continuous prior attainment score in each subject. The figure shows a very strong overall positive association. For example, in maths, the percentage of students making their target GCSE grade ranges from below 20% to above 80% as we move from the lowest to the highest KS2 scores. Thus, it is harder for low prior attainers to make expected progress than it is for high prior attainers. Low prior attainers are set relatively tough target GCSE grades while high prior attainers are set relatively easy target GCSE grades. Schools with higher prior attaining intakes will therefore do better on EP. Put differently, EP therefore under-adjusts for school differences in prior attainment and so is not a pure measure of progress in the way that CVA is. Looking more closely at Fig. 1, we also see that the overall positive association between the percentage of students making EP and their prior attainment takes on an illogical sawtooth (zig zag) shape with sharp discontinuities in the probability of making EP as we move from the top of one KS2 level to the bottom of the next. These discontinuities reveal that at these transitions, even students with effectively the same prior attainment are set very different educational challenges in terms of their target grades. This is clearly undesirable.

Our third statistical concern is that EP takes no account of students' socioeconomic and demographic characteristics and therefore will be biased in favour of schools which serve more advantaged student groups.

Our fourth statistical concern with EP is that it makes no attempt to quantify and communicate the statistical uncertainty in measuring school effects. There is no obvious way for users to establish whether measured differences between schools, or differences from national averages are meaningful, or whether they more likely reflect the variations of chance. Consider a school with 180 students where 70% make EP. The associated 95% Wald binomial confidence interval ranges from 63% to 77% and so the school has a 7% point margin of error which would be completely unacceptable in any survey or poll of public opinion. When we plot the 95% confidence interval for every school in the country (not shown), we see that over a third of schools cannot be distinguished from the national average in either English or mathematics.

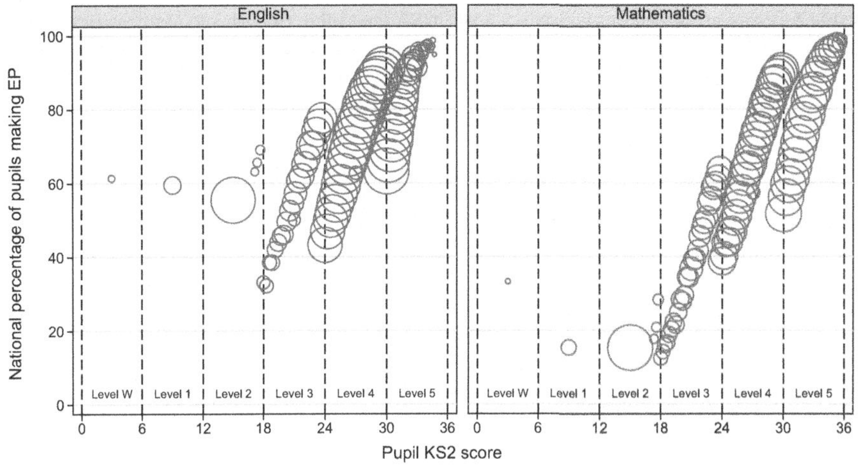

Fig. 1 National percentage of students making EP (expected progress) during secondary schooling against KS2 score in 2014, reported separately for English and mathematics. The magnitude of the hollow circles are proportional to the national number of students with that KS2 score. The dashed vertical lines denote the KS2 level thresholds. Level W = working towards level 1. For clarity, the plot is restricted to values of KS2 score for which there were at least 100 students nationally

5 Expected Progress Versus Contextual Value-Added

We have explained how CVA and EP, based on value-added and value-table methodologies, are fundamentally different measures of school progress. However, if the two measures lead to similar rankings then it could be argued that our arguments are largely academic. In this section we therefore analyse the 2010 data (3,056 schools) to contrast the two methods of calculating school progress empirically.

Table 2 reports Pearson correlations between the CVA, EP, 5 A*-C and KS2 APS (average point score across English and maths). We see that CVA and EP are only moderately positively correlated (correlations of 0.36 and 0.29 between CVA and EP in English and maths). Schools ranked high on EP are often ranked low on CVA and vice versa. The two measure are clearly measuring very different things. EP is much more highly correlated with 5 A*-C (correlations of 0.85 and 0.89) and is therefore closer to being a pure attainment measure of school performance than a pure progress measure. This is supported by the high correlations between EP KS2 APS (correlations of 0.64 and 0.67), whereas there is effectively no relationship between CVA KS2 APS (correlation of −0.02); a schools' success in EP is very much predetermined by how academic their students are at intake.

An interesting exercise is to consider how schools' ranks would likely change were the Government to revert back from EP to CVA and in particular, what types of schools would benefit or not by such a move. In Fig. 2, we plot the change in national rank against school mean KS2 APS. As expected, EP is strongly biased in favour of schools with high prior attaining intakes: schools with high prior attaining

Table 2 Pearson correlations between 5 A*-C, CVA, and EP in English and Mathematics in 2010

	5 A*-C	CVA	EP English	EP Maths	KS2 APS
5 A*-C	1				
CVA	0.24	1			
EP English	0.85	0.36	1		
EP Maths	0.89	0.29	0.77	1	
KS2 APS	0.87	−0.02	0.64	0.67	1

Number of schools = 3,056. 5 A*-C = percentage of students with five or more GCSEs (or equivalent qualifications) at grade A* to C; CVA = contextual value-added score; EP English = percentage of students making expected progress in English; EP English = percentage of students making expected progress in mathematics; KS2 APS = KS2 average point score

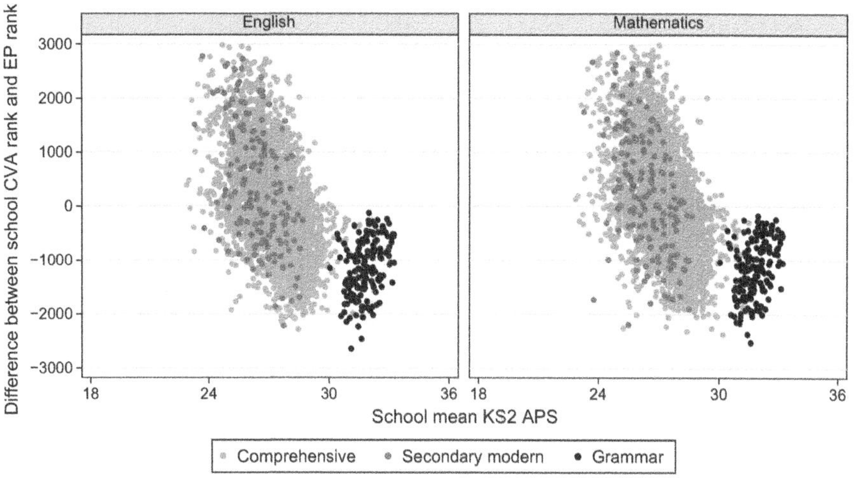

Fig. 2 Difference between school CVA and EP ranks against school mean KS2 average point score, based on 2010 school league table data, reported separately for English and Mathematics. KS2 levels map onto the KS2 point score scale as follows: [18, 24) = KS2 level 3 (i.e., low prior attainers); [24, 30) = KS2 level 4 (i.e., middle prior attainers); [30, 36) = KS2 level 5 (i.e., high prior attainers)

intakes would see very large drops in their national ranking were the Government to switch back from EP to CVA. The distinct cluster of schools which would particularly lose out with a return to CVA are 'grammar' schools, a small subset of around 160 schools which select students academically at intake and therefore have especially high school mean prior attainment. In grammar school areas, 'secondary modern' schools take the remaining students and so these schools therefore have especially low mean prior attainment. The point, however, is a more general one which is that CVA and EP are quite different school performance measures leading to substantially different rankings which will be systematically biased in favour or against particular types of schools.

6 Progress 8

In 2016, the Government will withdraw EP replacing it with P8, a new value-added based measure derived from a multiple linear regression model, a simplified version of which can be written as

$$y_{ij} = \mathbf{x}'_{ij}\beta + r_{ij}, \quad \bar{\hat{r}}_{.j} = \frac{1}{n_j}\sum_{i-1}^{n_j} \hat{r}_{ij} \tag{4}$$

where y_{ij} denotes the GCSE score of student i $(i = 1, ..., n_j)$ in school j $(j = 1, ..., J)$, $\mathbf{x}_{ij}$ denotes a flexible function of their KS2 score, r_{ij} denotes the student residual, and $\bar{\hat{r}}_{.j}$ denotes the predicted school value-added effect which is the P8 score.

By adjusting for a flexible function of student prior attainment, P8 should avoid the borderline effects and biases of EP. P8 scores will also once again be presented with 95% confidence intervals and therefore avoid that criticism of EP. However, P8 will continue to ignore school differences in the socioeconomic and demographic composition of their students.

7 Conclusion

The UK Government's reasons for withdrawing CVA, a value-added based measure, are questionable. CVA's successor, EP, a value-table based measure, appears fundamentally flawed. In particular, EP perversely incentivises schools' efforts on borderline students, it is severely dependent on prior attainment, it ignores school differences in students' backgrounds, and it fails to communicate statistical uncertainty. CVA, while by no means perfect, largely avoided these pitfalls. P8 is conceptually a return to the value-added based approach of CVA and should therefore also avoid these pitfalls, however, it will continue to ignore students' socioeconomic and demographic backgrounds and we think this fundamentally problematic in terms of holding schools accountable [3].

Acknowledgements This research was funded by UK Economic and Social Research Council grant ES/K000950/1.

References

1. Amrein-Beardsley, A.: Rethinking value-added models in education: critical perspectives on tests and assessment-based accountability. Routledge (2014)
2. Castellano, K.E., Ho, A.D.: A Practitioners Guide to Growth Models. Council of Chief State School Officers (2013)

3. Department for Education: Secondary school GCSE (and equivalent) performance tables 2010, Bristol, City of, 801. Department for Education, London (2011). http://www.education.gov.uk/schools/performance/archive/schools_10/pdf_10/801.pdf. Cited 6 Oct 2016
4. Department for Education: The importance of teaching: the schools white paper 2010. Department for Education, London (2010). https://www.gov.uk/government/uploads/system/uploads/attachment_data/file/175429/CM-7980.pdf. Cited 6 Oct 2016
5. Department for Education: Key Stage 2 to key stage 4 progress measures 2014. Department for Education, London (2015). http://www.education.gov.uk/schools/performance/2014/secondary_14/Guide_to_KS2-KS4_progress_measures_2014.pdf. Cited 6 Oct 2016
6. Department for Education: Compare school and college performance. Department for Education, London (2010). https://www.compare-school-performance.service.gov.uk/. Cited 6 Oct 2016
7. Goldstein, H.: Multilevel Statistical Models, 4th edn. Wiley, Chichester, UK (2011)
8. Leckie, G.: The complexity of school and neighbourhood effects and movements of pupils on school differences in models of educational achievement. J. Roy. Stat. Soc. Series A (Statistics in Society) **172**, 537–554 (2009)
9. Leckie, G., Goldstein, H.: The limitations of using school league tables to inform school choice. J. Roy. Stat. Soc. Series A (Statistics in Society) **172**, 835–851 (2009)
10. Leckie, G., Goldstein, H.: A note on 'The limitations of school league tables to inform school choice'. J. Roy. Stat. Soc. Series A (Statistics in Society) **174**, 833–836 (2011)
11. Leckie, G., Goldstein, H.: Understanding uncertainty in school league tables. Fiscal Stud. **32**, 207–224 (2011)
12. Leckie, G., Goldstein, H.: The evolution of school league tables in England 1992–2016: 'contextual value-added', 'expected progress' and 'progress 8'. Brit. Edu. Res. J. 43, 193–212 (2017)
13. Leckie, G., Charlton, C., Goldstein, H.: Communicating uncertainty in school value-added league tables. Centre for Multilevel Modelling, University of Bristol (2016). http://www.cmm.bris.ac.uk/interactive/uncertainty/. Cited 23 Mar 2017
14. My School: Australian Curriculum, Assessment and Reporting Authority, Sydney (2017). https://www.myschool.edu.au/. Cited 23 Mar 2017
15. Rasbash, J., Leckie, G., Pillinger, R., Jenkins, J.: Children's educational progress: partitioning family, school and area effects. J. Roy. Stat. Soc. Series A (Statistics in Society) **173**, 657–682 (2010)
16. School Value-added Information System: Education Bureau The Government of the Hong Kong Special Administrative Region, Hong Kong (2017). https://svais.edb.gov.hk/. Cited 23 Mar 2017
17. Tennessee Value-Added Assessment System: Tennessee Department for Education, Tennessee (2017). http://www.tn.gov/education/topic/tvaas/. Cited 23 Mar 2017
18. Leckie, G.: Avoiding bias when estimating the consistency and stability of value-added school effects using multilevel models. J. Educ. Behav. Stat. Forthcoming (2018). https://doi.org/10.3102/1076998618755351

Part VII
Economic and Financial Data Analysis

Indexing the Normalized Worthiness of Social Agents

Giulio D'Epifanio

Abstract A class of indexes is proposed to evaluate the "worthiness" of the performance of social agents (e.g. governors of health-care districts, schools, etc.), which is fully standardized on the conventional reference-framework specified by the policy-maker. An interdisciplinary attempt is made herein to integrate concepts and methods from different fields (management and political science, decision theory, statistics, economics, artificial intelligence, etc.). The performance is interpreted from the view of the policy-maker which pursues his overall-goal on a sequential planning of goals. The index is adapted on the data of the reference standard-agent, also normalized on the conventional behavior which has been specified by stakeholders through setting of a probabilistic model. Pseudo-Bayes tools are used into the normalization process.

Keywords Normalizing · Social performance index · Pseudo-Bayes
Standardizing

1 Social Performance Against Planning

1.1 The General Topic

This work deals the issue of designing and implementing an index for evaluating social agents (governors of a type of social service to citizens), with respect to the "social worthiness" of their performance (e.g. see [6]) from the view of the policy-maker (PM) which pursues his overall-goal on a sequential planning of goals. Within the general topic of the performance assessment of social agents, various visions are reported in a lot of papers. For a partial review see, for example, [8, 9]. However, the activity of evaluating and benchmarking is pragmatically conceived in this work from a cybernetic perspective, essentially with the aim of helping the PM to guide (through acts of deciding on incentives/penalties) behaviors of his social agents

G. D'Epifanio (✉)
Department of Political Science, University of the Study of Perugia, Perugia, Italy
e-mail: ggiuliodd@gmail.com

C. Perna et al. (eds.), *Studies in Theoretical and Applied Statistics*,
Springer Proceedings in Mathematics & Statistics 227,
https://doi.org/10.1007/978-3-319-73906-9_24

toward planned goals, rather than to describe or perhaps to explain them as in an econometric perspective. An interdisciplinary attempt is made to integrate concepts and methods scattered in various fields (management and political science, decision theory, statistics, economics, artificial intelligence, etc.).

1.2 Performance and Planning

The policy maker (PM) would evaluate social agents $A_1, A_2, \ldots, A_p$ (the governors of services to citizens), in a certain reference class D (e.g. a type of health-care service, of school, etc.), with respect to the "worthiness" (the intrinsic value) of their performance against a given planning of ordered goals:

$$O_0 \preceq O_1 \preceq O_2 \preceq \cdots \preceq O_l \preceq \cdots \preceq \cdots \preceq O_{L-1} \preceq O_L := O_{Full} \tag{1}$$

Here, assuming a certain reference framework $\mathcal{F}$, the PM has deployed (taking into account principles and criteria of social and economic interest) its ideal overall-goal O_{Full} (to be wished for each citizen) through a chain of $(L+1)$ increasingly more stringent goals with binary outcome (achieved/not achieved). The scale of the level of results, related to planning (1), $Y \in \{0, 1, \ldots, l, \ldots, L\}$ could be now defined through setting logical identification of proposition "*goal O_l is achieved*" with that of "*result-level Y is at least l*", briefly setting $O_l :\equiv \{Y \geq l\}$, for $l := 0, 1, \ldots, L$. Thus, for any citizen, the degree of achievement of the pursued ideal overall-goal is represented by the level of Y, starting from the lower level (this is associated to the "tautological" base-line goal $O_0 :\equiv (Y \geq 0)$ which is "achieved by default" for any citizen) up to the higher level which is associated to the achievement of the more desired goal $O_L :\equiv (Y \geq L)$. The social behavior of any agent A is that which will emerge from the behaviors of its subjected citizens against planning (1). In the assumed framework $\mathcal{F}$, social reference conditions might be considered also, by the PM, in order to justify evaluations of the agents performance which are related to the status x of the governed citizens (e.g. related to gender, age, etc.) on a finite space-state $x \in \{x_1, \ldots, x_R\}$. Then, for any social agent $A \in D$, the social performance will be associated to its "government capacity" to advance (from the starting base-line) its subjected citizens on planning (1), conditionally on the citizens status x on the established "reference social conditions" $\{x_1, \ldots, x_R\}$. Hence, the performance of agent $A \in D$ against planning (1) is intended described by the set $\{p_{|x}[A] := (p_{0|x}, p_{1|x}, \ldots, p_{l|x}, \ldots, p_{L|x})[A],\ x \in \{x_1, \ldots, x_R\}\ \}$ of the probability distributions of Y, on the set of the citizens that A governs, conditional on status $x := X \in \{x_1, \ldots, x_R\}$; here $p_{l|x}[A] := Pr\{Y = l|x\} > 0$.

1.3 Reference Data

An example of agents data, interpreting here distributions $p_{l|x}[A]$ above as empirical distributions, is reported in data Table 1.

1.4 What This Work Presents

In this work assuming an engineering view, it is proposed the design and the construction of an index for evaluating the performance worthiness (see [7–9]), even taking also into account "reference conditions" of the agents, standardized on a set of conventional specifications by the PM (or by other stakeholders) within the assumed design framework $\mathscr{F}$. The index borrows its formal structure from the Quiggin-Yaari's functional, used in the decision theory (e.g. see [2]); but, it is re-framed and re-interpreted herein. It is constructed on the "increases of the worthiness" (see [7] for details) through the planned goals on sequence (1), provided these increases have been primarily standardized on the design requirements specified by the PM. The standardization process proceeds in two stages. Firstly, the PM explicitly chooses a certain concrete instance A^* of standard-agent. At least roughly, such a standard-agent would represent a certain type of behavior for reference. Subsequently, the PM attempts to normalize such a concrete behavior upon an "ideal behavior". Such ideal behavior is that which adheres to the set of requirements, specified by the PM through setting constraints on parameters upon an assumed formal reference model. The conceptual aim is to identify that "ideal standard-agent" from the which the "normalized increases of the worthiness" on goals sequence (1) have to be extracted. Subsequently, these increases will enter the index structure.

To this purpose, an highly structured probabilistic model is used in the process of identification of the normalized "increases of the worthiness" on goals sequence (1). This model is essentially viewed as a conceptual tool to represent "what should be the normalized behavior" on the which the agents performance have to be conventionally interpreted. It serves to represent, within the reference design framework $\mathscr{F}$, the point of view assumed by the PM (or by other stakeholders) in a formal manner, thus including its vision and beliefs in a transparent way. A specific interpretation instance is established through setting of constraints on a proper (hyper-)parameters space. It is briefly outlined an example where the sequences of the "increases of the worthiness" on planning (1), interpreted through a complex probabilistic model, are interrelated through latent processes which are driven by states $\{x_1, \ldots, x_R\}$ by means of an (hyper-)parameters profile γ. Constraints specifications on parameters allow to include various types of working assumptions, requirements and perhaps beliefs of the PM inside the design. Technically, the container model is viewed as a class of distributions, which is hierarchically structured unless an (hyper-)parameters profile γ. Given specifications of constraints according to certain design requirements, (hyper-)parameters have to be regulated to adapt (as much as possible) the

Table 1 Example of performance data. The social agents and the standard agent

Agent A1	Performance level (Y)					Agent A2	Performance level (Y)					Agent A3	Performance level (Y)				
Status (X)	I	II	III	IV	V	Status (X)	I	II	III	IV	V	Status (X)	I	II	III	IV	V
x1	25	130	125	15	7	x1	30	142	153	11	3	x1	15	170	142	18	5
x2	13	95	132	18	7	x2	9	122	165	21	16	x2	12	105	155	25	16
x3	17	74	183	49	5	x3	21	85	137	74	35	x3	9	81	163	64	21
x4	13	28	115	39	21	x4	6	17	82	39	12	x4	15	37	114	42	31
x5	4	3	27	12	15	x5	3	5	21	17	7	x5	2	5	35	11	13

Agent A4	Performance level (Y)					Reference standard agent A^*	Performance level (Y)				
Status (X)	I	II	III	IV	V	Status (X)	I	II	III	IV	V
x1	27	130	174	19	12	x1	17	160	164	9	1
x2	21	107	137	29	15	x2	11	108	165	23	6
x3	7	64	181	44	13	x3	7	64	181	44	3
x4	9	27	115	37	31	x4	2	18	99	30	11
x5	1	5	29	15	8	x5	0	0	20	7	4

interpretative conventional model on the data associated to the specific concrete choice of standard agent A^*. To this technical aim, it is used a pseudo-Bayesian tool for the automatic identification of (hyper-)parameters γ from the data of agent A^*. Finally, it is recovered the sequence of the "increases of the worthiness" which will enter the index, normalized on the model specifications, conditionally on the social status x.

2 Indexing the Worthiness

Let A^* denote the standard agent (e.g. a recognized meaningful "best practice", from the PM's view), $\mathcal{P}^*$ the population of the citizens which A^* governs. The concept of "intrinsic worthiness" has been conceptually defined in ([7]) and operationalized herein. The criterion of intrinsic worthiness may be interpreted (see [7]), in particular, on a probabilistic setup.[1] We define the *"worthiness increase standardized on agent A^*"* between any two goals on planning (1) (i.e. between two any adjacent levels of $Y := l \in (0, 1, \dots, L)$), conditional on status $x := X \in \{x_1, \dots, x_R\}$, as the following variation:

$$\omega^*_{l|x} := \Delta_{l-1} Val_{|x} := Val_{|x}(O_l) - Val_{|x}(O_{l-1}) =$$
$$= \varphi_l(\frac{Pr\{Y = l-1|x;\ \mathcal{P}^*\}}{Pr\{Y \geq l-1|x;\ \mathcal{P}^*\}}) = \varphi_l(\frac{p^*_{l-1|x}}{p^*_{l-1|x} + p^*_{l|x} + \dots + p^*_{L|x}}) \geq 0 \qquad (2)$$
$$l := 1, \dots, L$$

conventionally setting $\omega^*_{0|x} := 0$ also. Here, $\varphi_l(.)$ denotes a continuous monotone function of the conditional probability rates $\tau_l := \frac{Pr\{Y=l-1|\,x; A^*\}}{Pr\{Y \geq l-1|\,x; A^*\}}$. Thus, the basic scale of Y can be re-quantified to that of $Y^* \in \{0, \omega^*_{1|x}, \omega^*_{1|x} + \omega^*_{2|x}, \dots, \omega^*_{1|x} + \omega^*_{2|x} + \dots + \omega^*_{L|x}\}$ which is standardized on agent A^*. The choice of $\varphi_l(.)$ confers some features to the scale (see [7]).

Example 1 Among various choices, it might be chosen (a) the function identity $\varphi_l(\tau_l) = \tau_l$ which will provide a non additive scale through the goals on planning (1); (b) $\varphi_l(\tau_l) := \log\{\frac{1}{1-\tau_l}\} = -\log Pr\{Y \geq l|\ Y \geq l-1;\ x\}$ which will provide an additive scale.

[1]Consider hierarchical chain of goals (1). Given that a certain goal O_{l-1} has been achieved in condition x, the greater the resistance, with reference to the evaluation framework, to also achieve the next pursued goal O_l, by continuing to improve, the greater the increment of value due to the intrinsic worthiness of who, in condition x, actually is able to achieve it. Thus, for any citizen "i" in the reference-status x, having achieved goal O_{l-1} on planning (1), the higher "the risk of failing the next goal O_l", referring such a risk to the population $\mathcal{P}^*$ of the citizens governed by A^* given x, the greater the "*increase of worthiness*", due to the worthiness of the agent which governs "i" *as if* "i" was in $\mathcal{P}^*$given x, whenever citizen "i" actually achieves goal O_l.

Formally interpreting such "worthiness increases" as "utility increases", the "rank dependent expected utility" (RDEU) (recalling the Yaari-Quiggin functionals, e.g. see [2], pp. 559) will provide the following conditional index:

$$A \in D,\ x :\mapsto W[A|x;\ A^*, \psi(.)] := \sum_{l:=1}^{L} \omega_{l|x}(A^*) \cdot \psi(Pr\{Y \geq l|x\}[A]) \quad (3)$$
$$= \sum_{l:=1}^{L} \varphi_l(\frac{Pr\{Y = l-1|\, x; A^*\}}{Pr\{Y \geq l-1|\, x; A^*\}}) \cdot \psi(Pr\{Y \geq l|x\}[A])$$

In general, parameters $Pr\{Y \geq l|x\}[A]$ may be intended as intrinsic quantities to be estimated by the data of agent A. However, for the sake of convenience in communication, from now up such parameters are intended as explicit empirical probabilities, associates to the actual data of agent A.

In RDEU theory (e.g. see [2]), $\psi(.)$ represents a weighting function (a continuous monotone function on (0, 1) such that $\psi(0^+) = 0$, $\psi(1^-) = 1$) which may be interpreted to take into account, for example, aversion of the PM to social risk. Therefore, while index inherits coherence with general principles of rationality in social choices, it may be adapted on the setting of specific interest for the PM. Of course, the version of index (3) resized on [0, 1] is the following:

$$W^*[A|\, x;\ A^*, \psi(.)] := \frac{W[A|x;\ A^*, \psi(.)] - W[A_{Worst}|x;\ A^*, \psi(.)]}{W[A_{Best}|x;\ A^*, \psi(.)] - W[A_{Worst}|x;\ A^*, \psi(.)]} \quad (4)$$

Here, A_{Worst} and A_{Best} denote the "ideally worst" and the "ideally best" social agent such that $W[A_{Worst}|x;\ A^*, \psi(.)] := 0$ and $W[A_{Best}|x;\ A^*, \psi(.)] = \omega^*_{1|x} + \omega^*_{2|x} + \cdots + \omega^*_{L|x}$, respectively, provided reference-status $x := X \in \{x_1, \ldots, x_R\}$ of the citizens.

Example 2 (Example 1, continued) Using choice (b), as in example (1), and letting $\psi(.)$ be the identity function, the correspondent instance of index (4) is the following [2]:

$$W^*[A|\,x;\, A^*, 1(.)] =$$
$$= k \cdot \sum_{l:=1}^{L} \log(Pr\{Y \geq l|\, Y \geq l-1,\, x; A^*\}) \cdot Pr\{Y \geq l|x\}[A] \qquad (5)$$

where constant $k := \frac{1}{\log(Pr\{Y \geq L|\, Y \geq 0, x; A^*\})}$.

Finally, the overall evaluation of any agent $A \in D$, on set $x \in \{x_1, \dots, x_R\}$ of reference conditions, is provided by index

$$A \longmapsto \sum_{r:=1}^{R} q_r \cdot W^*[A|\, x_r;\, A^*, \psi(.)] \qquad (6)$$

Here, $q_r \geq 0$, $\sum_{i:=1}^{R} q_r = 1$, are interpreted (at the light of some criterion, e.g. that of the different "political relevancy" of social reference-domains to the aim of the PM) as weights.

3 Model-Based Conditional-Scale of the Worthiness

Here, the process is briefly delineated which provides parameters $\omega^*_{|x} = (\omega^*_{1|x}, \dots, \omega^*_{L|x})$ entering indexes (3), (4) and (6). Such a process takes into input the data of standard agent A^*, e.g. those reported in Table 1. Then, it interprets the data of standard agent A^* on the formal framework represented through a probabilistic model, for example that represented below in Eqs. (7)–(11). Therefore, it can recovers the artificial data-table associated to the behavior of an "idealized standard agent" A^{**}, in this example that reported in Table 2 below. From this table, rates $\tau_l := \frac{Pr\{Y=l-1|\, x; A^{**}\}}{Pr\{Y \geq l-1|\, x; A^{**}\}}$ could be calculated which yield parameters $\omega^*_{|x}$ entering indexes (3), (4) and (6).

In this example, the reference model is structured as it follows.

The manifest outcome model :

$$Y_r := \{Y_{r0}, \dots, Y_{rL}\} \mid \psi_r \overset{ind.}{\underset{r:=1,\dots,R}{\sim}} Mult(y_{r0}, \dots, y_{irL}; \psi_{r0}, \psi_{r1}, \dots, \psi_{rL}, n_r) \qquad (7)$$

The normative model :

[2] Since $\frac{Pr\{Y \geq l|\, x; A^*\}}{Pr\{Y \geq l-1|\, x; A^*\}} = Pr\{Y \geq l|\, Y \geq l-1,\, x; A^*\}$ and $\frac{Pr\{Y \geq L|\, x; A^*\}}{Pr\{Y \geq 0|\, x; A^*\}} = Pr\{Y \geq L|\, Y \geq 0,\, x; A^*\}$, using conditional probabilities on chain (1), it follows index (5).

$$\psi_r := (\psi_{r0}, \psi_{r1}, \dots, \psi_{rL}) | \, m_r, a_r \underset{r:=1,\dots,R}{\overset{ind.}{\sim}} Dirichlet(\psi_{r0}, \dots, \psi_{rL}; m_r, a_r) \quad (8)$$

$$m_r := (m_{r0}, \, m_{r1}, \dots, m_{rL}), \; 0 < m_{rl} := E[\psi_{rl}] < 1 \quad (9)$$

$$\sum_{s:=0}^{L} m_{rs} = 1, \; a_r := w_r, \; w_r > 0$$

$$Constraints \; specification:$$

$$\begin{cases} v_{r1} := \frac{m_{r0}}{m_{r0}+m_{r1}} = \frac{e^{\eta_{r1}}}{1+e^{\eta_{r1}}} \\ \dots \\ v_{rl} := \frac{m_{r0}+\dots+m_{r(l-1)}}{m_{r0}+m_{r1}+\dots+m_{rl}} = \frac{e^{\eta_{irl}}}{1+e^{\eta_{rl}}} \\ \dots \\ v_{rL} := \frac{m_{r0}+\dots+m_{r(L-1)}}{m_{r0}+m_{r1}+\dots+m_{rL}} = \frac{e^{\eta_{rL}}}{1+e^{\eta_{rL}}} \end{cases} \quad (10)$$

$$Latent \; regression \; equations:$$

$$\eta_{rl} = \mu_0 + \sum_{s:=1}^{L} \delta_l \cdot I_{(s=l)} + \sum_{w:=2}^{R} \beta_{w(l-1)} \cdot I_{(X(r)=w)}; \; I_{(.)} \; bin. \; indic. fun. \quad (11)$$

$$r := 1, \dots, R := 5 \; (ref. \; soc. \; condit.); \; l := 1, \dots, L := 4 \; (scale \; level \; transit.)$$

On the rth stratum of the n_r individuals in the condition x_r, the manifest outcome $Y_r := (Y_{r0}, \dots, Y_{rL})$ is distributed as a multinomial, given expectation-profile $\psi_r := (\psi_{r0}, \psi_{r1}, \dots, \psi_{rL})$ which is distributed in turn according to the Dirichlet model, given complementary profile w which represents prior-vagueness into the model specification. The entire model is viewed as a structured class of distributions of the latent (underlying the manifest multinomial outcome Y_r, crossing the social strata recoded by $r := x \in \{1, 2, \dots, R := 5\}$) profile $\Psi := E[Y] := (\psi_1, \dots, \psi_r, \dots, \psi_R)$, parametrized by vectorial hyper-parameter $\gamma := (\mu_0, \delta, \beta)$. The hidden transitions $(l := 1, \dots, 4)$ on the chains of planning (1) are intended as regulated by a set of "latent worthiness parameters" v_{rl}, conditionally on strata $r \in \{1, \dots, R := 5\}$, which in turn are interrelated through a system of latent regression equations, unless a set of hyper-parameters $\gamma := (\mu_0, \delta, \beta)$. Thus, sub-profile (μ_0, δ) represents the common "background effects" (shared among the various conditional scales), the starting base-line μ_0 and the background profile $\delta := (\delta_1, \dots, \delta_l, \dots, \delta_L)$ of the latent level increments, respectively. Sub-profile β represents the "crossed effects" due to the interactions between social strata $r := 1, \dots, R := 5$ and result levels $l := 1, \dots, 4$.

In attempting to norm the behavior of standard social agent A^* on such a model, a value-profile γ^* is recursively searched, using the actual performance data y of A^* given social conditions x as reported in data Table 1, such that the multi-dimensional "residual from updating"

$$\mathbf{\Delta}_{y;x} T := \| \; Vec \; (\, E(\Psi \mid y, x; \gamma, \, w) - E(\Psi \mid x; \gamma, \, w) \,) \; \|$$

Table 2 Recovered model-based table, about the normalized behavior of the standard agent

Normalized standard agent A^*	performance level (Y)				
Status (X)	I	II	III	IV	V
x1	13.594267	164.4959028	161.21639	10.10709	1.586352
x2	13.0176318	105.9409114	163.5118	25.05272	5.47693
x3	6.9416901	59.9375878	187.47076	38.14704	6.502919
x4	3.3570336	17.6207492	98.87564	31.31299	8.833582
x5	0.5863062	0.9039199	17.50527	9.04469	2.959813

is minimized at value-profile γ^* (see also [5]). Here, $E(\Psi \mid y, x; \gamma, w)$ denotes the formal predictive expectation of full parameter profile $\Psi := (\psi_1, \ldots, \psi_R)$ which is updated by outcome y, whereas $E(\Psi \mid x; \gamma, w)$ is his non updated counterpart, over the design-point x.

The rationale of this approach is based on a general "minimum information principle" (e.g. see [3], [4]):

> the less a prior representation of knowledge is updated by current data, having specified the constraints in the model, the more intrinsically it already was accounted for by the *intrinsic information* added by such data

Here, this principle is implemented on a pseudo-Bayesian setting (e.g. see [1]). The computational process used iterated projections on a differential geometric manifold (notes are reported below) up to an equilibrium configuration γ^* is reached, from the which the normalized ideal behavior is recovered, as reported in Table 2. Generally, associated to the structured probabilistic model, the main geometric manifold (as well also its geometric sub-manifolds, for a sketch see Fig. 1) it is constructed to be regular in the differential geometric sense (an example to tackle even non regular manifolds into a regular frame is provided in [5]), so that his dimension coincides with the number of the free hyper-parameters (recursively top-down, the dimension of any geometric regular sub-manifold coincides with the number of the free sub-hyper-parameters which describe that sub-manifold), analogous to the generalized coordinates in the systems theoretical mechanics. Consequently, derived parameters $\omega^*_{|x} = (\omega^*_{1|x}, \ldots, \omega^*_{L|x})$ can be calculated which will enter index (3)–(4). For example on status x_1, recalling the specific choices of example (1), $\omega^*_{|x_1} = (0.007316, 0.1238, 0.4989, 0.3699)$.

In concluding, the aim here has been that of re-formatting the behavior of the reference standard agent A^*, normalized on the reference model conventionally assumed by the stakeholder. However, incidentally, it could be formally noticed (see [4] for details) that, provided that the standard agent is a real actual agent and that the conventional model is actually right also (i.e. at least "adequate" to represent the actual behavior), the method would be also consistent and efficient in a classic usual sense.

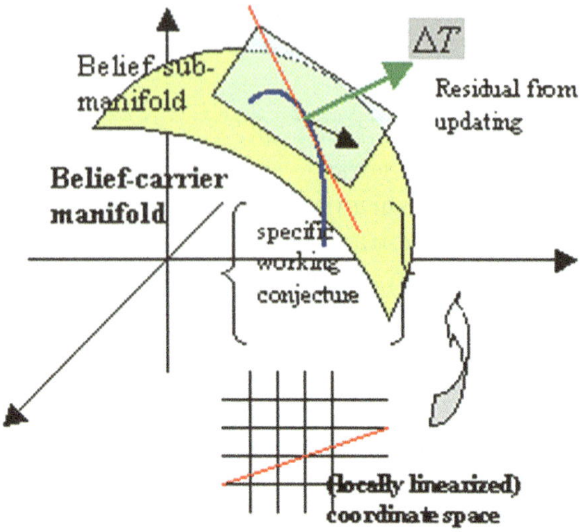

Fig. 1 Sketch: differential geometric manifolds

3.1 Note on the Computational Process

Let $\mathbf{P}(\gamma, w) := < \frac{\partial}{\partial(\gamma,w)}, [\frac{\partial}{\partial(\gamma,w)}]^t >^{-1} \cdot [\frac{\partial}{\partial(\gamma,w)}]$ denote the coordinate projector of the full variation $\mathbf{\Delta}_{y;x}T(\gamma, w)$ upon the tangent space of the sub-manifold at $p(\gamma, w)$. Here, $[\frac{\partial}{\partial(\gamma,w)}]$ denotes the basic coordinate (row-)vector system, $< . >$ the usual inner product. The operator

$$(\gamma, w) \longmapsto \mathbf{P}(\gamma, w)[\mathcal{W}^{-1/2}\mathbf{\Delta}_{y;x}](\gamma, w)$$

is a vector field which yields a vector over the tangent space at coordinate-point (γ, w). This vector field induces a dynamic over the coordinate space, which yields the following iterative (dynamical) process:

$$(\gamma, w)^{(q+1)} = (\gamma, w)^{(q)} + \rho \cdot \mathbf{P}((\gamma, w)^{(q)})[\mathcal{W}^{-1/2}\mathbf{\Delta}_{y;x}](m, \Sigma)((\gamma, w)^{(q)}).$$

Here, $\mathcal{W}$ represents some proper weighting system matrix, ρ denotes the step-length. Provided this process converges (due to the non-linearity, ρ should be sufficiently small to assure convergence), the convergence-point would satisfy the orthogonal equation:

$$\mathbf{P}(\gamma, w)[\mathcal{W}^{-1/2}\mathbf{\Delta}_{Y;x}(\gamma, w)] = 0$$

The convergence-point would be a "constrained fixed point" (CFP, see [3, 4]) solutions, by checking that progressively the process reduces distances.

4 Results and Discussion

From an interdisciplinary methodological point of view, the results of this work concern the proposal of a structured approach, driven by specifications of the policy-maker/stakeholder of interest, in order to develop a wide class of indexes for evaluating the "worthiness" of the performance of social agents, beyond a mere economic perspective. Incidentally, various concepts have been re-formulated from different disciplinary fields, attempting to make operative the meaning of the "worthiness", normalized on a reference setup for the standardization from the point of view of the policy-maker/stakeholder of interest. It has been briefly showed as the reference setup might also use sophisticated structured probabilistic models, in a no-conventional pseudo-Bayesian modeling perspective, which could be geometrically represented. Perhaps, a future advancing might consider links to the "systems theory" and socio-cybernetics, in social-economics fields where complex social planning could be related to a network of goals.

References

1. Casella, G., Robert, C.P.: Monte Carlo Statistical Methods (third printing). Springer, New York (2002)
2. Chateauneuf, A., Cohen, M., Meilijson, I.: Four notions of mean-preserving increase in risk, risk attitudes and applications to the rank-dependent expected utility model. J. Math. Econ. **40**, 547–571 (2004)
3. D'Epifanio, G.: Notes on a recursive procedure for point estimation. Test **5**(1), 1–24 (1996)
4. D'Epifanio, G.: Properties of a Fixed Point Method. Annales de L'ISUP Paris, vol. XXXXIII, Fasc. 2–3, pp. 69–83 (1999)
5. D'Epifanio, G.: Data dependent prior modeling and estimation in contingency tables. In: Vichi, M., et al. (eds.) Studies in Classification Data Analysis and Knowledge Organization. Springer, New York (2005)
6. D'Epifanio, G.: Implicit social scaling. From an institutional perspective. Soc. Ind. Res. **94**, 203–212 (2009)
7. D'Epifanio, G.: Sviluppo di un Indice Multi-attributo per la Valutazione del Merito. In: Criteri e indicatori per misurare l'efficacia delle attività universitarie, vol. I, p. 279, CLEUP, Padova. http://www.ec.unipg.it/DEFS/depifanio.html?lang=it (2011)
8. Franco-Santos, M., Lucianetti, L., Bourne, M.: Contemporary performance measurement system: a review of their consequences and a framework for research. Manag. Account. Res. **23**(2), 79–119 (2012)
9. Royal Statisticall Society: Performance indicators: good, bad, and ugly. J. R. Stat. Soc. A **168**(Part 1), 1–27 (2005)

Financial Crises and Their Impacts: Data Gaps and Innovation in Statistical Production

Emanuele Baldacci

Abstract Financial crises damage output and social cohesion. Lack of timely and accurate data makes it more difficult to assess risks' build up. Information gaps can also limit the ability to respond to crises. This calls for better data to monitor economic and financial risks. Several measures taken by the international official statistics community address information needs. These include efforts to fill the data gaps, ensure policy relevance of key indicators, and measure the "unmeasured" complex dimensions of economy and society. Harnessing new data sources and promoting innovation in statistical production processes are key to improving timeliness and adequacy of statistical information services. Nowcasting and predictive analytics can enhance the provision of early warnings about crises.

Keywords Financial crises • Modernisation of official statistics
Big data

1 Background

There have been many financial crises throughout history, from the United States banking crash of 1929 to the recent global financial meltdown.

Although crises have differed across time in terms of their key triggers and contagion mechanisms (e.g., from balance sheet crises and fiscal sustainability shocks to financial sector risks), they have in common similar impacts. Typically, these range from short-term damages to output growth and social cohesion, to the long-term fall out on capital and labour markets [1]. Natural capital is also affected by economic consequences of financial turmoil. Crises also cause structural changes in economic agents' behaviours, including on households' consumption profiles, labour market participation, labour demand and government policies. Understanding what causes financial crises and their effects is, therefore, critical to limit

---

E. Baldacci (✉)
European Commission, Luxembourg, Luxembourg
e-mail: emanuele.baldacci@ec.europa.eu

C. Perna et al. (eds.), *Studies in Theoretical and Applied Statistics*,
Springer Proceedings in Mathematics & Statistics 227,
https://doi.org/10.1007/978-3-319-73906-9_25

economic and social harm from these shocks and hamper loss of human, physical and natural assets.

Lack of timely and accurate data makes it more difficult to send early warning signals about financial risks' build up, identify upcoming (micro-macro) economic imbalances, and ultimately assess financial crises' effects. Lessons from the recent global crisis show that information gaps can take different shapes. These could include lack of relevant data, non-availability of pertinent indicators and failure to assess the extent of interdependencies across sectors and agents.

These needs call for official statistical production to better monitor economic and financial risks; assessing risk interdependencies and links between micro and macro dimensions; improving the communication and consumption channels of official statistics; and ensuring cross-country comparability of data.

In response to the global crisis, the international statistical community has undertaken several initiatives. These aim to address data needs and strengthen the ability to assess crises' risks and the economic and social implications of financial turmoil. Actions include efforts to fill the data gaps, ensure policy relevance of key indicators, and measuring the "unmeasured" in areas such as wellbeing, risk dependencies, resilience, and sustainability.

Section 2 of this paper briefly discusses the transmission mechanisms and the key impacts of financial crises on economic and social dimensions, reflecting key findings arising from a rich and well-documented literature. Section 3 focuses on the key statistical information gaps that emerged from the recent global crisis, both in terms of the inability to assess risks accumulated over time and to predict their consequences. Section 4 describes the initiatives undertaken by the international statistical community to address data gaps and to innovate statistical production processes in order to deliver better information services; it aims also at presenting the European experience of the ESS Vision 2020 portfolio of projects to modernize statistical production systems. In Sect. 5, the paper focuses in particular on the role of new data sources to help strengthen the timeliness of statistics. This could support early identification of crises' risks and assess their implications. Section 6 indicates the next steps for research in this area.

2 Financial Crises Effects

Financial crises had disruptive effects on economic, social and environmental indicators for many decades [3].

Crises have taken different shapes. During the 1970s, financial crises were mostly triggered by excessive absorption of the domestic market from external sources. In particular, these crises hit countries with strong dependencies from imports for key raw materials and limited export competitiveness.

During the 1980s, several financial crises, which originated in the government sector, hit the economies of Latin America and had widespread consequences in other regions. These shocks arose from the excessive accumulation of public debt financed through international credits from the banking sector. Fiscal sustainability issues led to debt repudiation, high inflation and sharp currency depreciations.

The financial crises of the 1990s hit mostly economies in Asia and Eastern Europe. They stemmed from balance sheet vulnerabilities in the private sector, with strong accumulation of external currency denominated debt. Currency and maturity mismatches in corporate and banking sector balance sheets, led to sharp recessions with large accumulation of public debt. This, in turn, slowed down the path to economic and social recovery.

The most recent global financial crisis started in the United States financial sector in 2007. At the beginning, it hit the housing sector, with an excessive credit build up in the high-risk subprime mortgage sector. However, due to risk links among sectors and instruments, it quickly became a banking crisis with severe losses. This, in turn, transmitted sharp shock waves to the economic system through credit crunch, hitting trade and investment and fostering risk aversion. The need for the governments to step into limit economic damages and the fall in output affecting revenue fostered an increase in public debt. The sovereign debt crisis then hit back the economy, with higher interest rates and pro-cyclical fiscal policies.

Its long-term consequences are still visible today, as global economic growth remains below pre-crisis levels, long-term and youth unemployment remain high in several countries. At the same time, public debt has increased sharply both in mature and emerging economies with scope for crowding out effects.

Damages produced by these crises are not limited to short-term sharp declines in economic output, usually accompanied by worsening conditions in the labour market and an increase in poverty incidence. Medium-term implications of financial crises can be substantial. While output growth typically recovers after the end of financial shocks, potential output levels and potential output growth are harmed by the structural productivity decline. This is a consequence of the human and physical capital destruction caused by the crisis.

Transient unemployment and poverty can easily become entrenched, leading to higher structural poverty incidence and long-term unemployment spells. This, in turn, can lead to underinvestment in human capital [9], which further affects productivity and long-term output growth perspectives.

Negative implications of crises go, however, beyond economic and labour market effects, spreading to other sectors of society. During crises, households spending on preventive health can be compressed as a result of a fall in income, with harmful consequences for population living conditions and stronger demand on public services. The latter are typically under strain due to budget constraints and may not be able to absorb increased demand.

Natural capital can also suffer from the economic fall out of financial crises. With risk appetite in the financial industry severely curtailed by the economic shock, investment in more risky environmental friendly production systems in manufacturing could fall. Public spending on environmental protection measures may also be restrained by tight budget constraints.

Economic agents' behaviours could also change in response to crises. Risk-taking is reduced by lack of adequate supply of financial services and risk aversion by the financial industry. Households' consumption models may also change in response to the fall in income. This can lead to a reduction in assets, postponement of childbearing decisions, and a structural reduction in savings buffers.

Corporate behaviours also change in response to shocks with a variety of impacts. Corporate investment tends to fall during crises, while a selection process takes place in its aftermath. Companies that have adapted better to the opportunities and risks emerging from the new economic and social environment have a higher probability of surviving. The financial industry is typically massively affected by these shocks, with a process of bank consolidation and risk segregation taking place and tighter regulation being introduced. Risk taking by the banking sector and the capital markets also tend to suffer at least in the short term, which leads to less financing for innovation.

As briefly described above, socio-economic implications of financial crises can be massive and widespread, affect the macro as well the micro level of the economy and society, and lead to changes in behaviours by different agents. Understanding the causes of financial crises is therefore important to assess their likelihood ex ante and prevent the accumulation of excessive risks in the economy. Assessing the transmission channels of financial crises is also important to understand their impacts and the consequences on different sectors of economy and society.

3 Statistical Information Gaps

In the aftermath of the recent global crisis, statistical information gaps emerged in several areas.

These gaps are critical, as they limit the ability of decision makers to have a complete picture of the risks accumulated in the economic system before the crisis. It They also cloud the understanding of the channels through which the shocks propagate across sectors and regions. Key gaps were identified in the following four areas.

First, **balance sheet vulnerabilities** were not sufficiently covered by the statistical information systems. In particular, data about maturity, currency, and instruments mismatches both in the public and in the private sector were insufficient to draw a full map of risks and their interdependencies. This is particularly the case in financial systems where capital market intermediaries play a significant role, along with the more traditional banking sector.

The second area where statistical information was found to be less than adequate is the **assessment of the underlying strength of the economy**. Focus on headline economic growth indicators, was providing misleading signals in cases where the economic engine was heavily dependent from historically unprecedented performances in some sectors (e.g., financial sector, housing investment, commodity export). As economic trends in these sectors reverted to "normal" conditions and "super cycles" of commodity prices ended, headline output growth and other key macroeconomic indicators started to signal a less than healthy condition.

Fiscal risks have also been underestimated before the crisis. The focus on headline fiscal indicators, as opposed to cyclically adjusted targets, gave a sanguine picture of government finance's health, reflecting a higher-than-normal trend in economic output and revenue. Public debt roll over risks were also masked by high-risk appetite of financial markets, which compressed credit risk premiums on sovereign papers.

Financial interdependencies was another area were data and analysis gaps emerged. Risks spilled over from capital markets to the banking sector and hit rapidly sovereigns in many of the economies that were affected by the global crisis. Lack of full understanding of risk cross correlation across markets was in part linked to insufficient analysis of risk patters and risk propagation. This is related, in turn, to limited knowledge about the structure of credit markets and of the links between financial sector and macroeconomic stability.

Finally, despite abundant statistical information on social statistics, assessment of social disparities and their impact on **households' behaviour** [4] was limited. First, links between macroeconomic and microeconomic variables was weak. Second, lack of comprehensive data hampered the understanding of the relationships between income, consumption and wealth. Third, focus on consumption and income dimensions of households well-being did not allow to capture the more complex interdependencies in individuals' responses to economic and social shocks. Finally, the link between quality of life and environmental sustainability was not established.

These data and analysis gaps reflected the limited availability of statistical information systems, pertinent policy-oriented indicators, and sound conceptual framework to inform decisions in the run up to crises and in their aftermath.

In order to respond to these challenges, several initiatives have been undertaken by the international statistical community. These actions were complemented by efforts from researchers to use better data in order to assess early warning signals of crises and crises' effects.

Efforts to address statistical information gaps have focused on three key areas: dealing with data gaps; strengthening statistical indicators; work on more solid statistical frameworks.

Several actions aimed to address data gaps. The G-20 Data Gap Initiative focuses on making comparable data available across G-20 countries in areas that have resulted particularly relevant in the context of financial crises assessment. The initiative fosters better coordination among statistical producers of economic and financial data. It strengthens the monitoring of risks in the financial sector at country level through financial soundness indicators. It also provides more statistical information on international network connections, including financial linkages and cross-border financial flows. Finally, it includes information on vulnerability to external shocks, through better balance sheet and flow of funds data and asset prices. The key outcome of this initiative is a set of comparable Principle Global Indicators (www.principalglobalindicators.org).

Other initiatives to fill data gaps have led in Europe to more timely dissemination of key economic indicators, such as GDP [8] and the Harmonised Index of Consumer Prices (HICP), through flash estimates based on coincident and leading indicators. Efforts also focused on strengthening the information on asset prices, in particular in the housing sector, and to enrich statistical information on households' social conditions.

The international statistical community has also focused on developing indicators for policy decisions. Recently, the United Nations endorsed the Sustainable Development Goals, which are supposed to guide governments in designing and implementing sound public policies, with holistic development targets in mind.

In Europe, the Macroeconomic Imbalance Procedure scoreboard identifies imbalances in EU countries beyond the fiscal sustainability dimensions (http://ec.europa.eu/economy_finance/economic_governance/macroeconomic_imbalance_procedure/mip_scoreboard/index_en.htm). The scoreboard focuses on external imbalances, including external position, competitiveness, export performance, and internal imbalances (public and private debt, asset markets, financial sector liabilities and unemployment) to complement indicators of fiscal health. During the crisis, excessive debt and lack of external competitiveness were associated with higher exposure to the risk of being hit by the shock and to suffer more deeply from its consequences.

New indicators have also developed in the aftermath of the crisis to measure progress towards structural reforms in key areas of the economy (e.g., Europe 2020 indicators). This reflects the fact that during the recent global crisis, economies with stronger economic institutions and solid performance on structural reforms were more able than others to withstand the effects of shocks.

Early warning indicators can also be helpful in signalling ex ante the risk of a crisis (e.g., in the case of the growth at risk index used by the International Monetary Fund) or to identify source of potential stress to the fiscal sector (e.g. the fiscal stress index).

Finally, decision-oriented indicators have increasingly been adopted in policy-making. For example, measures of the output gap complement information on headline GDP growth to measure the underlying strength of the economy. Indicators of structural unemployment are also used to supplement information about labour market health not affected by cyclical patterns in the economy. These efforts aim to limit the risk of underestimating the build-up of economic risks in periods of rising asset prices or unprecedented growth in selected growth engines of the economy. The use of these indicators can support decisions in terms of the right macroeconomic policy stance in order to avoid pro-cyclicality.

One of the key lessons from the recent global crisis is that statistical systems need to provide a better understanding of the links between macroeconomic trends and individual behaviours, the distribution of income consumption and wealth, the role of cash and in kind transfers across households and their consequences for aggregate economic variables and social cohesion.

These information need to be conceptually linked to other quality of life variables, including well-being perception by individuals, as well as environmental

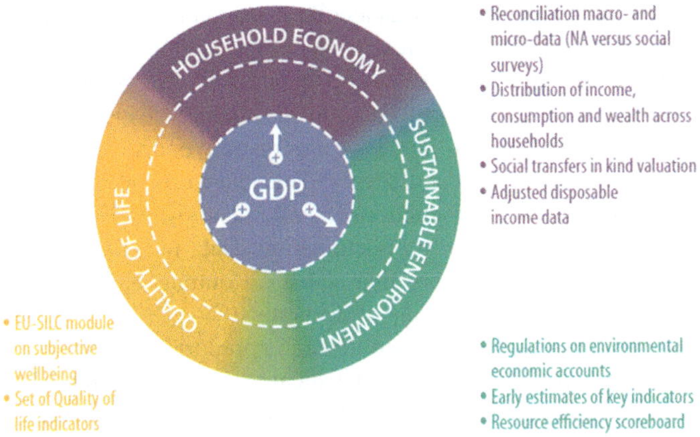

Fig. 1 Beyond GDP: conceptual framework. *Source* Eurostat

sustainability (Fig. 1). Continuous efforts have been made in recent years to strengthen the capacity of statistical organisations to disseminate information on these integrated dimensions, based on a sound conceptual framework. The availability of better data to capture complex dimensions that go beyond GDP can provide an additional source of knowledge to assess crises implications. It also helps assess social cohesion vulnerabilities that could pose risks for economic and financial stability.

4 Modernization of Official Statistics

In addition to addressing statistical information gaps, official statistics organisations have undertaken significant steps to modernise production processes in recent years.

This is motivated by the need to harness technological progress and in particular, the "data revolution", achieve efficiency to lower budget costs and soften respondents burden, and provide better statistical products and services to cater to users' demands.

While innovation in statistical production processes is not directly linked to data needs arising from the financial crisis, the ability of statistical organizations to deliver timely, pertinent and accurate information in response to demands from stakeholders requires a change in the way the statistical "factory" works. The change in the statistical production engine that is required for this quality leap in official statistics has four main components.

First, the raw materials used in the production process (i.e., the **data sources**) are increasingly based on the reuse of existing information (e.g., administrative) rather than its direct collection through surveys. This reflects high budget costs and response burden of traditional surveys. Innovation in survey collection modes, with increasing use of computer-assisted data collection, can only in part respond to the need to have more integrated data for statistical production.

Reliance on administrative sources, supplemented by direct data collection, leads to the second major ongoing transformation in official statistics production: **data integration**. Direct linkage of massive administrative data sets and statistical matching among these sources and survey data, increases dramatically the information sources for statistical production. It could also lower the time to market of statistical products, from design to dissemination. Data integration should rely on sound quality assessment to ensure that statistical outputs are comparable across countries and fit for purpose in terms of accuracy.

Integrated data need to be processed through appropriate statistical **methods**. In the multi-source environment of statistical production, model-based and algorithm-based estimation methods are required [2]. Models are already used in the production of official statistics in selected areas, such as the production of statistics for small areas, seasonal adjustment and flash estimates of macroeconomic variables. However, the change envisaged by the new "data factory" calls for a more widespread use of models in statistical production, which in turn requires good quality frameworks and methodological guidelines for model design, testing and implementation.

The final component of this transformation is the **dissemination** of statistics. Users are mostly driving this change on the characteristics of statistical outputs. They request fast-changing, on-demand statistics, tailored to their needs. This means that statistical organizations have to adjust the way they produce statistical information, to be able to deliver statistics as a service, going beyond statistical products.

Delivering services implies building platforms for data consumption, which are interactive for users and guided by strong metadata systems that enable integration of information into knowledge and storytelling (Fig. 2).

In the European Union, the European Statistical System (ESS) has endorsed a transformation program based on the above principles. A portfolio of projects, whose main purpose is more efficient and effective production of European statistics, supports the ESS Vision 2020 (http://ec.europa.eu/eurostat/web/ess/about-us/ess-vision-2020). This takes into account users' needs and leveraging technologies to innovate the way statistical information is produced (Fig. 3). The key components of the portfolio are: (i) the use of multiple sources and methods to process them in statistical production; (ii) the development of platforms and applications for data processing, which rely on data reuse; (iii) new information products and services on the corporate sector and intra-EU trade; (iv) new data consumption and dissemination service for end users.

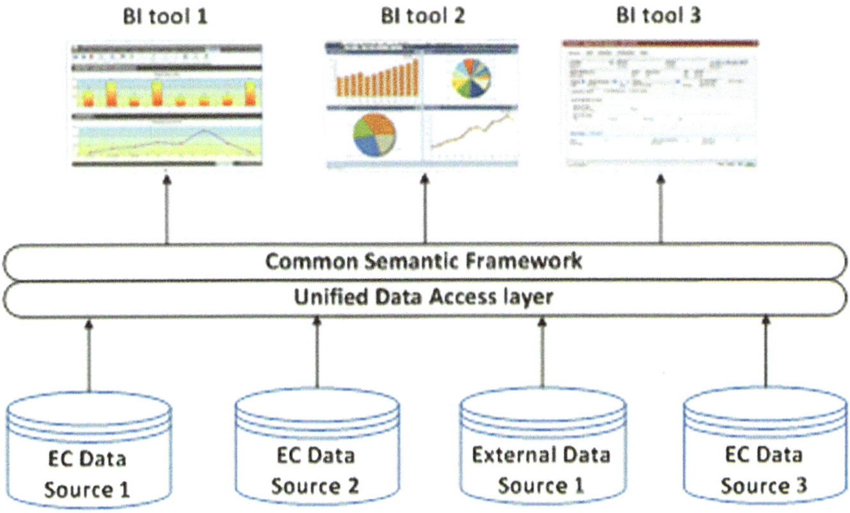

Fig. 2 A new statistical production architecture. *Source* Eurostat

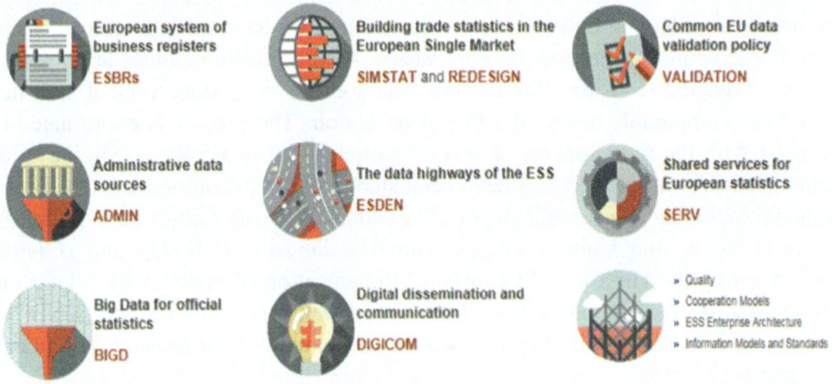

Fig. 3 The ESS Vision 2020 implementation portfolio

5 Harnessing New Data Sources

Recent developments in information technologies led to an increase in the volume, the frequency and the types of data, which can be accessed and potentially used for statistical production.

Big data can be a significant source of supplemental information in official statistics, helping measure concepts that are difficult to assess with traditional sources and dramatically increasing the timeliness of statistics. Harnessing these data sources is, however, not without challenges, as the lack of structure in the data

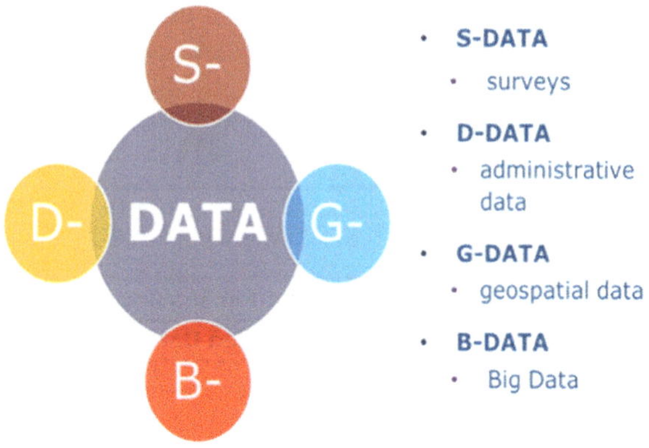

Fig. 4 Multi-source statistical production. *Source* Eurostat

and limited knowledge of the data generating process could affect the quality of the statistical estimates.

Notwithstanding the difficulties, big data are increasingly emerging as a key component of the multi-source statistical production environment described above (Fig. 4). Several pilots have been launched on the use of big data sources for the production of official statistics. These include activities undertaken at national level by statistical offices in collaboration with research centers and international activities. For example, in Europe the ESS Vision 2020 portfolio includes an important project dedicated to using different big data sources to produce official statistics, which are comparable across the European Union. The project is coordinated by Eurostat with the participation of several national statistical offices. Similar international collaboration activities have been started under the umbrella of the United Nations, including the establishment of a Global Working Group on big data for statistics led by the United Nations Statistical Division (UNSD), and activities carried out by the High Level Group on Modernization of Statistical Products and Process led by the United Nations Economic Commission for Europe (UNECE). These activities also entail the collaboration with a network of researchers and data scientists.

The use of big data could be particularly important for addressing some of the information gaps emerged during the recent financial crisis. In particular, new data sources can be used as stand-alone data or supplement information provided by traditional sources and models in two areas: (i) nowcasting and predictive analytics; and (ii) risk spillovers assessment. Both areas are critical to develop early warning systems to help detect risks' build up and identify vulnerabilities leading to potential shocks for the economy.

Big data have been used in several applications of nowcasting economic series. Results, however, have been mixed. While there is evidence that indicators based on web activity, for example, can improve the quality of unemployment, sales and output predictions, these results are conditional on the quality of the models and the indicators chosen. Also parameters' stability tends to be an issue, with changes in the underlying big data series affecting the predictive performance.

Recent empirical evidence based on studies carried out by Eurostat shows that nowcasting models using big data as a source of supplementary information can significantly outperform models based on traditional data sources. In particular, the use of online search results to build statistical variables with coincident and leading information about macroeconomic variables, such as inflation, retail sales growth and unemployment, can help improve the quality of nowcasting and short-term prediction models.

However, care should be taken in the choice of the appropriate statistical model. Sparse regression models (such as LASSO) tend to work better for very large data sets, but data reduction techniques (such as Partial Least Squares) are more helpful with multiple indicators. In general, the findings of several experiments show that dimensionality reduction methods improve nowcasting, but pooled models work best [7]. A data driven strategy is therefore deemed to be the preferred approach, which entails a model rotation based on forecasting performance at different intervals.

Other areas of application of different big data sources seems equally promising on the basis of initial experimental findings, resulting from work carried out at Eurostat and in collaboration with the ESS partners. For example, webscraping data on companies can be used to track the extent of factory less production and could integrate statistical information on activities based on company establishment data. Smart meters can help track real time electricity consumption and temporary vacant dwellings, which could be inputs into nowcasting models for aggregate demand. Automatic identification systems for ships can lead to estimates of trade flows on the basis of vessel identification data, Finally, mobile phone data offer information on individuals' mobility and real time population.

Big data sources can also help explore the amount of risk dependencies and spillovers, which are important dimensions of early warning systems, to help detect crises' occurrence. For example, credit risk premia have been used to assess the risk correlations across regions, instruments and markets and can complement traditional space-state models that explore financial sector risk transmission [5].

Notwithstanding encouraging early results from ongoing pilots, the structural use of big data in a multi-source production environment for official statistics requires a comprehensive business plan [6]. The ESS has developed a big data roadmap and action plan to help address the different dimensions underlying the use of big data in official statistics beyond experiments and pilots. These include the following priorities:

- secure high-quality data sources, which are accessible, relatively stable over time and whose provenance is well documented;
- deal with data privacy and data confidentiality issues, based on existing data protection regulations and statistical legislation;
- build adequate skills and partnerships both inside and outside the official statistics community;

- provide adequate computing platforms and related processing services that enable the use of large streams of data;
- continue to invest in methodological and quality issues, related to the use of big data in the context of a multi-source production environment.

6 Findings and Future Research Lines

Financial crises are costly for the economy and damaging for the social fabric, this is why governments try to prevent their occurrence if possible and otherwise mitigate the impacts.

However, crises are complex to predict in the absence of solid early warning systems. These should be based on comprehensive statistical information systems and accurate analytical models, which take into account crises triggers and transmission channels.

Data gaps have emerged as a key legacy of the recent global financial crisis, which the international statistical community has been tackling by strengthening statistical information on risks spillovers, wellbeing and inequality. New indicators dashboards on vulnerabilities and statistical frameworks to assess quality of life and environment complement these measures.

Addressing data gaps, although important, is not a unique answer to pressing information and knowledge needs. Innovations in statistical production can further help support the quest for data-driven policies, that allow lowering the risk of costly crises in the future. This requires a change in the statistical "factory", from data collection to dissemination. It calls for a shorter time to market of statistical products to respond to users' needs and the development of statistical services centred on data analytics.

The European Statistical System (ESS) has been at the forefront of these initiatives with the ESS Vision 2020 program, aiming at improving the statistical production, by leveraging new data sources and methods to deliver better information services.

The data revolution offers a new opportunity to address the pending challenges. A large amount of new sources are available through digital processes. Harnessing new data sources for the production of official statistics requires adequate methods to ensure output quality. New statistical production processes based on multi-source data mesh ups are also required and offer the opportunity to provide better and more flexible data analytics services to stakeholders.

Data analysis techniques can be used to produce timely information and nowcasting of relevant indicators, thereby contributing to better understanding of crises risks. Based on lessons learned from pilot results, the use of appropriate statistical methods to process the data is critical. Models that use a combination of sources that include non-traditional ones (big data), tend to outperform traditional models.

However, information about data structures and properties is paramount to avoid spurious causality errors and misspecifications. Strengthening economic and financial statistical information system is critical to addressing knowledge gaps.

Acknowledgements An earlier version of the draft has benefitted greatly from inputs by Dario Buono and exchange of views with other colleagues at Eurostat, which are gratefully acknowledged here.

References

1. Baldacci, E., de Mello, L., Inchauste, G.: Financial crises, poverty and income distribution. Financ. Dev. **39**(2) (2002)
2. Citro, C.: From multiple modes for surveys to multiple data sources for estimates. J. Off. Stat. **30**(3, September), 381–442 (2014)
3. Classens, S., Kose, M.A., Laeven, L., Valencia, F.: Financial Crises: Causes, Consequences and Policy Responses. International Monetary Fund, Washington, DC (2013)
4. Deaton, A.: Economics and Consumer Behaviour. Princeton Press (1980)
5. Dell'Erba, S., Baldacci, E., Poghosyan, T.: Spatial spillovers in emerging market spreads. Empir. Econ. **45**(2), 735–756 (2013)
6. Eurostat: ESS Big Data Action Plan and Roadmap (2014). https://ec.europa.eu/eurostat/cros/content/big-data_en
7. Eurostat: Big Data and Macroeconomic Nowcasting, by Kapetanios and Marcellino (2016). http://ec.europa.eu/eurostat/cros/sites/crosportal/files/item_2.1_big_data_and_macroeconomic_nowcasting_short_0.pdf
8. Eurostat: Euro Area and European Union GDP Flash Estimates at 30 Days, Statistical Working Paper Collection (2016)
9. Radermacher, W.: Do we need natural capital accounts for measuring the performance of societies towards sustainable development, and if so, which ones? EURONA 1/2015

European Welfare Systems in Official Statistics: National and Local Levels

Alessandra Coli and Barbara Pacini

Abstract In the last decades, European welfare systems have undergone continuous reforms in the light of financial pressures. Monitoring changes requires to consider several dimensions of welfare systems, such as the composition of risks and needs covered, the rules for accessing benefits or the type of social benefits delivered. Finally, it is relevant to take into account the geographical area where beneficiaries live, since in some countries local governments are assigned managing and, sometimes, legislative competencies on social protection areas. This paper aims at exploring official statistics on European welfare systems, by focusing on social benefits. The objective is assessing if available statistics allow one to compare the level and the kind of social benefits delivered across European countries both at national and sub-national levels. We focus on the Italian case to provide some examples.

Keywords Official statistics · Social protection systems · Local welfare

1 Introduction

In the last decades, European countries have undertaken innovation processes of their welfare systems, essentially through a reshaping of the role of public actors along with a reduction of financial resources.

Several aspects need to be considered to detect changes in welfare systems such as the composition of risks and needs covered, the rules for accessing benefits or the type of social benefits delivered. The level of decentralization of a welfare system represents a further key dimension to be considered since, in some countries, local governments are in charge of administrative and managerial responsibilities as well as of legislative power or political control. For example, the Italian Constitution (art. 117 as amended in 2001) provides a list of matters on which State and Regions (Nuts

A. Coli (✉) · B. Pacini
Dipartimento di Scienze Politiche, Università di Pisa,
Via Serafini, 3, 56126 Pisa, Italy
e-mail: alessandra.coli1@unipi.it

C. Perna et al. (eds.), *Studies in Theoretical and Applied Statistics*,
Springer Proceedings in Mathematics & Statistics 227,
https://doi.org/10.1007/978-3-319-73906-9_26

2 level, [7]) have concurrent competencies. Among these, we find health protection, education, and complementary and supplementary social security. Furthermore, in several countries, municipalities play an important role in the field of social assistence (disability, old age, poverty etc.). As a result, the place of residence may represent a crucial dimension to understand the actual kind of social protection received by citizens.

The awareness of the importance of comprehensive, up to date, comparable and accessible data on social protection has impelled international official statistics to promote the stocktaking of existing social protection international data and indicators. [3] describes the on-going process and presents a mapping of data and indicators. These pertain to most of the above-mentioned dimensions of welfare systems, however neither decentralization nor the need of territorial data are mentioned explicitly.

Here, we aim to examine information given by European official statistics on social protection benefits both at national and local levels, and to investigate the real comparability in terms of social domains, classifications adopted, aggregation levels, timing and coverage. We discuss the coherence of data coming from different data sources pointing out common features and main differences. As an example, we use different sources to provide some empirical evidence on territorial disparities in Europe and, particularly, in Italy.

The paper is structured as follows. Section 2 presents the definition of social protection shared at international level and discusses social protection characteristics, which should be taken into account to properly monitor changes of welfare systems in time. Section 3 describes available European official statistics on social benefits whereas Sect. 4 points out their differences, similarities and main gaps. Some descriptive analyses are provided on territorial disparities in the provision of social protection benefits in Europe and in Italy. Section 5 illustrates conclusions.

2 Social Protection in Official Statistics: Main Concepts

A universally accepted definition of the scope of social protection does not exist. Therefore, European official statisticians established a definition of social protection considering the needs of both producers and users of statistics. Such definition was proposed along with the development of Esspros (European system of integrated social protection statistics), a framework created in the late 1970's by Eurostat and European Union member states to allow international comparison among administrative national data on social protection [5]. According to Esspros, social protection is defined as encompassing "all interventions from public or private bodies directed to relieve households and individuals of the burden of a defined set of risks and needs, provided that there is neither a simultaneous reciprocal nor an individual arrangement involved" [5, p. 9]. Furthermore, Esspros traces the boundaries of social protection domain, making a list of the risks/needs covered, namely: sickness/health care, disability, old age, survivors, family/children, unemployment, housing and social exclusion. This means that only interventions falling within one of these areas can be labelled as social protection activity. For years Esspros definition has represented a yardstick in the field of social protection statistics.

Monitoring the evolution of welfare systems implies looking at changes of several aspects of social protection. The literature on the identification and clustering of welfare regimes has proposed a number of characteristics (dimensions) to be considered [4, 9, 10]. The type of actors delivering services, the composition of risks and needs covered, the quota of people covered and the rules for accessing benefits (e.g. whether they are means-tested or not) are among the most recurrent dimensions. However, other aspects seem particularly relevant to characterize a welfare system model, like the kind of economic transaction through which benefits are delivered: monetary transfers (benefits in cash), direct provisions of goods and services (benefits in kind) or tax breaks. Also, it is relevant to point out whether the funding of social protection is public (general government contributions from taxes) or private (social contributions from employers or from protected persons) and to identify the institutional sector that pays benefits.

The level of decentralization of welfare system represents a further key dimension since in some countries local governments are assigned managing and, sometimes, legislative competencies on social protection areas. To explore this issue, it is necessary to go beyond the examination of national legal systems, and consider the actual implementation of constitutional provisions by decentralized entities [2]. Indeed, effects of local social policies can be assessed only analysing sub-national data and indicators.

3 European Official Statistics on Social Protection Benefits

Four main data sources provide internationally comparable statistics on social protection expenditures and receipts: Esspros and National Accounts (NAs) by Eurostat, Socx (Social expenditure database) by Oecd and SSI (Social Security Inquiry) by ILO.

The objective of SSI is mainly to address the lack of (comparable) social protection statistics outside the Oecd world (see ILO website). In this respect, the inquiry adopts a systematic approach compatible with existing statistical frameworks for Oecd countries. The SSI database mainly incorporates Esspros and Socx statistics when dealing with social protections benefits of European countries. Esspros, in turn, is a NAs satellite account [6]. Therefore, the four data sources share significant common concepts and supply statistics on common thematic areas.

National statistical offices disseminate also micro data on the supply and use of social protection services. However, these statistics seldom permit sound comparisons among countries. Data are fragmented and often inconsistent, their availability and quality vary across countries. This depends on the fact that social protection programmes are carried out by a multitude of actors (public, private or non-profit institutions) at different level of government (central, local) and a systematic and shared data gathering methodology is lacking. In the following subsections, we examine the main data sources for the analysis and comparison of social benefits across European countries.

3.1 National Accounts (NAs)

NAs record social protection expenditure under the "Social benefits category". According to the European System of Accounts (ESA 2010, [6]) "Social benefits are transfers to households, in cash or in kind, intended to relieve them from the financial burden of a number of risk or needs, made through collectively organized schemes, or outside such schemes by General Government (GG) units and by Non-Profit Institutions Serving Households (NPISHs); they include payments from general government to producers which individually benefit households and which are made in the context of social risks or needs" (ESA 2010, 4.83). Risks or needs covered are the following: sickness, invalidity/disability, occupational accident or disease, old age, survivors, maternity, family, promotion of employment, unemployment, housing, education, general neediness (ESA 2010, 4.84). In particular, NAs distinguish two categories: "Social benefits other than social transfers in kind" and "Social transfers in kind". The first covers social transfers benefiting households (retirement pensions, unemployment allowances, family and maternity allowances, sick-leave per diem allowances) and these are recorded in the Secondary distribution of income account. The second includes expenditures of GG and NPISHs on the provisions of various individual services (healthcare, education etc.) but also the reimbursement of purchases of goods and services such as medical consultations and medicines, as well as housing allowances; these transactions are recorded in the Redistribution of income account. "Social benefits other than social transfers in kind" and "Social transfers in kind" contribute to determine the amount of two significant NAs balancing items, namely "Disposable income" and "Adjusted disposable income". Disposable income equals gross primary income minus current monetary transfers paid (e.g. taxes on income and wealth or social contributions), plus monetary transfers received (among which social benefits in cash). Disposable income shows how much can be consumed without running down assets or incurring liabilities. However, it is worth reminding that Disposable income is not appropriate for comparing people's material well-being across countries with different welfare systems. In fact, depending on the type of social protection systems, a relevant share of social protection is allocated through social transfers in kind. To face this problem, System of National Accounts 1993 (SNA 93, United Nations [11]) introduced the Adjusted disposable income, which is equal to Disposable income plus social transfers in kind.

NAs allow one to distinguish various typologies of social benefits. Particularly, "Social benefits other than social transfers in kind" break down into three categories, namely "Social insurance benefits in cash", "Other social insurance benefits" and "Social assistance benefits". The first category includes benefits paid out by social security plans organized by government and by private pension plans in return for prior contributions (ESA 20101, 4.103). The second refers to benefits payable by employers in the contest of other employment social insurance schemes (ESA 20101, 4.104). The third category identifies benefits provided without any previous contribution of beneficiaries (ESA 2010 4.105). Within "social transfers in kind", NAs distinguish individual goods and services provided directly to the beneficiaries by non-market producers (i.e. by GG or NPISHs) from individual goods and services provided directly by market produces on behalf of GG or NPISHs (ESA 2010, 4.109).

NAs classify the different kinds of social benefits according to the different types of paying sectors, i.e. by institutional sectors and by the sub-sectors thereof (e.g. Private/Public within Non-financial corporations or Central/Local governments within General Government).

Finally, NAs supply two pieces of information concerning decentralization and territorial data. The former concerns social benefits (namely "Social benefits other than social transfers in kind" and "Social transfers in kind provided by market producers") paid by Local government; the latter is the distribution of "Social benefits other than social transfers in kind" received by Households (HH), at Nuts 2 level.

3.2 The European System of Integrated Social Protection Statistics (Esspros)

Esspros records the accounting of social protection schemes distinguishing receipts (sources of financing) from expenditures (uses of financing). Two supplementary modules contain statistical information on pensions' beneficiaries and on net social benefits [8].

Receipts are analysed according to the nature of the payment (general government contributions, employers' social contributions and contributions paid by protected people) as well as to the kind of paying institutional sector. Institutional sectors correspond exactly to the NAs' ones. Expenditures include social benefits but also administration costs (costs charged to the scheme for its management and administration) and other miscellaneous costs.

Social benefits are further analysed by function and by type. The function identifies the primary purpose for which social protection is provided [5, 109], i.e. the risk or need covered. The type of benefit refers to the form in which the protection is provided. In particular, Esspros distinguishes between benefits paid in cash (further detailing between those paid at regular intervals or in the form of a lump sum) and benefits in kind [5, 110–115]. Finally, Social benefits are broken down between means-tested and non means-tested benefits. Means-tested benefits are conditional on the beneficiary's income and/or wealth falling below a specified level [5, 116, 117].

Esspros database does not contain details at the local level. However, it allows one to detect the part of social protection receipts coming from "State and local government".

3.3 Social Expenditure Database (Socx)

The Oecd database was developed in the 1990 s as a tool for monitoring trends in aggregate social expenditure and analysing changes in its composition. Oecd defines social expenditures as "the provision by public and private institutions of benefits to, and financial contributions targeted at, households and individuals in order to provide support during circumstances which adversely affect their welfare, provided that the provision of the benefits and financial contributions constitutes neither a direct payment for a particular good or service nor an individual contract or transfer" [1, p. 90].

Oecd distinguishes nine social different policy areas, which only approximately correspond to the risks/needs specified by NAs and Esspros: Old age, Survivors, Incapacity-related benefits, Health, family, Active labour market programmes, Unemployment, Housing, and Other social policy areas. Social expenditure comprises cash benefits, direct in-kind provision of goods and services (benefits in kind), and tax breaks with social purposes.

Social benefits are classified as public when general government (central, state or local governments, social security funds) controls the relevant financial flows. All social benefits not provided by general government are considered private. Private social benefits further break down into two sub-categories: mandatory private social expenditure, which includes social support stipulated by legislation but operated through the private sector (e.g. direct sickness benefits paid by employers); voluntary private social expenditure, which includes benefits accruing from privately operated programmes. Socx mainly incorporates Esspros data when dealing with European countries, with the exception of health and active labour market programmes data, which come from thematic Oecd databases.

Oecd publishes also an estimate of net social expenditure, which considers the effects of tax systems on social protection. Broadly speaking, this happens through direct taxation of benefit income, indirect taxation of consumption by benefit recipients and tax breaks for social purposes [1]. This effect can be considerable and vary across countries.

3.4 EU Statistics on Income and Living Conditions (Eusilc)

Eusilc is the reference source for comparative statistics on income distribution and social inclusion in the European Union. The reference population includes all private households and their current members residing in the territory of the countries at the time of data collection. Eusilc collects information on social benefits received by households and their members. Social benefits are defined (in accordance with Esspros) as "current transfers received during the income reference period by households intended to relieve them from the financial burden of a number of risk or needs, made through collectively organised schemes, or outside such schemes by government units and NPISHs". Areas covered are the following: unemployment benefits, old-age benefits, survivor benefits, sickness benefits, disability benefits, and education related allowances. Eusilc benefits do not cover benefits in kind, with the only exception of housing benefits. Finally, Eurostat does not deliver estimates of benefits by function based on Eusilc data, neither at the national nor at the sub-national level.

4 Combining Information from Different Data Sources

Differences among data sources are due to two main reasons. The first reason concerns the different boundaries of the social domain, i.e. the distinction between social spending and not-social spending. The second relates to the breakdown of social expenditure among functions. NAs and Esspros have undoubtedly a more homogeneous base and comparable data although some differences are present. A major difference is that NAs include education in the social domain while Esspros does not. Furthermore, social benefits within Esspros cover both current and capital transfers whereas the definition offered by NAs refers to current transfers only. Finally, NAs in kind transfers also cover transfers, which do not have a social protection objective. For example, they include expenditures on sport, cultural and recreational activities [5, p. 65]. Esspros statistics undoubtedly provide a richer analysis of social protection accounting than NAs. However, NAs have the advantage of directly linking changes in social protection expenditure to changes in households disposable income. The scope of Socx is arguably larger than that of NAs and Esspros. Socxs expenditure also considers lost revenues due to tax breaks with social purposes [1, p. 110]. Furthermore, differently from NAs and Esspros, Socx includes all spending on public health or labour market programmes like investment in medical facilities, preventive health initiatives or health education and training, not only expenditures that can be "allocated" to individuals or families (individual consumptions). Like Esspros and differently from NAs, Socx does not include education within the social domain (except pre-primary education, which is recorded under the family policy area). All Esspros social protection benefits are included in Socx with the only exception of some expenditures for disability, sickness and unemployment that are directly taken from thematic Oecd databases.

Table 1 attempts to give an overview of available information on social protection benefits at the macro and micro level (limited to Eusilc).

The first column makes a list of main social benefits categorizations, which can be detected in at least one of the analysed data sources. The following columns are headed to the data sources described in Sect. 3.

Table 1 Social benefits (paid or received) in EU official statistics

Social benefits paid or received	NAs	Esspros	Socx	Eusilc
By Needs/risk covered		paid	paid	received
Means-tested/ Non-means tested		paid		
Cash/in-kind	paid	paid	paid	
Public/private			paid	
Gross/net		paid	paid	
By paying sector	received	received		
By geographic area	received			received
By households' typologies				received

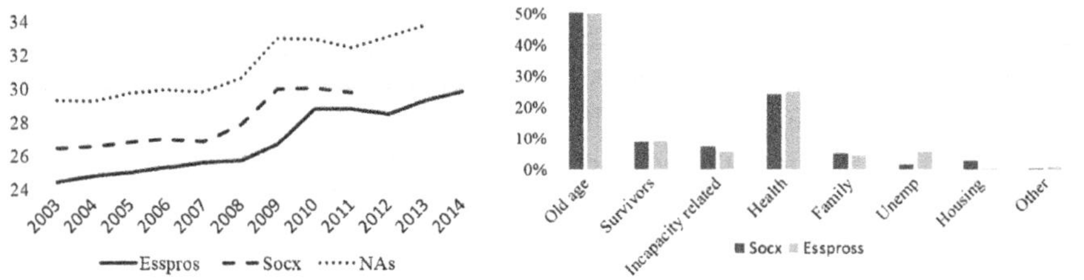

Fig. 1 Gross social protection expenditure (% GDP) and Social benefits by function (shares), left and right panel, respectively—Socx Esspros and NAs data—Italy 2011

Full cells indicate that the data source (column heading) contains the information described in the row heading. Furthermore, the cell specifies whether social benefits are recorded as received or paid transfers. We observe that, in some cases, more than one data source gives the same kind of information. In the following, we present some analysis on overlapped topics for Italy. Figure 1, left panel, shows gross social protection expenditure in percentage of GDP from 2003 to 2014, computed on the basis of NAs, Esspros or Socx data. NAs percentages are higher than Socx percentages, which in turn overcome Esspros values. Net social expenditure shows similar levels in Espross and Socx databases, about 25–26 points of GDP according to the latest estimates. Figure 1 (right panel) compares Esspros and Socx data on gross expenditures by type of risk/need covered.

Figure 2 compares cash benefits (left panel) and benefits in kind (right panel) both expressed as percentages of GDP, stemming from Esspros, Socx and NAs data sources. Percentages show some significant differences, however changes in time show similar patterns.

Our exploring of available official statistics on social benefits has shown a general shortage of data for measuring the level of decentralization of a welfare

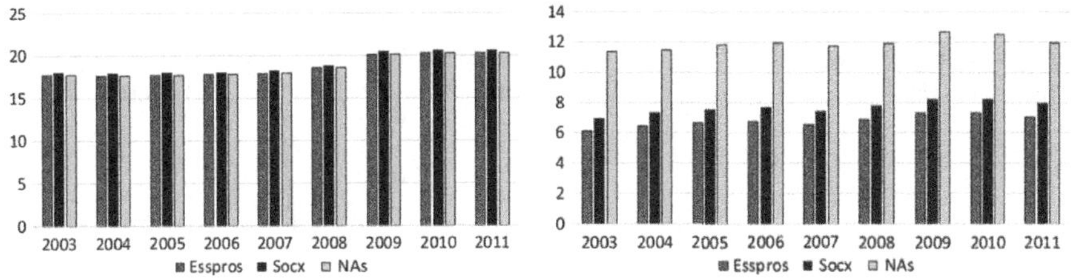

Fig. 2 Cash benefits, and Benefits in kind (% GDP), left and right panel, respectively—Socx Esspross and NAs data—Italy, years 2003–2011

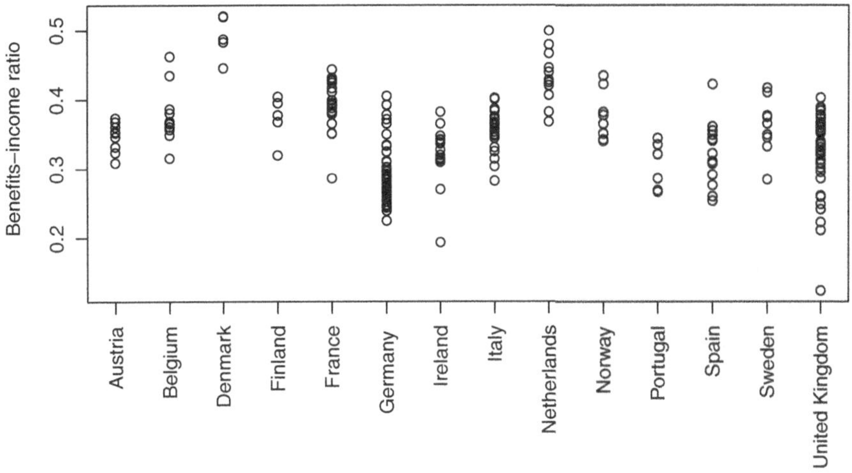

Fig. 3 Share of Households disposable income covered by cash benefits by regions (Nuts 2), in a selection of European countries—NAs data, year 2013

system. Figure 3 shows disparities among a selection of European countries, based on the only indicator available, namely the amount of cash benefits received by households, by region (see Sect. 3.1). In particular, the figure shows the share of disposable income covered by cash benefits in each region (dot) of each country.

National statistical offices disseminate statistics on local social protection activity. Istat, for example, regularly carries out a survey on municipal social protection expenditure, which gives an interesting insight into the heterogeneity of social expenditure distribution on the Italian territory. Figure 4 shows how municipalities of each region share differently their resources among family, old age and disability functions. From an in depth analysis, which is not the focus of the present paper and it is not reported here due to space constraints, we can state that disparities in social protection provision generally do not agree with the local population needs and the specific demographic dynamics (see, for example [12]).

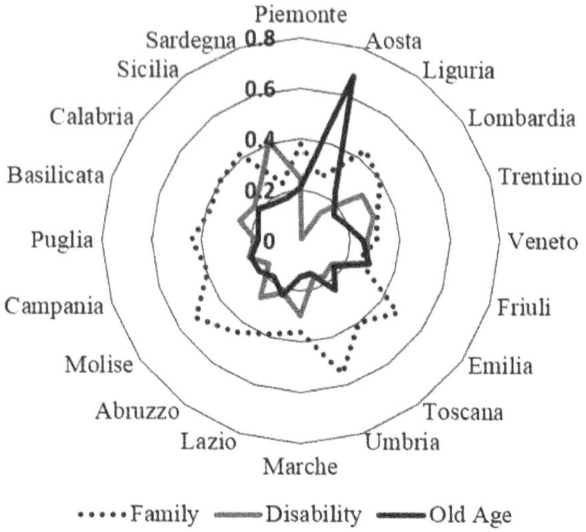

Fig. 4 Municipal social expenditure by function and region (Nuts2)—Italy, 2012

5 Concluding Remarks

The main objective of this study was to map comparable official statistics on social protection benefits at national and local levels for European countries. To this end, we explored Eurostat and Oecd main data sources on social protection accounting to provide a complete picture of data and indicators theoretically available, pointing out differences and matches among the different data sources. According to our analysis, European welfare states can be compared only at the country level and not at the local level, at least for what concerns the financial aspects. The distribution of cash benefits by region (Nuts 2) from National Accounts represents the only official data on social protection benefits delivered at the local level. Furthermore, currently disseminated Eusilc micro data allow one to estimate cash benefits only for large geographical areas (Nuts 1). National statistical offices supply more information on social protection at the local level. For example, Istat for example disseminates data and indicators on municipal social protection expenditures at the province level. The analysis shown in Sect. 4 shows significant territorial disparities. Further research is required to understand whether these disparities are the successful responsiveness of local governments to the specific needs of local populations, or, conversely, a sign of inequity and territorial injustice.

References

1. Adema, W., Fron, P., Ladaique, M.: Is the European Welfare State Really More Expensive?: Indicators on Social Spending, 1980-2012; and a Manual to the Oecd Social Expenditure Database (Socx), OECD Social, Employment and Migration Working Papers, No. 124, Oecd Publishing (2011). https://doi.org/10.1787/5kg2d2d4pbf0-en
2. Addis, P., Coli, A., Pacini, B.: Welfare state and local government: the impact of decentralization on well-being. In: 34th IARIW General Conference, Dresden, Germany, 21–27 Aug 2016
3. Bonnet, F., Tessier, L.: Mapping international social protection statistics and indicators. ESS Paper Series (SECSOC)-ESS 38-ILO (2013)
4. Esping-Andersen, G.: The Three Worlds of Welfare Capitalism. Polity Press, Cambridge (1990)
5. Eurostat: The European system of integrated social protection statistics. Manual (2011)
6. Eurostat: European system of accounts (ESA 2010). Manual (2013)
7. Eurostat: NUTS Nomenclature of Territorial Units for Statistics, by regional level (2013)
8. Eurostat: Statistics Explained (2016). http://epp.eurostat.ec.europa.eu/statisticsexplained/
9. Ferrera, M., Fargion, V., Jessoula, M.: Alle radici del Welfare all'italiana Origini e futuro di un modello sociale squilibrato. Saggi e ricerche, collana storica della banca dItalia. Marsilio (2012)
10. Titmuss, R.: Social policy. An Introduction, London, Allen & Unwin (1974)
11. United Nations: System of national accounts 1993. Manual (1993)
12. Venturi, S.: Disparities in local social protection systems from a demographic perspective. The population ageing. In: Policy, Welfare and Financial Resources: The Impact of Crisis on Territories. Pisa University Press (2017)

Financial Variables Analysis by Inequality Decomposition

Michele Costa

Abstract This paper illustrates the use of the methods related to inequality decomposition for the analysis of financial variables. By means of the overlapping component and of the inequality between it is possible to detect and to assess the main factors determining the cross section assets variability.

Keywords Inequality decomposition · Financial variables · Gini index

1 Introduction

Advantages of inequality measurement for the analysis of financial variables have been only partially exploited. Literature on the methods for inequality measurement explores many different aspects, it addresses the relation with the measurement of risk [1], but it does not exhaust all topics related to financial variables. Our first aim is to stress the usefulness of inequality measurement for the analysis of financial variables and, in order to support our point, we propose three main motivations.

First, we recall that the focus of inequality indexes is represented by the tails of the distribution, concerning, on the left side, the poverty measurement, and, on the right side, the inequality issues. Also in assets returns analysis, the tails play naturally a key role, with the debate on risk measurement focused on the left side of the distribution. Methods and tools for poverty analysis could, therefore, provide interesting insights on risk evaluation.

A second property of inequality indexes which could be helpful in assets returns analysis, refers to the degree of asymmetry of the distribution. In the inequality measurement, the presence of asymmetry precludes the use of the standard statistical methods, based on the normality assumption, and requires specific indicators, such as the Gini index. Also asset returns can exhibit relevant levels of asymmetry, particularly during financial crises, or speculative bubbles. When assets returns

M. Costa (✉)
Department of Economics, University of Bologna, Bologna, Italy
e-mail: michele.costa@unibo.it

C. Perna et al. (eds.), *Studies in Theoretical and Applied Statistics*,
Springer Proceedings in Mathematics & Statistics 227,
https://doi.org/10.1007/978-3-319-73906-9_27

distribution departs from the Gaussian, and shows an asymmetric form, inequality methods provide a natural solution for financial studies.

The third aspect of inequality measurement of interest for the analysis of financial variables refers to the decomposition by subgroups of total inequality. Inequality decomposition allows to detect the factors underlying the inequality, that is, the factors determining the global result. Following a similar pattern, by dividing asset returns into subgroups, inequality decomposition allows to detect which characteristics (such as risk level, firm size, business sector, etc.) are most relevant into assets cross section variability.

Focusing on the relevance of this last point, our purpose is to exploit the advantages of the methods for inequality decomposition into the analysis of financial variables. To the best of our knowledge, this work is the first attempt to analyze financial variables by means of inequality decomposition. We contribute to the literature on inequality measurement by adding the analysis of financial variables as a new dimension of interest. We also contribute to the analysis of financial variables by introducing inequality decomposition as a non parametric method for detecting the main determinants of assets cross section variability.

2 Inequality Decomposition and Financial Variables

Inequality analysis starts from a set of n observations on some variable of interest Y (such as income, consumption or wealth), which can be disaggregated into k subgroups. For a population disaggregated into k subgroups of size n_j, with $\sum_{j=0}^{k} n_j = n$, one of the most important inequality measure, the Gini index, can be expressed as

$$G = \frac{1}{n\bar{y}^2} \sum_{j=1}^{k} \sum_{h=1}^{k} \sum_{i=1}^{n_j} \sum_{r=1}^{n_h} |y_{ji} - y_{hr}| \tag{1}$$

where $\bar{y}$ is the arithmetic mean of Y in the overall population, y_{ji} is the value of Y in the i-th unit of the j-th subgroup and, accordingly, y_{hr} is the value of Y in the r-th unit of the h-th subgroup. For a detailed analysis of G see, e.g., [2, 3].

Moving from the traditional inequality analysis to the study of financial variables, the n observed units (which usually can be individuals, households, regions or also countries) become n listed companies, while the variable of interest is the mean asset return of these companies.

Within inequality analyses, equidistribution represents the situation where each unit possesses the same quantity of Y, while we have maximum concentration when $n - 1$ units have 0 and only one unit possesses the total amount of Y. In the case of financial variables, we get equidistribution if all companies have the same return. Conversely, maximum concentration corresponds to the case where $n - 1$ companies have zero return and only one company gets the total return.

One of the main goals of inequality analyses refers to inequality decomposition (see, e.g., [4–7])which allows to research the sources of inequality, that is the relevant determinants of poverty condition. In the framework of financial variables, we can use the same methods in order to identify the characteristics able to explain the assets cross section variability.

Among the many methods, which allow to decompose the Gini index, we use the decomposition proposed by Dagum [8], where the differences $|y_{ji} - y_{hr}|$ in (1) are assigned to G_w, the component of inequality within subgroups, when $j = h$, to G_b, the component of inequality between subgroups, when $j \neq h$, $\bar{y}_j \geq \bar{y}_h$, $y_{ji} \geq y_{hr}$, and to G_t, the component of overlapping, when $j \neq h$, $\bar{y}_j \geq \bar{y}_h$, $y_{ji} < y_{hr}$. Globally we have $G = G_w + G_b + G_t$.

G_w evaluates the contribution to total inequality related to the variability existing within the k subgroups, while $(G_b + G_t)$ captures the contribution to total inequality related to the differences among the k subgroups. An important feature of the Dagum's decomposition is to explicitly consider the role of overlapping units, thus allowing to separately evaluate the contribution to total inequality related to the overlapping between the subgroups. For a detailed description of the Dagum's decomposition see [8–10].

Overlapping subgroups play a relevant role in inequality decomposition (see, e.g., [11–14]). High levels of overlapping indicate that the factor used to obtain the subgroups only slightly contributes to total inequality. Conversely, low levels of overlapping suggest a stronger contribution. Overlapping can be extremely helpful also in portfolio analysis, since it allows to detect assets with particular characteristics. It could be the case, for instance, of assets with low standard deviation, but with high mean return. Furthermore, high levels of inequality between subgroups indicate strong differences between subgroups, thus stressing the relevance of the factor used to get the subgroups.

It is possible to analyse over time the contribution of G_b and G_t by means of the ratios G_b/G and G_t/G. Together with inequality between subgroups, overlapping analysis provides straightforward information on inequality structure.

3 A Case Study

In order to illustrate the usefulness of inequality methods for the analysis of financial variables, we propose a case study related to the 30 companies belonging to the Dow Jones index. By analyzing monthly data from 1990 to 2015, observations are represented by the mean returns calculated over a 12-months period. Figure 7 illustrates, for example, the frequency distribution of the mean returns of the 30 Dow Jones companies for the interval Jan–Dec 2015.

The observed data depicted in Fig. 1 represent an intermediate situation between the two extreme cases corresponding, respectively, to absence of variability or equidistribution (all companies have the same return, see Fig. 2) and to maximum

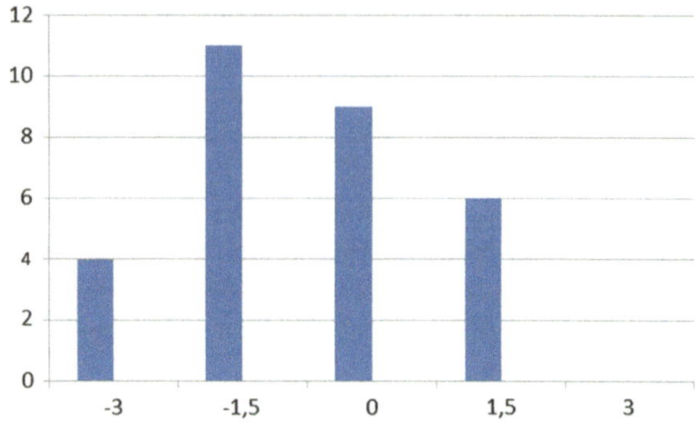

Fig. 1 Frequency distribution of the mean returns of the 30 Dow Jones companies, observed data Jan–Dec 2015

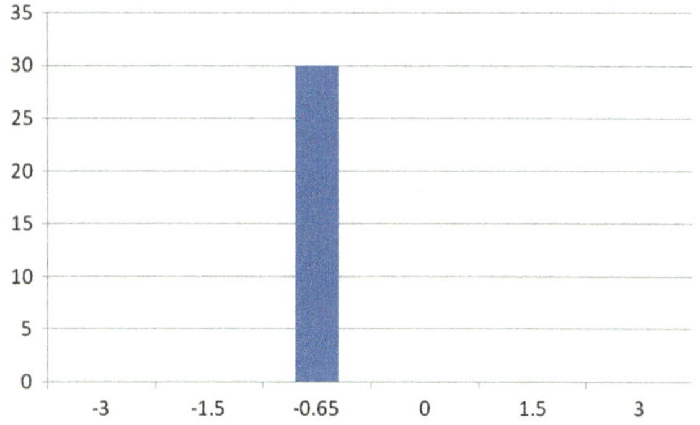

Fig. 2 Frequency distribution of the mean returns of the 30 Dow Jones companies, scenario of equidistribution

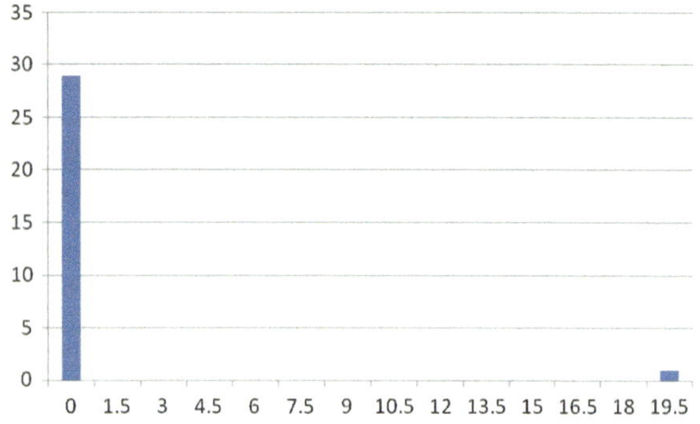

Fig. 3 Frequency distribution of the mean returns of the 30 Dow Jones companies, scenario of maximum concentration

variability or maximum concentration ($n - 1$ companies have 0 return and only one company gets the total return, see Fig. 3).

Our purpose is to analyze the variability existing within the observed data by disaggregating the n observations into k subgroups and by evaluating the contribution to total inequality given by the three components G_w, G_b and G_t.

In the following k = 2 subgroups are selected on the basis of low/high standard deviation calculated on the same 12-months period used to obtain the mean returns. Figure 4 depicts, for the interval Jan–Dec 2015, the frequency distribution of the 2

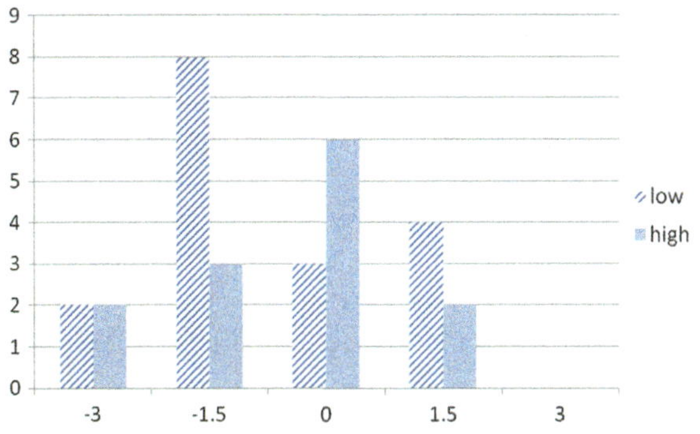

Fig. 4 Frequency distribution of the mean returns of the 30 Dow Jones companies classified into low/high volatility subgroups, observed data Jan–Dec 2015

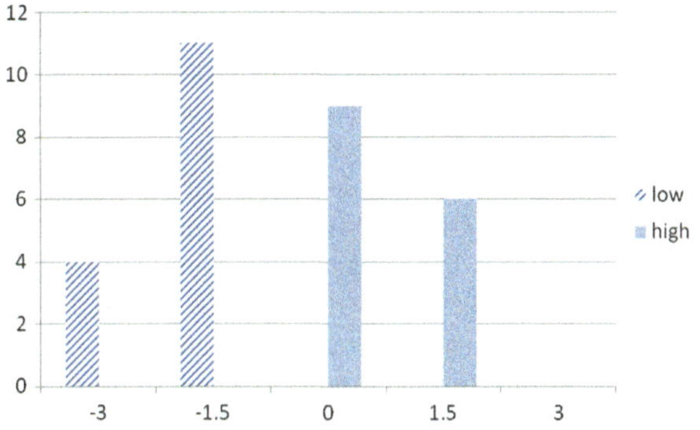

Fig. 5 Frequency distribution of the mean returns of the 30 Dow Jones companies, classified into low/high volatility subgroups, scenario of no overlapping

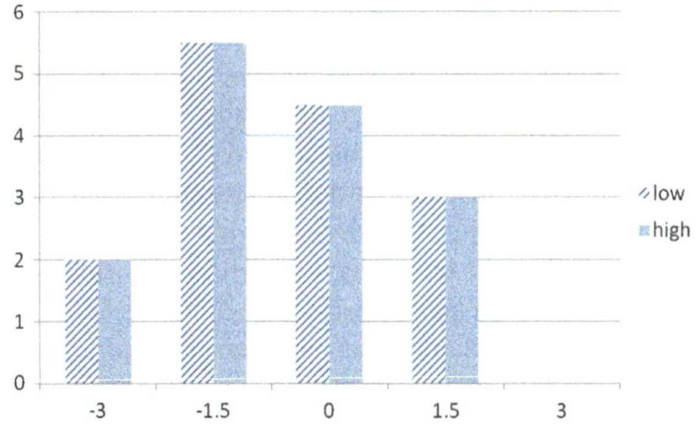

Fig. 6 Frequency distribution of the mean returns of the 30 Dow Jones companies, classified into low/high volatility subgroups, scenario of perfect overlapping

subgroups constituted by the 30 mean returns classified according to low/high level of standard deviation.

Observed data illustrated in Fig. 4 still represent an intermediate situation between two extreme cases: real data show a partial degree of overlapping between the two subgroups, that is neither absence of overlapping nor complete overlapping.

The case of non-overlapping subgroups is illustrated in Fig. 5, where the two subgroups are completely separate. Conversely, Fig. 6 depicts the opposite case, where the two subgroups are perfectly overlapping.

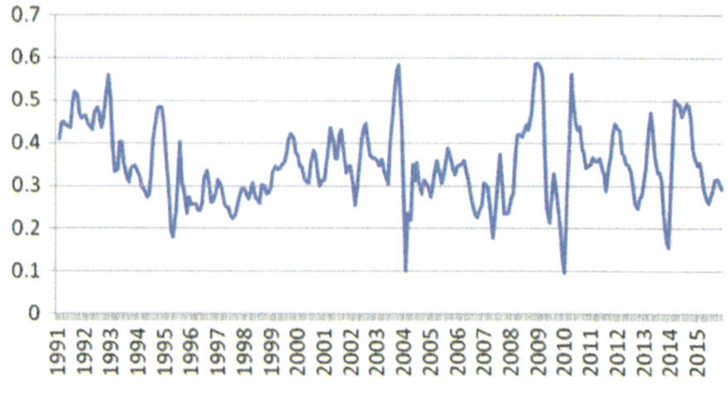

Fig. 7 Results of the decomposition of the Gini index: ratio G_b/G from 1990 to 2015 for the 30 Dow Jones companies mean returns divided into log/high volatility subgroups

By referring to standard portfolio theory, it is immediate to associate the case of no overlapping illustrated in Fig. 5 to the mean-variance analysis, where low (high) standard deviation implies low (high) mean return. Absence of overlapping suggests that the variability of the mean returns is strongly affected by the standard deviation. On the contrary, complete overlapping depicted in Fig. 6 indicates the absence of a relation between standard deviation and mean return.

The Gini index decomposition calculated on the 30 mean returns of our example, (Jan–Dec 2015), disaggregated into the 2 subgroups low/high standard deviation, yields $G_w/G = 0.50$, $G_b/G = 0.29$, $G_t/G = 0.21$, thus indicating an intermediate degree of importance for overlapping and inequality between components.

By repeating the same procedure for all 12-months periods within the interval 1990–2015, we obtain a series of 301 Gini indexes which can be decomposed in order to analyze the variability existing among the mean returns.

Our main results are represented by the ratios G_b/G and G_t/G, illustrated in Fig. 7 and in Fig. 8, respectively. High levels of G_b/G indicate that the risk (measured by standard deviation) strongly affects the mean returns, while high levels of G_t/G suggest a low influence of standard deviation on mean return. The analysis of Figs. 7 and 8 provides striaghtforward and useful information, since it allows to assess how the importance of risk evolves over time. For example, by comparing 2014 to 2015, it is possible to detect a decrease of G_b/G, which implies a weaker effect of standard deviation on mean return, and an increase of G_t/G, which also suggests a lower influence of standard deviation on mean return. Therefore, from 2014 to 2015 our results indicate an important change into the relation between standard deviation and mean return.

As for the standard deviation, the ratios G_b/G and G_t/G allow to evaluate the role of any other possible determinant of the assets cross section variability. By adding to the analysis more inequality factors and by comparing the related results, inequality decomposition is able to provide a relevant and complete information set on financial variables behaviour.

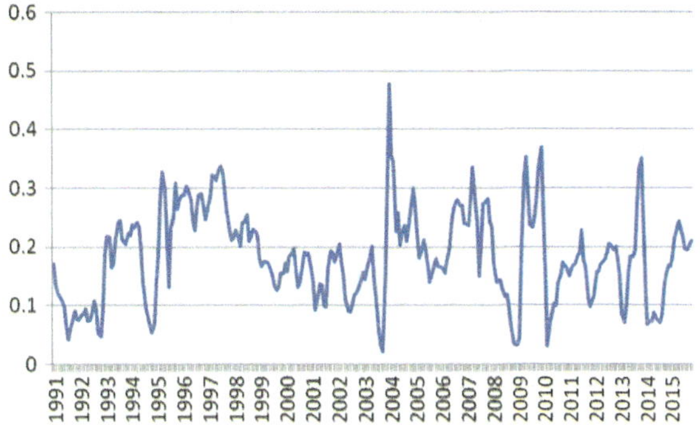

Fig. 8 Results of the decomposition of the Gini index: ratio G_t/G from 1990 to 2015 for the 30 Dow Jones companies mean returns divided into log/high volatility subgroups

4 Conclusion

The paper introduces inequality decomposition as a method for the analysis of financial variables.

The inequality between subgroups, together with the overlapping component, provides powerful insights on the main determinants of total inequality. In the framework of financial variables, they allow to detect the main factors explaining assets cross section variability. By means of the ratios G_b/G and G_t/G, it is also possible to assess the importance of the relevant factors and to evaluate their evolution over time.

To the extent of our knowledge, inequality decomposition is still unapplied in financial variables analysis and we believe that the method developed in the paper can be successfully applied to different research areas in the field of financial markets.

References

1. Breitmeyer, C., Hakenes, H., Pfingsten, A.: From poverty measurement to the measurement of downside risk. Math. Soc. Sci. **47**, 327–348 (2004)
2. Dagum, C.: Gini ratio. In: The New Palgrave Dictionary of Economics. Mac Millian Press, London (1987)
3. Giorgi, G.M.: Gini's scientific work: an evergreen. Metron **63**, 299–315 (2005)
4. Dagum, C., Zenga, M.: Income and Wealth Distribution. Inequality and Poverty. Springer, Berlin (1990)
5. Frosini, B.V.: Approximation and decomposition of Gini, Pietra-Ricci and Theil inequality measures. Empiric. Econom. **43**, 175–197 (2012)
6. Giorgi, G.M.: The Gini inequality index decomposition. An evolutionary study. In Deutsch, J., Silber, J.: The measurement of individual well-being and group inequalities. Routledge, London (2011)
7. Yitzhaki, S., Lerman, R.: Income stratification and income inequality. Rev. Income Wealth **37**, 313–329 (1991)
8. Dagum, C.: A new decomposition of the Gini income inequality ratio. Empiric. Econom. **22**, 515–531 (1997)

9. Costa, M.: Transvariation and inequality between subpopulations in the Dagum's Gini index decomposition. Metron **67**, 134–120 (2009)
10. Costa, M.: Gini index decomposition for the case of two subgroups. Commun. Stat. Simulat. Computat. **37**, 631–644 (2008)
11. Costa, M.: Overlapping component and inequality decomposition: a simulation study for the Gini index. Metron **74**, 193–205 (2016)
12. Deutsch, J., Silber, J.: Ginis Transvariazione and the measurement of distance between distributions. Empiric. Econom. **22**, 547–554 (1997)
13. Frick, J., Goebel, J., Schechtman, E., Wagner, G., Yitzhaki, S.: Using analysis of Gini for detecting whether two subsamples represent the same universe: the German socio-economic panel study experience. Sociol. Methods Res. **34**, 427–468 (2006)
14. Yitzhaki, S.: Economic distance and overlapping distributions. J. Econometr. **61**, 147–159 (1994)

Part VIII
Sustainable Development: Theory, Measures and Applications

A Novel Perspective in the Analysis of Sustainability, Inclusion and Smartness of Growth Through Europe 2020 Indicators

Elena Grimaccia and Tommaso Rondinella

Abstract The comparison of different territorial areas according to multiple factors raises the challenge of representing synthetically the complexity of multidimensional phenomena, such as the targets of growth promoted by the Europe 2020 strategy. We considered data for 10 years in order to highlight the evolution of the similarities and dissimilarities of the 28 European countries in the whole period. The analysis is centred on a technique which combines cluster analysis with the use of a composite indicator, thus permitting to identify Countries both according to their structural characteristics and to their overall performance. We also look at convergence processes among countries and link our results to GDP growth to better qualify countries patterns of development.

Keywords Complexity · Composite indicators · Cluster analysis Europe 2020 indicators

1 Introduction

In 2010, with the expiration of the Lisbon Strategy for Growth and Jobs, the governments of the European Union launched the Europe 2020 initiative [1], a set of guidelines of action in order to fix mid-term political economy targets which extended economic growth to a few aspects which should characterize the European model: namely smartness, sustainability and inclusion. The broad policy objectives have been declined in quantitative and measurable targets, making governments "accountable" to the citizens and to the Commission. The targets are measured by eight headline indicators, concerning research and innovation, employment, education, poverty, climate change, renewable sources and energy efficiency.

E. Grimaccia (✉) · T. Rondinella
ISTAT, Rome, Italy
e-mail: elgrimac@istat.it

T. Rondinella
e-mail: rondinella@istat.it

C. Perna et al. (eds.), *Studies in Theoretical and Applied Statistics*,
Springer Proceedings in Mathematics & Statistics 227,
https://doi.org/10.1007/978-3-319-73906-9_28

These guidelines aim at increasing European competitiveness maintaining a social market economy and improving resource efficiency.

The comparison of different territorial areas according to multiple factors raises the challenge of representing synthetically the complexity of multidimensional phenomena. In particular, when comparing different areas (Countries in our specific case), the attention can be either on the performance shown by Countries, usually evaluated through rankings, or on the structural models which characterize different compositions of the various elements.

We present here a technique which combines cluster analysis with the use of a composite indicator, thus permitting to identify Countries both according to their structural characteristics and to their overall performance. The combined use of Cluster analysis and composite indicators allows us to represent the level of progress combined with the different patterns of growth which characterise different groups of countries. We considered data for ten years, before and after the economic crisis, in order to highlight the evolution of the similarities and dissimilarities of the 28 European countries in the considered period.

Finally, the building of composite indicators for the monitoring of sustainability, smartness, and inclusion, as well as for the overall initiative, describes the growth model of European countries according to different characterizations, looking at the association of Europe 2020 indicators with GDP growth paths. In this way, we are able to see which countries chose a smart, inclusive and/or sustainable growth, and which have not.

2 Conceptual Framework: The Europe 2020 Strategy

In June 2010, the European Council adopted the "Europe 2020 Strategy", put forward by the European Commission [1]. It defines three priorities for growth in the European countries:

- Smart growth: developing an economy based on knowledge and innovation;
- Inclusive growth: fostering a high employment economy, delivering social and territorial cohesion.
- Sustainable growth: promoting a more resources efficient, greener and more competitive economy.

The broad policy objectives (smartness, inclusiveness and sustainability) were declined in measurable, and thus well-defined, numerical targets, making governments "accountable" to the citizens and to the Commission.

Targets have been fixed for each country and indicator and the monitoring system is foreseen within the European Semester of economic policy coordination. This implies that Government must not only monitor the dynamic of the indicators but have also to make explicit the measures for achieving the goals [1].

"Smart growth" is measured by the proportion of Early leavers from education and training, and should be under 10%; the quote of persons (25–34 years old) with Tertiary educational attainment should be at least 40%, and the percentage of Gross domestic expenditure on research and development (R&D) on GDP should reach 3% in the EU as a whole.

"Inclusive growth" is measured by the Employment rate, with a target of 75% of the population aged 20–64 should be employed and the number of People at risk of poverty or social exclusion should decrease by 20 million from 2008 in total in EU.

"Sustainable growth" is measured by the Greenhouse gas emissions should be reduced by 20% compared to 1990, the Share of renewable energy sources in final energy consumption, to be increased to 20%, and the Energy efficiency should be improved by 20%.

Europe 2020 is the most advanced institutionalization of a multidimensional set of progress indicators. It is perhaps not a complete set of indicators for the evaluation of the progress of societies-literature shows that the choice of domains and indicators is usually much broader-, but it is the formal realization of European institutions that GDP cannot be the only target indicator, and that growth measurement has to be equipped with quality measures that take into account equity and sustainability. Not by chance Europe 2020 indicators have been included in the European "GDP and beyond" initiative [3] and present the methodological and accessibility characteristics, such as simplicity and reliability, needed to monitoring societal well-being. The initiative has also the merit of being based on a reduced number of goals and indicators fostering its transparency and intelligibility. The indicators come out from a political debate within the European Council, thus granting a certain political legitimacy, targets have been fixed for each country and indicator and the monitoring system is foreseen within the European Semester of economic policy coordination. This implies that Government must not only monitor the dynamic of the indicators but have also to make explicit the measures for achieving the goals.

Nevertheless, the strategy cannot represent a whole success story, since does not seem to really represent the priority for Europe and European Government and citizens. Maybe due to the tough economic and financial crisis, the attention has actually shifted towards the monitoring of the "macroeconomic imbalances" [2] which is the framework that member countries are actually asked to follow. There is no broad agreement over the relevance of the indicators for Europe 2020. An agenda for "smart, inclusive and sustainable growth" is not necessarily one for progress, well-being or sustainable development, but certainly it is a "beyond GDP one". Maybe its full legitimacy is invalidated by the lack of a public debate to support its definition, the strategy it is not sufficiently influential [8, 9].

3 Convergence Across Countries: Empirical Evidence

All “smartness” indicators show an overall improvement over time and a continuous process of convergence measured by the Coefficient of variation (CV).[1] In the considered period, EU Member States performance became more similar, for what concern education and research.

Regarding Inclusion indicators, the number of People at risk of poverty or social exclusion (RPSE) converges over time even if the indicator strongly worsen through the crisis. The convergence process is due to the rapid fall of poverty in Bulgaria, Romania, Latvia, Poland and Lithuania between 2005 and 2008.

Employment rate is a much more uniform indicator across countries, showing a very low value of the CV, yet after the burst of the crisis it rapidly increased from 7.9 to 9.4% because of the drop in rates in Croatia, Bulgaria, Greece and Spain, and the recovery in Sweden where it reaches nearly 80%.

Environmental indicators show different trends: since 2008 GHG decreased by 9% in the UE28, following very diverse patterns between +2.9% in Malta and −18.6 in Denmark. Yet the CV did not change significantly.

The share of renewable sources on energy consumption increased in all countries. The overall effect is a progressive and intense convergence among European countries.

Energy consumption per capita has been reduced through the years from 3.5 to 3.1 TOEpc since 2004 in the EU28. Yet some convergence appears especially in the first years of the period thanks to the reduction in Luxemburg, Sweden and Belgium.

Overall, the 28 Countries show a general process of convergence over time in the indicators, in particular those regarding education and research, risk of poverty and use of renewable sources (Table 1).

4 Cluster Analysis

We have applied a cluster analysis to the 28 EU Member States, with the aim to identify the groups of Countries with similar performance with regard to the eight Europe 2020 Indicators. We used a Hierarchical clustering. As measure of dissimilarity, we chose an Euclidean distance and the linkage criterion used was the decrease in variance for the cluster being merged (Ward’s criterion), because it gave the best results and it is the most common in the literature.[2]

---

[1]The ratio of the standard deviation to the mean, a standardized measure of dispersion.

[2]In Ward’s minimum-variance method, the distance between two clusters is the ANOVA sum of squares between the two clusters added up over all the variables. At each generation, the within-cluster sum of squares is minimized over all partitions obtainable by merging two clusters from the previous generation Ward’s method joins clusters to maximize the likelihood at each level of the hierarchy [6].

Table 1 Convergence of Europe 2020 indicators (coefficient of variation and percentage change) —2004–2013

	2004	2005	2006	2007	2008	2009	2010	2011	2012	2013	2013–2004
School leave	0,63	0,59	0,59	0,59	0,57	0,57	0,54	0,51	0,48	0,48	−0,16
Tertiary enrolment	0,37	0,37	0,37	0,35	0,33	0,32	0,30	0,28	0,28	0,25	−0,12
R&D	0,67	0,66	0,64	0,62	0,63	0,63	0,60	0,58	0,56	0,56	−0,11
Employment	0,09	0,08	0,08	0,08	0,08	0,08	0,08	0,09	0,09	0,10	0,01
Risk of poverty	0,45	0,44	0,42	0,41	0,33	0,34	0,34	0,33	0,33	0,32	−0,13
Greenhouse gasses	0,32	0,33	0,32	0,32	0,32	0,35	0,32	0,32	0,33	0,33	0,02
Renewables	0,89	0,88	0,87	0,83	0,81	0,77	0,73	0,72	0,69	0,66	−0,23
Energy efficiency	0,48	0,47	0,46	0,45	0,44	0,43	0,46	0,45	0,44	0,45	−0,04

Source Authors' elaboration on Eurostat data

The analysis of the eight indicators in each group permitted to identify their most relevant characteristics. In general, more homogenous groups form at an earlier stage and are better characterized, while those with higher internal dissimilarity present less defined communalities (Fig. 1).

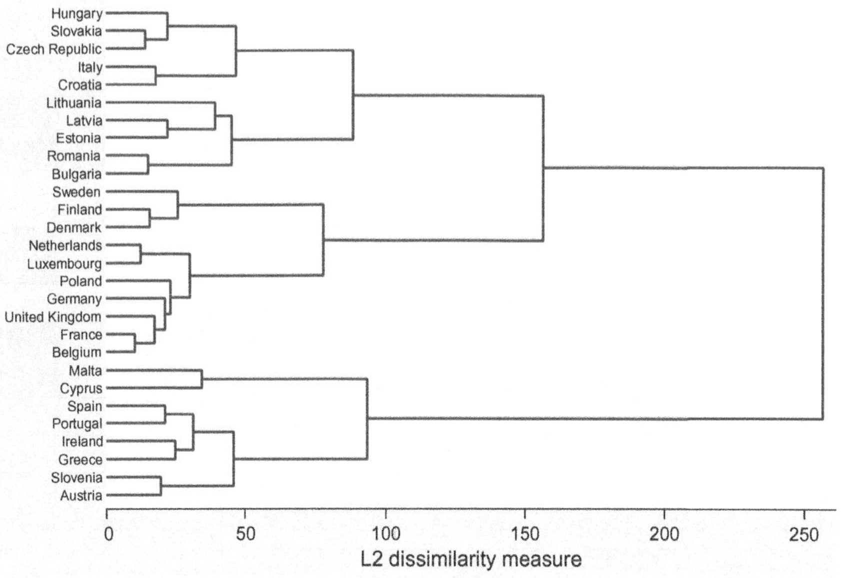

Fig. 1 Dendrogram for 2012 cluster analysis on Europe 2020 indicators (Ward method, Euclidean distance)

The first group to form has been the "Nordic Countries" (Sweden, Finland and Denmark), that present the highest level of education in the population and research expenditure (smartness), the highest level of employment and lowest of poverty (inclusion), and the best performance in terms of environmental sustainability. The group of Central Europe Countries is characterised by a high level of Research and Development expenditure, good performance of the education system, a good level of employment and inclusion, but a moderate level of emission.

Mediterranean Countries present a more problematic situation, with lower levels of education in the population, low level of expenditure in R&D, quite low employment and problems of inclusion. Austria and Slovenia present a peculiar pattern, being part of the first group most of the time, but with worse performance in tertiary education and GHG emission, and also less school drop-outs.

The Eastern Europe Group presents good levels of performance in the smartness sub group of indicators but a lower level of inclusion and, mostly, they can be identified by a lower attention on energy saving.

The group of Baltic Republic together with Romania and Bulgaria presents a low percentage of R&D expenditure, a high level of poverty but a good performance on sustainability, with low emissions and high percentage of the use of renewable sources.

We repeated the analysis for each year between 2004 and 2013 and have identified seven groups of Countries every year, fixing a value of 55 for the dissimilarity measure (Fig. 2).

	2004	2005	2006	2007	2008	2009	2010	2011	2012	2013	General	
Finland	1	1	1	1	1	1	1	1	1	1	Finland	1
Sweden	1	1	1	1	1	1	1	1	1	1	Sweden	1
Denmark	2	1	1	1	1	1	1	1	1	1	Denmark	1
Austria	1	4	4	1	4	1	1	1	6	1	Austria	4
Slovenia	1	4	4	1	4	1	1	1	6	6	Slovenia	4
France	2	2	2	2	2	2	2	2	2	4	France	2
Belgium	2	2	2	2	2	2	2	2	2	2	Belgium	2
Netherlands	2	2	2	2	2	2	2	2	2	2	Netherlands	2
Luxembourg	2	2	2	2	2	2	2	2	2	2	Luxembourg	2
UK	2	2	3	2	2	2	3	3	2	3	UK	4
Germany	2	2	3	3	3	2	3	3	2	2	Germany	4
Poland	3	3	3	3	3	2	3	2	2	4	Poland	4
Czech Rep	3	3	3	3	3	3	3	3	3	3	Czech Rep	3
Hungary	3	3	3	3	3	3	3	3	3	3	Hungary	3
Slovakia	3	3	3	3	3	3	3	3	3	3	Slovakia	3
Croatia	3	3	4	3	3	2	3	6	3	4	Croatia	4
Italy	4	4	4	3	4	2	6	6	3	4	Italy	4
Greece	4	4	4	4	4	6	6	6	6	6	Greece	6
Ireland	4	4	4	4	4	6	2	2	6	6	Ireland	6
Portugal	6	6	6	4	6	6	6	6	6	6	Portugal	6
Spain	6	6	6	6	6	6	6	6	6	6	Spain	6
Cyprus	6	6	6	6	6	7	7	7	7	7	Cyprus	7
Malta	6	6	6	6	6	7	7	7	7	7	Malta	7
Estonia	5	5	5	5	5	5	5	5	5	5	Estonia	5
Latvia	5	5	5	5	5	5	5	5	5	5	Latvia	5
Lithuania	5	5	5	5	5	5	5	5	5	5,5	Lithuania	5
Bulgaria	5,5	5,5	5,5	5,5	5,5	5,5	5,5	5,5	5,5	5,5	Bulgaria	5,5
Romania	5,5	5,5	5,5	5,5	5,5	5,5	5,5	5,5	5,5	5,5	Romania	5,5

Note:	1	2	3	4	6	7	5	5,5
	Nordic	France Benelux	Central Europe	Other	Mediterranean	MT+CY	Baltic	RO+BG

Fig. 2 Groups of cluster analysis on Europe 2020 indicators (cut L2 = 55), years 2004–2013

Through the years, the analysis presents quite stable results in the identification of groups. In the "Nordic Countries" group, Finland, Sweden and Denmark are present for the whole period. We can say that the best performing countries present also a stability in the Europe 2020 indicators trough the economic crisis. The "Central Europe" group is formed by Czech Republic, Slovakia and Hungary for the whole decade, with Poland before the crisis. The Mediterranean Countries group presents a different composition before and after the crisis, because Cyprus and Malta constitute a different group after 2008, but they still present quite high level of similarity, and similar values of the Europe 2020 indicators. The last two clearly identified clusters are those of Baltic countries and Romania and Bulgaria. They are all characterized by very low GHG emissions and large use of renewable sources. Moreover (with the exception of Estonia), they also present low R&D expenditure and high levels of poverty. Within this picture, Romania and Bulgaria can be grouped separately for the much worse results of smartness and inclusiveness indicators.

5 Composite Indicators

In order to assess the performance of Member States according to Europe 2020's indicators we opted for the building of a composite indicator for each one of the three dimensions (inclusiveness, smartness and sustainability) as well as for the overall initiative.

The methodology we adopted is inspired by the work done by Mazziotta and Pareto within the BES initiative [4] for building a composite index able to maintain comparability both in space and time. Most standardization techniques [7] allow comparisons between units (countries) but since the standardization reference is changing every year, the comparison over time shows only relative changes with respect to other units. The Mazziotta-Pareto method, a min-max based on an initial reference value, is able to monitor also time trends as a typical based index.

Differently to the MPI [5] we opted for perfect substitutability among indicators for using a more transparent measure and therefore we have not applied their correction based on the internal variability in one country's indicators.[3] Using this

[3]The standardization is expressed as:

$$r_{ij} = \frac{(x_{ij} - Min_x)}{(Max_x - Min_x)} 60 + 70 \qquad (1)$$

where x_{ij} is the value of the variable x for the unit j at time i; Min_x and Max_x are the "goalposts" for the variable j. If the variable has negative polarity (in our case school leave, poverty, greenhouse gasses and energy consumption), then the complement of (1) with respect to 200 is computed.

normalization, every variable will assume value equals to 100 for EU28 in 2004, and a value relative to such reference for each country and each year allowing both space and time comparison.

We built composite indexes of Smartness, Inclusiveness and Sustainability, as simple average of normalised values of the relevant indicators. A composite index for Europe 2020 is the simple average of the three. Should we have built it as the average of all the eight considered indicators, inclusiveness, represented only by two indicators, would be underweighted.

Trends of the eight indicators in recent years show, after the economic crisis in 2008, the pattern toward the target of reducing the Risk of poverty and social exclusion, and Employment show a reversal.

The results of composite indicators, as known, are characterized by a number of pros and cons [7]. They provide for an effective quantitative synthesis of multiple variables allowing for comparison and ranking among units, yet for their interpretation it is always necessary to go back to initial data, especially when the index comprises very different phenomena. Therefore, a similar value can describe very different situations, typically for mid-range units.

Looking at the results from the application of the composite indexes over time, in the EU28 the smartness index shows a continuous increase of +8% in the whole considered decade thanks to the positive contribution of all the indicators. Similarly, also the sustainability index increases by 7%, but it is determined by the rising use of renewable sources, and a substantial stability of GHG emissions and energy consumption. Inclusiveness index, instead, rose only by 2% with a dynamic of fall and recovery through the crisis and a number of countries experiencing a very weak or no recovery for both employment and poverty measures.

The overall Europe 2020 index rises by 6% during the decade, and looking at country results we find Sweden, Finland and Denmark on the top and Malta at the bottom. Poland and Bulgaria show an over 10% increase, while Croatia, Greece and Ireland increased the index only by 3%.

$$\begin{cases} Min_x = Ref_x - \Delta \\ Max_x = Ref_x + \Delta \end{cases} \quad \Delta\Delta(Sup_{A_x} - Inf_{A_x})/2$$

The goalposts are calculated as an interval $\pm\Delta$ around a reference value for the variable x, where Δ is the midrange of the dataset A_x comprising all units and times taken into account for the variable x [5].

In our case, the reference value considered is the one for European Union 28 in year 2004, while A_x comprises the values for the 28 countries in years 2004–2013.

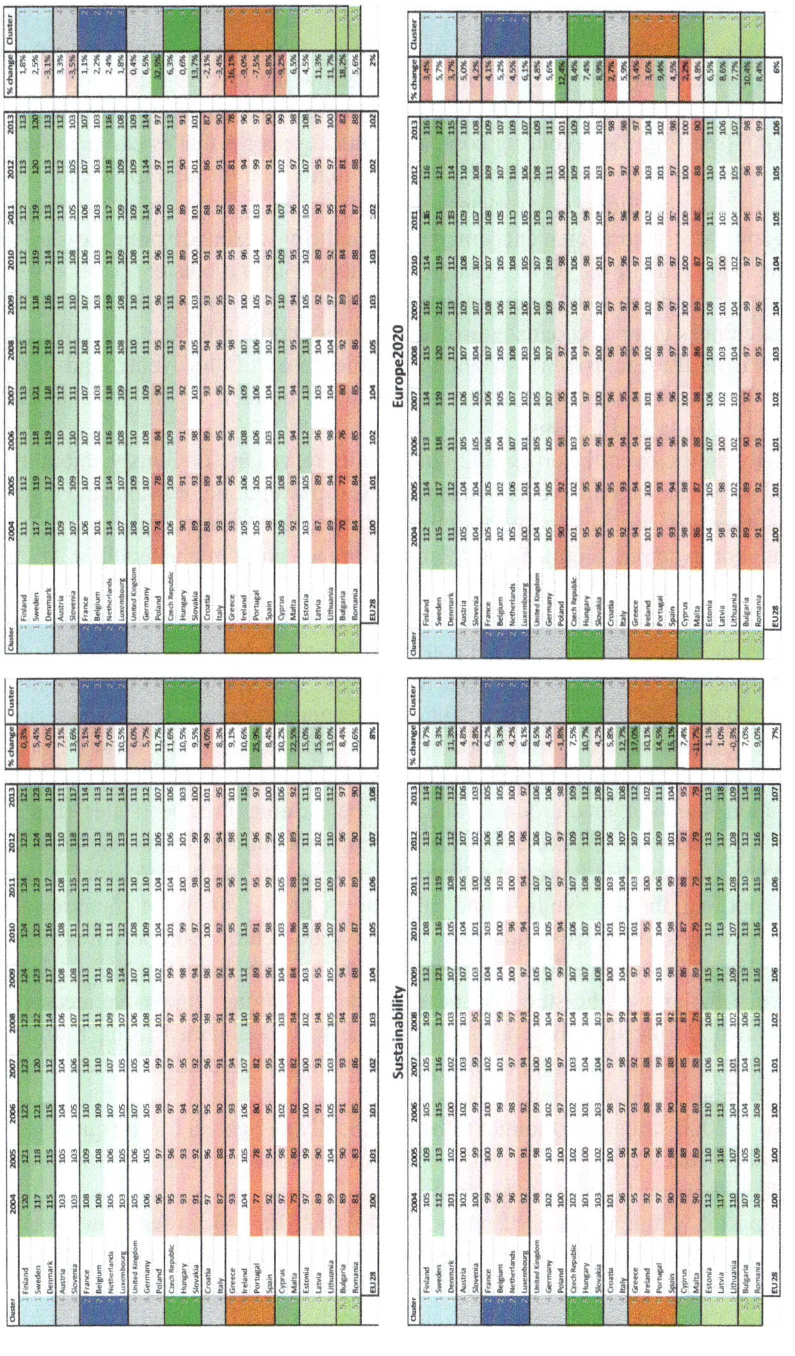

Fig. 3 Joint visualization of clusters and composite indexes scores (EU28 = 100, scores and percentage values; colour: from dark red to dark green on the scores' table and on the % change column)

6 Joint Analysis of Clusters and Composite Indicator

We propose a form of visualization of synthetic data which merges both cluster analysis and a composite measure in order to show the level and the evolution of countries while still focussing on their structural similarities as represented by the clustering (see Fig. 3). Colouring composite indexes scores makes easier to see the relative position of each country as well as it progression through the years. It also helps to show the different characterization of the clusters and how the indexes' scores and the structural elements emerged from the clustering do not necessarily go together with sometimes very relevant differences within a cluster.

However, performances within clusters tend to be quite homogenous, but relevant differences can be recognized in the Smartness domain for the cases of Ireland or Latvia, in the Inclusiveness domain for the cases of the Netherlands or the Central European cluster, and in the Sustainability domain for the Mediterranean cluster and Malta and Cyprus.

Finally, in the overall Europe 2020 index, differences are obviously smoothened but still some relevant differences emerge for Czech Republic or Malta and Cyprus. On the other way, we can find very similar values for countries which have different profiles. For example, Germany and Estonia, France and Czech Republic, UK and Slovenia, or Poland and Cyprus.

As an analysis of robustness of our method, we have also reversed the procedure, applying the cluster analysis to the countries, based on the synthetic indicators of sustainability, inclusion and smartness and then to the synthetic measure of all the Europe 2020 indicators in 2012. The results appear quite robust. The cluster analysis, carried out using the three synthetic indicators, shows (at 40% of dissimilarity) very close results with the cluster with the eight original indicators. The application shows a first group of the best performing member states (Finland, Sweden and Denmark), a second of Central Europe countries including the United Kingdom, then the Mediterranean countries and so on. Also the cluster analysis carried out with the one synthetic indicator shows similar groups of countries, such as the best performing Scandinavian countries, the central Europe countries (including UK), the worst performing member states (such as Italy, Greece and Spain).

Even if the aim of the study is not to estimate a unique conclusive measure of development in the EU, the consistency of the results obtained using a different order in the procedure can support the validity of the representation of the European union member states with our method.

Europe 2020 strategy is thought to be specifically an agenda for the qualification of Member States economic growth. It is thus relevant to associate growth paths with their "beyond GDP" characterization.

During the whole considered period 2004–2013 (Fig. 4), faster growing countries (mainly new member states) showed lower levels of R&D and tertiary education, higher poverty levels and a smaller effort for energy saving. Yet, they better performed in increasing employment and reducing poverty and worse performing on the three green indicators. On the other hand, as predictable, the Mediterranean

Cluster		2004	2005	2006	2007	2008	2009	2010	2011	2012	2013	2004-2013	Cluster
1	Finland	4,1	2,9	4,4	5,3	0,3	-8,5	3,4	2,8	-1,0	-1,4	1,2	1
1	Sweden	4,2	3,2	4,3	3,3	-0,6	-5,0	6,6	2,9	0,9	1,6	2,1	1
1	Denmark	2,3	2,4	3,4	1,6	-0,8	-5,7	1,4	1,1	-0,4	0,4	0,6	1
4	Austria	2,6	2,4	3,7	3,7	1,4	-3,8	1,8	2,8	0,9	0,3	1,6	4
4	Slovenia	4,4	4,0	5,8	7,0	3,4	-7,9	1,3	0,7	-2,5	-1,1	1,5	4
2	France	2,5	1,8	2,5	2,3	-0,1	-3,1	1,7	2,0	0,0	0,2	1,0	2
2	Belgium	3,3	1,8	2,7	2,9	1,0	-2,8	2,3	1,8	-0,1	0,2	1,3	2
2	Netherlands	2,2	2,0	3,4	3,9	1,8	-3,7	1,5	0,9	-1,2	-0,8	1,0	2
2	Luxembourg	4,4	5,3	4,9	6,6	-0,7	-5,6	3,1	1,9	-0,2	2,1	2,2	2
4	United Kingdom	3,2	3,2	2,8	3,4	-0,8	-5,2	1,7	1,1	0,3	1,7	1,1	4
4	Germany	1,2	0,7	3,7	3,3	1,1	-5,1	4,0	3,3	0,7	0,4	1,3	4
4	Poland	5,3	3,6	6,2	6,8	5,1	1,6	3,9	4,5	2,0	1,6	4,1	4
3	Czech Republic	4,7	6,8	7,0	5,7	3,1	-4,5	2,5	1,8	-1,0	-0,9	2,5	3
3	Hungary	4,8	4,0	3,9	0,1	0,9	-6,8	1,1	1,6	-1,7	1,1	0,9	3
3	Slovakia	5,1	6,7	8,3	10,5	5,8	-4,9	4,4	3,0	1,8	0,9	4,2	3
4	Croatia	4,1	4,3	4,9	5,1	2,1	-6,9	-2,3	-0,2	-2,2	-0,9	0,8	4
4	Italy	1,7	0,9	2,2	1,7	-1,2	-5,5	1,7	0,4	-2,4	-1,9	- 0,2	4
6	Greece	4,4	2,3	5,5	3,5	-0,2	-3,1	-4,9	-7,1	-7,0	-3,9	- 1,1	6
6	Ireland	4,2	6,1	5,5	5,0	-2,2	-6,4	-1,1	2,2	0,2	-0,3	1,3	6
6	Portugal	1,6	0,8	1,4	2,4	0,0	-2,9	1,9	-1,3	-3,2	-1,4	- 0,1	6
6	Spain	3,3	3,6	4,1	3,5	0,9	-3,8	-0,2	0,1	-1,6	-1,2	0,9	6
7	Cyprus	4,2	3,9	4,1	5,1	3,6	-1,9	1,3	0,4	-2,4	-5,4	1,3	7
7	Malta	-0,3	3,6	2,6	4,1	3,9	-2,8	4,2	1,5	0,8	2,6	2,0	7
5	Estonia	6,3	8,9	10,1	7,5	-4,2	-14,1	2,6	9,6	3,9	0,8	3,1	5
5	Latvia	8,8	10,1	11,0	10,0	-2,8	-17,7	-1,3	5,3	5,2	4,1	3,3	5
5	Lithuania	7,4	7,8	7,8	9,8	2,9	-14,8	1,6	6,0	3,7	3,3	3,6	5
5,5	Bulgaria	6,7	6,4	6,5	6,4	6,2	-5,5	0,4	1,8	0,6	0,9	3,0	5,5
5,5	Romania	8,5	4,2	7,9	6,3	7,3	-6,6	-1,1	2,3	0,6	3,5	3,3	5,5
	UE28	2,6	2,2	3,4	3,2	0,4	-4,5	2	1,6	-0,4	0,1	3,29	

Fig. 4 Joint visualization of clusters and GDP growth rates (percentage values; colour: from dark red to dark green on the whole table and on the 2004–2013 column)

Cluster		GDP	SMART	INCLUSIVE	SUSTAINABLE	E2020
1	Finland	1,2	0%	2%	9%	3,4%
1	Sweden	2,1	5%	3%	9%	5,7%
1	Denmark	0,6	4%	-3%	11%	3,7%
4	Austria	1,6	7%	3%	5%	5,0%
4	Slovenia	1,5	14%	-4%	3%	4,2%
2	France	1,0	5%	1%	6%	4,1%
2	Belgium	1,3	4%	2%	9%	5,2%
2	Netherlands	1,0	7%	2%	4%	4,5%
2	Luxembourg	2,2	10%	2%	6%	6,1%
4	United Kingdom	1,1	6%	0%	9%	4,8%
4	Germany	1,3	6%	7%	5%	5,6%
4	Poland	4,1	12%	33%	-2%	12,4%
3	Czech Republic	2,5	12%	6%	8%	8,4%
3	Hungary	0,9	10%	1%	11%	7,4%
3	Slovakia	4,2	9%	14%	4%	8,9%
4	Croatia	0,8	4%	-2%	6%	2,7%
4	Italy	- 0,2	8%	-3%	13%	5,9%
6	Greece	- 1,1	9%	-16%	17%	3,4%
6	Ireland	1,3	11%	-9%	10%	3,6%
6	Portugal	- 0,1	26%	-7%	14%	9,4%
6	Spain	0,9	8%	-9%	15%	4,5%
7	Cyprus	1,3	10%	-9%	7%	2,2%
7	Malta	2,0	23%	6%	-12%	4,8%
5	Estonia	3,1	15%	5%	1%	6,5%
5	Latvia	3,3	16%	11%	1%	8,6%
5	Lithuania	3,6	13%	12%	0%	7,7%
5,5	Bulgaria	3,0	8%	18%	7%	10,4%
5,5	Romania	3,3	11%	6%	9%	8,4%
		3,3	8%	2%	7%	5,9%

Fig. 5 Joint visualization of clusters, GDP growth rates and Europe 2020 composite indexes growth rates (percentage values; colour: from dark red to dark green on each column)

countries, which most suffered of the economic crisis, worsened their inclusiveness index (mainly employment levels) and increased the sustainability one (partly for the GHG reduction). We could also find among the slowest growing countries some of the best performing ones in qualitative terms, such as the Nordic countries, which show a very low increase in the Europe 2020 index.

In general, looking at the dynamics of the phenomena, we find that GDP growth is very much associated with increases in Europe 2020 index, with fast growing countries showing also the major improvements (Fig. 5).

7 Conclusions

Europe 2020 indicators allow us to analyse the different characteristics and patterns followed by the European countries, pointing out differences and similarities. They worked well in describing the effects of the crisis and they seem robust and sensible to changes. Member States show different results with respect to the eight indicators allowing for the identification of similarities characterizing different "patterns of development", the most evident being "Nordic countries", "France and Benelux", "Mediterranean countries", "Malta and Cyprus", "Central Europe", "Baltic countries" and "Bulgaria and Romania". Yet, results and trends for some indicators may deeply differ even within the groups. Such differences have sometimes strong influence when one tries to synthesize the complexity through a composite indicator. Composite indicators, as a matter of fact, hide the qualitative composition of original indicators leading to similar results for territories which may have a very different composition of a close final result. On the other hand, they allow the synthetic representation of the overall performance as well as of the trends countries have been following over time. The joint representation of clustering and composite indexes allows the association between overall performance and different models of development for a more comprehensive synthetic understanding of complex multidimensional phenomena.

Europe 2020 indicators worked well in describing the effects of the economic crisis and they seem robust and sensible to changes. It takes into account the economic issue but it goes "beyond GDP", however very often GDP changes reflect in Europe 2020 changes.

The European Union shows an overall convergence with progresses in sustainability and smartness indicators but with difficulties in the inclusion indicators (especially in Mediterranean countries). Faster growing countries (mainly new member states) showed lower levels of R&D and tertiary education, higher poverty levels and a smaller effort for energy saving. Yet, they better performed in increasing employment and reducing poverty and worse performing on the three green indicators. We can say about European GDP growth that where there have been more growth, it hasn't been green.

In conclusion, EU28 as a whole shows a progress in sustainability and smartness indicators but it failed in the inclusion process. Nevertheless, Europe 2020 indicators allow us to analyse the different characteristics and patterns followed by the European countries, pointing out differences and similarities in their growth models.

References

1. European Commission: Europe 2020 A Strategy for Smart, Sustainable and Inclusive Growth, COM (2010) 2020 Final. Brussels (2010)
2. European Union: Regulation (EU) No 1176/2011 of the European Parliament and of the Council of 16 November 2011 on the Prevention and Correction of Macroeconomic Imbalances (2011)
3. European Commission: Communication from the commission to the council and the European parliament GDP and beyond: measuring progress in a changing world, (COM/2009/0433 final) (2009)
4. Istat: BES 2015. Il benessere equo e sostenibile in Italia, Istat, Rome (2015)
5. Mazziotta, M., Pareto, A.: Non-compensatory composite indices for measuring changes over time: a comparative study. In: CESS 2014 Conference of European Statistics Stakeholders (2014)
6. Milligan, G.W.: An examination of the effect of six types of error perturbation on fifteen clustering algorithms. Psychometrika **45**(3), 325–342 (1980)
7. OECD, JRC.: Handbook on constructing composite indicators: methodology and user guide. OECD Publishing, Paris (2008)
8. Rondinella, T.: Policy use of progress indicators. In: Rondinella, T., Signore, M., Fazio, D., Calza, M.G., Righi, A. (eds.) Map on Policy Use of Progress Indicators, e-Frame—European Framework for Measuring Progress, EU FP7 "e-Frame" Project, Deliverable 11.1, pp. 8–17 (2014). www.eframeproject.eu
9. Rondinella, T., Segre, E., Zola, D.: Participative processes for measuring progress: deliberation, consultation and the role of civil society. Soc. Indic. Res. 1–24 (2015). https://doi.org/10.1007/s11205-015-1207-z

The Italian Population Behaviours Toward Environmental Sustainability: A Study from Istat Surveys

Isabella Mingo, Valentina Talucci and Paola Ungaro

Abstract The interest of the Scientific Community in environmental protection issues aiming at guaranteeing future sustainability is constantly increasing. For this reason, the environmental social sciences, in recent years, are treating the interrelationships between population and environment. In Italy, an informative contribution comes from both Istat traditional Multipurpose Survey "Aspects of daily life" and Istat more recent Survey "Energy Consumption of Households". The aims of the paper are: (1) to propose synthetic measures of PECB and PEEB (respectively, Pro-Environmental Curtailment and Efficiency Behaviours), by a methodology which facilitates the replicability of the analysis over time, on the basis of the upcoming Istat surveys on these topics; (2) to analyse the determinants of pro-environmental behaviours of Italian citizens, by deepening the direction of the relationships with socio-demographic and other relevant characteristics, using a multivariate data analysis approach.

Keywords Environmental sustainability · Istat surveys · Multivariate analysis

1 On Environmental Sustainability and Behaviors

In 1987, the World Commission on Environment and Development provided a definition that represents the starting point of this paper: "*Humanity has the ability to make development sustainable to ensure that it meets the needs of the present without compromising the ability of future generations to meet their own needs*" [1].

I. Mingo (✉)
Sapienza University of Rome, Rome, Italy
e-mail: isabella.mingo@uniroma1.it

V. Talucci (✉) · P. Ungaro (✉)
ISTAT, Rome, Italy
e-mail: talucci@istat.it

P. Ungaro
e-mail: ungaro@istat.it

C. Perna et al. (eds.), *Studies in Theoretical and Applied Statistics*,
Springer Proceedings in Mathematics & Statistics 227,
https://doi.org/10.1007/978-3-319-73906-9_29

The core of this definition is the concept of "intergenerationality", which means to enforce economic, social and environmental factors that are fair and sustainable for present generations and especially for future ones. The environmental sustainability is therefore one of the three crucial dimensions of sustainability and is meant as the capability to preserve over time ecosystems' fundamental characteristics and to safeguard natural assets. Sustainability is a dynamic process, which involves the whole society, the public opinion, the economic and productive systems, media and policy makers.

In particular, an increasing number of researches highlight the relevance of population behaviours that minimize ecological harm and support natural resource defence [2–4]. These behaviours can refer to private dimension of daily life, as well as social ones, with regard to activism and social participation such as civic engagement, social environmentalism, financial support to environmental actions.

This paper is focused on private dimension, considering as a pillar the "conservation lifestyle behaviours" [5] which includes many daily activities such as recycling, riding a bicycle, etc.

With regard to different impact on environmental sustainability, these actions can be categorized in two different types: "curtailment behaviours" and "efficiency behaviours" [6, 7]. The first are those activities which reduce the harm to the environment (e.g. turning off the light when it is not needed, separating waste, etc.); the second category includes the adoption of more efficient technologies in order to contain the waste of resources (replacing equipment, installing house thermal insulation, etc.).

In this study we adopt a broad definition of pro-environmental behaviours considering both these types of activities.

The approach is "exploratory" and the main goals are the following:

- Identifying "how many" and "which" pro-environmental behaviours are adopted by Italian population;
- Deepening the regional differences and socio-demographic determinants of these activities.

2 Statistical Source and Data: Istat Sample Surveys

Istat population[1] sample surveys allow to obtain key data useful to study and measure the subject of this paper. In particular, the traditional Multipurpose Survey "Aspects of daily life" and the most recent survey "Energy Consumption of Households" provide much information in order to study the relationship between population and environment, with a particular focus on energy issues.

[1]In this paper the expression "Italian population" is referred to residents in Italy (Italian or foreigners).

The Multipurpose Survey[2] "Aspects of daily life" (ADL), carried out for the first time in 1993 and then annually, is the main source for social indicators related to the daily life of Italian population. In the editions of 1998 and 2012, an ad hoc module investigating behaviours, opinions, concerns towards environment was introduced. The successive editions present some of those questions and new ones in order to monitor the population contribution to environmental sustainability, mostly with regard to "curtailment behaviours".

In 2014, 71.4% of the population pays attention not to waste electricity, while 67% not to waste water. About 80 people on 100 avoid littering on the street, while smaller is the share of individuals who avoid adopting noisy driving behaviour which strongly contribute to noise pollution (about 44%). Moreover, about 20% of citizens avoid the consumption of disposable products, and the same percentage prefers means of transport sustainable and alternative to private motor vehicles. Behaviours of food consumption are finalized to environmental protection, but also to personal health care: 35.5% of citizens read food labels before buying, 18% buy food products at km 0 and only 9% biological products [8].

The European Commission is devoting an increasing attention to the issue of energy sustainability[3] [9]. The survey[4] on energy consumption of households gives for the first time in Italy timely statistical information on the energy behaviour of resident households. The survey has been carried out in 2013 and the aim is to estimate energy consumption for intended use and source of energy (gas, electricity, fuel oil, biomass, etc.); moreover, it allows to get information about "efficiency behaviours". The interviewed households say they had carried out, over the past five years, investment in money to reduce energy costs. We consider "efficiency behaviours" the followed items: (a) the replacement of one or more household appliances (10.6%); (b) the replacement of the boiler with one more efficient (9.3%); (c) the replacement of doors, windows and frames (8.4%); (d) residual % of e.g. the application of thermostats and thermostatic valves, the installation of renewable energy system or replace one or more individual units with more efficient equipment; [10].

[2]The survey technique is PAPI (*Paper and Pencil Interview*), it involves about 24 thousand families and 50 thousand individuals, the questionnaires are two: one family that deepens socio-economic dimensions of the family, housing, accessibility services, etc.; and one individual that concerns topics such as opinions and attitudes, health, media, leisure and security, etc. The thematic can be fixed, rotary and modular according to the frequency with which they are subjected to respondents.

[3]One of the three objectives of Europe 20-20-20, aimed at reducing climate change, is focused on the reduction of 20% of primary energy demand by increasing energy efficiency (Directive 2012/27UE).

[4]The survey technique is CATI, on a sample of 20,000 households representative at regional level, with stratification of approximately 8,000 Italian municipalities for demographic size and altitude zone.

3 Method and Results

Operationally, the analysis of pro-environmental behaviours and its relationship with the objective and subjective conditions of the Italian population takes into account information concerning individuals, households and housing conditions.

Statistical data analysis requires two steps: (1) the construction of synthetic measures of PECB and PEEB (respectively, Pro-Environmental Curtailment and Pro-Environmental Efficiency Behaviours), by a methodology which facilitates the replicability of the analysis over time, on the basis of the upcoming Istat surveys on these topics; (2) identifying the relations between these indices and other variables by regression model.

3.1 The Curtailment Behaviours of Italian Population: Synthetic Measure and Determinants

The first synthetic measure (PECB, Pro-Environmental Curtailment Behaviours) is computed using ADL micro-data. We considered seven variables y_j expressed by a Likert scale type (1 = Usually—4 = Never): (1) buying biological food and products; (2) buying local food and products; (3) not throwing papers down the street; (4) being careful not to wastewater; (5) being careful not to waste electricity; (6) not adopting noisy driving; (7) choosing sustainable and alternative transports.

The PECB index for the i-th individual (aged 14 and over) is obtained counting the answers "1 = Usually" to each of the variables considered,[5] in order to stress a stronger pro-environmental disposition of the population.

$$PECB_i = \sum_{j=1}^{7} y_{ij} \quad \forall y_j = 1$$

PECB index varies from 0 to 7 (mean = 3.08; σ = 1.57; median 3.00; Q_1 = 2; Q_3 = 4) and it aims to quantify an individual's capability to adopt "environmentally sustainable" conduct. Only the 0.9% of Italians usually adopts all the seven pro-environmental curtailment behaviors while for the 7.1% these behaviors are never adopted.

At regional level, PECB index varies from 2.78 in Campania to 3.41 in Bolzano and it presents different variability of individual pro-environmental behaviours within regions, as shown by regional coefficients of variation which range from

[5]The synthesis of elementary indicators was also carried out through Categorical Principal Components Analysis (CATPCA), obtaining a synthetic measure correlated with additive PECB (r = 0.69 and rho = 0.75). We decided to consider the simplest index to allow its replicability over the time.

43.71 in Val d'Aosta to 60.16 in Calabria. To adopt PECB index as the dependent variable in regression model, it has been transformed into a binary variable ($PECB_D$: low 55.4%/high 44.6%) using the mean as a criterion.[6] The predictors variables are: socio-demographic characteristics (age, gender, educational level, employment condition); opinion on environmental matters (relevance of nature reserve); subjective wellbeing (satisfaction for the health and for the environment); region and municipality's dimension. Another predictor is an additive index of subjective environmental concerns (EC), computing how many of the following concerns (max 5) have been expressed by each individual: greenhouse effect and ozone hole; extinction of plant and animal species, climate change, production and disposal of waste, noise, air, soil and water pollution, hydrogeological risk, natural man-made disasters, destruction of forests, electromagnetic pollution, degradation of the landscape, depletion of natural resources).[7] The EC index varies from 0 to 5 (mean = 3.94; median = 5; $Q_1 = 3$; $Q_3 = 5$).

A logistic regression model has been applied [11] because the dependent variable $PECB_D$ is dichotomous, so that many of the key assumptions of linear regression (linearity, normality, homoscedasticity) are not requested.

The final model, in which non-significant variables were eliminated,[8] correctly classified 61.5% of individuals. Table 1 reports the estimated parameters (odds ratios and significance) that identify those aspects that influence $PECB_D$ assuming other conditions remaining unchanged. The analysis of the odds ratios highlighted that EC index has significant and positive effect on the high level of $PECB_D$: the propensity to pro-environmental behaviour increases of about 20% for each concern comparing to those who did not indicate any. This result is concordant with the effect of the opinion about the relevance of natural reserve, which enhances the chance of high level of $PCBE_D$ of about 65%, revealing a greater overall sensitivity to environmental issues for the subjects with higher scores of $PCBE_D$ and with the satisfaction for own health (about 38%). Even socio-demographic characteristics have significant positive effect on the high level of $PEBC_D$, as shown by odds ratios which grow with the age and with the level of education. People 65–74 years aged increase the propensity for a high level of $PEBC_D$ by 3.65 times compared to the young people aged 14–24; similarly, high educated individuals tend to perform a higher number of curtailment behaviours, since their propensity is 1.78 times compared with those who have a low level of education. Besides the model shows that women have a greater propensity (about 20%) to engage in conduct conform to environmental sustainability than men. With regard to characteristics of the territorial context, the different domain of residence has a significant effect: compared

[6]Despite PECB index is an ordinal variable, ordinal logistic regression has not been applied because the proportional odds assumption did not hold for our data as evidenced by the test of parallel lines (Chi-Square statistic sig. < 0.000). So a less restrictive model has been applied.

[7]Descriptive statistics of individual indicators are not presented here [8]. Each individual could indicate up 5 concerns.

[8]The following variables were eliminated: employment condition, satisfaction for the environmental (air, water, noise) state of the residence area.

Table 1 Logistic model: factors affecting the pro-environmental curtailment behaviours ($PECB_D$)—odds ratios

Predictors	Values	Sign.	Exp (B)
Gender (*reference = Male*)	Female	0.000	1.198
Age (*reference = 14–24*)	over 74	0.000	2.604
	65–74	0.000	3.651
	55–64	0.000	3.291
	45–54	0.000	2.640
	35–44	0.000	2.310
	25–34	0.000	1.576
Educational level (*reference = Low*)	High	0.000	1.772
	Upper	0.000	1.380
	Middle	0.014	1.091
	Low		
Domain (*reference = Municipalities with less than 10.001 inhabitants)*	Centres and peripheries of metropolitan areas	0.000	0.809
	Municipalities with more than 50.000 inhabitants	0.078	0.944
	Municipalities with 10.001–50.000 inhabitants	0.002	0.917
Region (*reference = Sicilia*)	Piemonte	0.000	1.425
	Valle d'Aosta	0.015	1.224
	Lombardia	0.075	1.105
	Bolzano	0.000	1.829
	Trento	0.000	1.303
	Veneto	0.000	1.408
	Friuli	0.004	1.223
	Liguria	0.033	1.161
	Emilia	0.000	1.688
	Toscana	0.002	1.213
	Umbria	0.000	1.479
	Marche	0.191	1.094
	Lazio	0.000	1.263
	Abruzzo	0.095	1.118
	Molise	0.000	1.355
	Campania	0.138	0.917
	Puglia	0.000	1.331
	Basilicata	0.667	1.032
	Calabria	0.002	1.222
	Sardegna	0.000	1.889
Health satisfaction (*reference = No*)	Yes	0.000	1.382
Natural reserve (*reference = No*)	Yes	0.000	1.653

The reference categories is $PECB_D$ = low level
Final Model: −2LL = 31017,94, Chi square = 3274,824, df = 36, Sig = 0,000

with individuals resident in small municipalities, people who live in a metropolitan area have a chance less than 19% to have a high level of $PECB_D$ measure. Finally, considering the region of residence, odds ratios show some significant positive effects. Compared to the region of reference (Sicilia), the highest effects regard Sardegna and Bolzano: living in these regions increases the chance of taking pro-environmental behaviours over than 80%. The effect of living in Emilia (about +70%) and Piemonte, Umbria e Veneto (about +40%) are high and significant. Conversely, for other regions, no significant difference emerges (Lombardia, Marche, Abruzzo, Campania, Basilicata).

3.2 The Efficiency Behaviors of Italian Population: Synthetic Measure and Determinants

The second synthetic measure (PEEB, Pro-Environmental Efficiency Behaviours), computed using data from energy consumption of households survey, quantifies the propensity of Italian households to implement energy efficiency behaviours. It has been computed considering households who have carried out in the last 5 years at least an economic investment to reduce expenses for heating and/or electricity, by selecting only some of the possible types of investment.

The p_j variables, related to energy saving investments, are: (1) central heating system or single heating devices replacement; (2) installation of renewable energy systems for heating or electricity production; (3) application of devices for heating and electricity consumption regulation; (4) thermal insulation; (5) replacement of doors and windows; (6) replacement of appliances with more efficient ones.

$$\mathrm{PEEB_i} = 1 \ \text{if} \ \sum_{j=1}^{9} p_{ij} \geq 1$$

$PEEB_i$ varies between 0 to 1, where 0 = absence of any investment and 1 = presence of at least one investment. This synthetic measure allows both a simple and immediate interpretation of the phenomenon and comparisons over time.

At the national level, 27.7% of households adopt energy efficiency behaviours. The regions showing the greatest propensity of households to make investments for energy savings are in Northern Italy. In particular, Liguria (about 34%), Trento (33.5%), Valle d'Aosta (32%), Lombardia (about 32%), Piemonte (32%) and Emilia Romagna (31%). The efficiency and energy saving choices are still contained in the Center and the South areas, with minimum levels in Sicily (17.3%), Abruzzo (23%) and Puglia (23%).

A logistic regression model has been applied, where the dependent variable is PEEB and variables considered as predictors are: socio-demographic characteristics (type of household, level of education and employment condition); the territorial characteristics (region, municipality's dimension, climate zone); the characteristics of the house (type of dwelling, year of construction, housing tenure (property/rent/free), house size, change of electricity and/or gas supplier).

The results of Table 2 confirm that in the North-east (Lombardia and the Autonomous Province of Trento) the propensity to invest in energy efficiency is higher than in southern Italy (assuming Sicily as reference). On the contrary, in the South the risk to belong to a household "not energetically sustainable" increases. Considering socio-demographic characteristics, comparing with single person aged 65 or more, large (with children) and young families show a higher propensity to make sustainable energy choices, probably in connection with a more dynamic approach and more relevant needs of economic saving due to the family size. As already observed for $PECB_D$, educational level represents an incentive, as respondents with higher education are more likely to make investments for energy savings. Finally, an interesting indication comes from households that, in the last 5 years, have changed electricity and/or gas supplier, who perform more frequently interventions for home energy efficiency, as a sign of more dynamism and propensity to change. Regarding the characteristics of the house, home ownership increases the propensity to make investments in energy efficiency comparing to living in rented houses, because the owner can also benefit from tax incentives as a

Table 2 Logistic model: factors affecting the pro-environmental efficiency behaviours (PEEB)—odds ratios

Predictors	Values	Sign.	Exp(B)
Region *(reference = Sicilia)*	Piemonte	0.090	1.763
	Valle d'Aosta	0.144	1.833
	Lombardia	0.000	2.012
	Bolzano	0.009	1.916
	Trento	0.000	2.167
	Veneto	0.001	1.907
	Friuli Venezia Giulia	0.284	1.711
	Liguria	0.001	1.864
	Emilia Romagna	0.108	1.742
	Toscana	2.131	1.423
	Umbria	1.697	1.494
	Marche	6.178	1.302
	Lazio	1.082	1.468
	Abruzzo	2.906	1.155
	Molise	6.710	1.294
	Campania	6.749	1.251
	Puglia	4.422	1.318
	Basilicata	2.817	1.425
	Calabria	4.590	1.310
	Sardegna	2.210	1.411

(continued)

Table 2 (continued)

Predictors	Values	Sign.	Exp(B)
Kind of households	Other types	0.000	1.835
(reference = Single person more than 65 years)	Couple with 1 child	0.000	2.124
	Couple with 2 children	0.000	2.014
	Couple with 3 + children	0.000	2.504
	Childless couple with p.r. 65 years	0.000	1.741
	Childless couple with p.r. less than 65	0.000	1.958
	Single parents	0.000	2.003
	Single person less than 65 years	0.003	1.589
Educational level *(reference = Lower)*	Middle	0.000	1.356
	High	0.000	1.648
	Upper	0.000	1.783
Dwelling typology *(reference = Apartment in small building (> 10)*	Apartment in small building (10–27)	0.008	1.261
	Apartment in large building (28 +)	1.173	1.298
	Single family house	0.000	1.376
	Semi-detached house	1.558	1.151
Year of construction *(reference = After 1990)*	1950–1969	0.000	2.017
	1970–1989	0.000	2.003
	Don't know	0.119	1.592
	Before 1950	0.000	1.924
Housing tenure *(reference = Rent)*	Free	0.106	1.716
	Property	0.000	1.942
	Rent	0.007	1.861
House size (Mq) *(reference = less 60mq)*	61 to 70 mq	0.905	1.334
	71 to 80 mq	0.289	1.361
	81 to 90 mq	0.002	1.465
	91 to 100 mq	0.000	1.540
	Over 100 mq	0.000	1.697
Switching suppliers	Yes	0.000	1.409
The reference categories is PEEB = absence of any investment			
Final Model: −2LL = 24586.814, Chi square = 9234909, df = 61, Sig = 0.000			

result of such interventions. Compared to apartments in small buildings (less than 10 apartments), single family dwellings and apartments in large buildings are more often subject of investments, the first for a wider freedom of action of single families comparing with condominium, the latter given the opportunity to amortize the costs over a larger number of families than those who live in small buildings. The larger the size of the dwelling the stronger the propensity to invest, while the period of construction outlines a split in two between newer homes (built after 1990), which are virtuous from the energy point of view, and older homes that require investments for modernization.

4 Conclusion

The analysis showed that environmental sustainability is an issue that the Italian population has begun to be conscious of through appropriate behaviours. They express concern over proximate and long-terms environmental threats and have attention to the protection of natural resources.

Considering the set of curtailment behaviours, synthesized from the $PECB_D$ index, the results showed that subjective environmental concerns, the age and the level of education are significant and positive predictors. Furthermore, studying the distribution of the population by gender, women are more active in adopting environmental behaviours, especially in the context of the choices of food consumption (purchase of organic and zero km products).

About the pro-environmental efficiency behaviours (PEEB), over the last five years, 27 out of 100 families have invested money to reduce energy costs. Socio-demographic characteristics, housing and territorial features are significant predictors of these investments.

At regional level, the results indicate significant differences as showed in the scatter plot between PECB and PEEB ($r = 0{,}574$ and $rho = 0{,}626$) centred on the national average (Fig. 1).

In the first quadrant there are the "virtuous regions" (Trento, Liguria, Valle d'Aosta, Piemonte, Veneto, Emilia Romagna, Umbria) with values higher than the national mean, for both PEEB and PECB; in the second one there is Lombardia where families and population privilege "efficiency behaviours" than "curtailment"

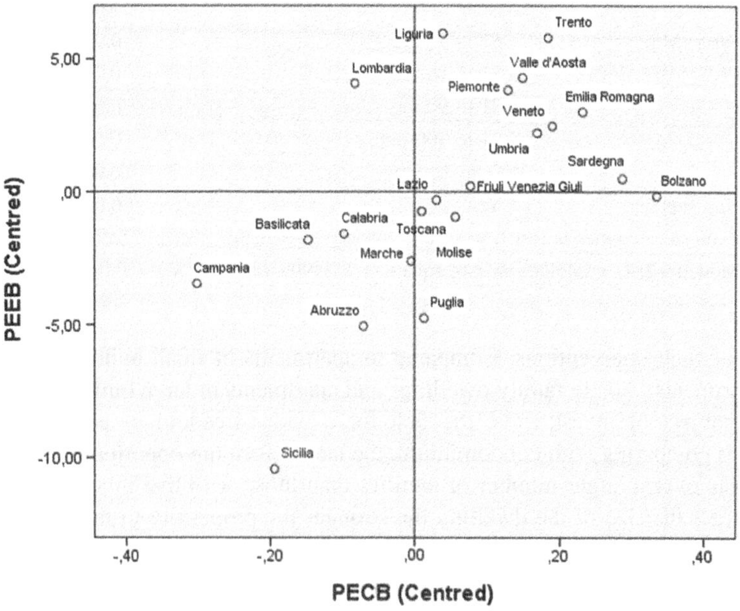

Fig. 1 Scatter plot between the two synthetic measures

ones; in the third one there are regions (Sicilia, Campania, Abruzzo, Basilicata, Calabria) which have low values of both types of environmental behaviours synthetic measures; in the fourth quadrant there are regions that privilege "curtailment behaviours" to "efficiency behaviours" (Bolzano, Molise, Lazio, Toscana e Puglia).

These results suggest the need for information and communication strategies aimed to promote an adoption of "virtuous" behaviours in the population, with a focus to households and individuals resident in the Southern part of Italy. On the other hand it's important to increase social participation and environmental awareness for all the social categories.

References

1. Brundtland, G.H., et al.: Our Common Future, Brundtland Report (1987)
2. Ardoin, N., Heimlich, J., Braus, J., Merrick, C.: Influencing Conservation Action: What Research Says About Environmental Literacy, Behavior, and Conservation Results. National Audubon Society, New York (2013)
3. Ehrlich, P.R., Kennedy, D.: Millennium assessment of human behavior. Science **309**, 562–563 (2005)
4. Gardner, G., Stern, P.C.: Environmental Problems and Human Behavior, 2nd edn. Allyn & Bacon, Boston (2002)
5. Larson, L.R., Stedman, R.C., Cooper, C.B., Decker, D.J.: Understanding the multi-dimensional structure of pro-environmental behavior. J. Environment. Psychol. **43**, 112–124 (2015)
6. Fornara, F., Pattitoni, P., Mura, M., Strazzera, E.: Predicting intention to improve household energy efficiency: the role of value-belief-norm theory, normative and informational influence, and specific attitude. J. Environment. Psychol. **45**, 1–10 (2016)
7. Stern, P.C., Gardner, G.T.: Psychological research and energy policy. Am. Psychol. **36**, 329–342 (1981)
8. Istat: Popolazione e ambiente preoccupazioni e comportamenti dei cittadini in campo ambientale, Statistiche Report, 22 dicembre 2015 (2015)
9. European Commission: EUROPA 2020 Una strategia per una crescita intelligente, sostenibile e inclusiva, Bruxelles (2010)
10. ISTAT: I consumi energetici delle famiglie, Statistiche Report, 15 dicembre 2014 (2014)
11. Hosmer, D., Lemeshow, S.: Applied Logistic Regression. Wiley, New York (1989)

Estimating the at Risk of Poverty Rate Before and After Social Transfers at Provincial Level in Italy

Caterina Giusti and Stefano Marchetti

Abstract Considering the local areas where citizens live is fundamental to investigate deprivation and social exclusion, particularly in a period of increasing financial difficulties and reduction of public funding. In this work we estimate the at risk of poverty rate of Italian households before and after social transfers at provincial level. To obtain these estimates we use data coming from the EU-SILC 2013 survey and data coming from the population census and administrative archives in a small area estimation framework, since the design of EU-SILC survey does not allow for reliable direct estimation at provincial level. Our results, besides indicating the essential role of social transfers in the reduction of the at risk of poverty rate, allow a sub-national analysis of the phenomenon of interest that would be lost by using traditional statistical techniques.

Keywords Area level models · Small area estimation · EU-SILC

1 Introduction

In time of increasing financial difficulties and reduction of public funding, welfare systems are subject to several transformations which may impact the quality and homogeneity of the offered services. To understand the impact of these changes on several crucial needs that may give riser to social protection, such as health care, old-age and unemployment, it is fundamental to consider the local areas where households live. Indeed, social protection benefits and allowances are often managed by local administrations, such as Regions and Provinces in Italy.

In this paper we are interested in evaluating the impact of social transfers on poverty at local level by using data from the EU-SILC survey. The EU-SILC (European Survey on Income and Living Conditions) is an annual European survey devoted to collect comparable longitudinal and cross-sectional data on income,

C. Giusti (✉) · S. Marchetti
Department of Economics and Management, Via Ridolfi, 10, 56124 Pisa, Italy
e-mail: caterina.giusti@unipi.it

C. Perna et al. (eds.), *Studies in Theoretical and Applied Statistics*,
Springer Proceedings in Mathematics & Statistics 227,
https://doi.org/10.1007/978-3-319-73906-9_30

poverty, social exclusion and living conditions. In particular, EU-SILC data allow to compute the households' at risk of poverty rate (ARPR[1])—a measure of the incidence of poverty—both excluding ("before") and including ("after") in the household income cash social transfers, such as, for example, unemployment benefits, old-age benefits and sickness benefits. Indeed, the reduction in the ARPR due to social transfers, calculated as the percentage difference between the ARPR before and after social transfers, has been indicated as dashboard indicator on the effectiveness of social protection systems by the EC Social Protection Committee [5].

In recent years several authors have tried to investigate the impact of social transfers in the reduction of poverty using EU-SILC data. For example, Longford and Nicodemo [11] studied the effectiveness of several European social welfare systems by investigating the reduction in the poverty gap—a measure of intensity of poverty—resulted by cash social transfers. Fabrizi et al. [6] analyzed the anti-poverty effects of social cash transfers using a micro-economic approach by estimating the conditional probability for a household to receive social transfers given that the household is poor and the consequent conditional probability that a family moves out of poverty given that it receives the social transfers. Moving from this study, Baldini et al. [1] adopted a microeconomic approach to evaluate the ability of several typologies of European welfare systems to reach the poor, using also a longitudinal definition of poverty.

As done in the papers described above, data from the EU-SILC can be used to compute reliable indicators (e.g. the ARPR) at national level or for the macro-regions corresponding to the NUTS 1 definition in the nomenclature of territorial units for statistics of the EU. If one is interested in computing the indicators of interest at a more detailed geographical level—for example at LAU 1 or 2 levels (e.g. Provinces and Municipalities in Italy)—there is need to resort to appropriate methodologies, since the sample size is usually too small (or even equal to zero) to obtain reliable direct estimates (estimates that use only the information coming from the EU-SILC survey).

Our research question is thus the following: can small area methodologies be used with EU-SILC data to compute the ARPR in Italian provinces before and after the social transfers? Small area methodologies have been already used to compute poverty indicators using data from the EU-SILC survey in several studies (see for example [9]). Since the availability of unit-level data with a fine territorial reference is not trivial (e.g. availability of the province of residence of Italian households included in the EU-SILC sample), small area models defined at the area level can represent a flexible method to obtain reliable local estimates.

The rest of the paper is organized as follows. In Sect. 2 the small area methodologies employed are described. The data used in the application and the results are presented in Sect. 3, conclusions are in Sect. 4.

[1]The at risk of poverty rate (ARPR) is also known as Head Count Ratio (HCR).

2 A Short Review of the SAE Methods Used in the Application

Data obtained from surveys are usually used to estimate characteristics for subsets of the survey population. If the sample from a subset is small, then a traditional design-based estimator may have unacceptably large variance. These subsets are commonly known as *small areas*. Different methods have been discussed in literature; here the availability of the data limited us to the area-level methods [7].

Let assume that there are m small areas of interest and that θ_i represents the population characteristic of interest in area i, such as a mean or proportion. A survey provides a direct estimator $\hat{\theta}_i^{dir}$ of θ_i for some or all of the small areas. A p-vector $\mathbf{X}_i$ contains the auxiliary data sources of population characteristics for area i. Borrowing strength from auxiliary data it is possible to reduce the mean squared error of the direct estimates. Let assume that the auxiliary variables $\mathbf{X}_i$ are known exactly. An area-level model (FH) can be expressed as follows

$$\hat{\theta}_i^{dir} = \mathbf{X}_i^T \beta + u_i + e_i \quad i = 1, \ldots, m, \tag{1}$$

where $u_i \overset{iid}{\sim} N(0, \sigma_u^2), i = 1, \ldots, m$ are the model errors and $e_i \overset{ind}{\sim} N(0, \psi_i), i = 1 \ldots, m$ are the design errors, with e_i independent from u_j for all i and j. It is assumed that the quantity of interest in area i is $\theta_i = \mathbf{X}_i^T \beta + u_i$.

Under the normality assumption the best linear unbiased predictor (BLUP) of θ_i is

$$\tilde{\theta}_i^{FH} = \gamma_i \hat{\theta}_i^{dir} + (1 - \gamma_i)\mathbf{X}_i^T \tilde{\beta}, \quad \gamma_i = \frac{\sigma_u^2}{\sigma_u^2 + \psi_i}. \tag{2}$$

The predictor $\tilde{\theta}_i^{FH}$ is a convex combination of the direct estimator $\hat{\theta}_i^{dir}$ and the predicted value $\mathbf{X}_i^T \tilde{\beta}$ from the regression model where $\tilde{\beta} = \left\{\sum_{i=1}^m \mathbf{X}_i \mathbf{X}_i^T (\psi_i + \sigma_u^2)^{-1}\right\}^{-1}$ $\left\{\sum_{i=1}^m \mathbf{X}_i \hat{\theta}_i (\psi_i + \sigma_u^2)\right\}$. The extent to which it depends on the two terms is determined by the relative sizes of the model error variance σ_u^2 and the sampling error variance ψ_i.

According to the theory of small area estimation [13], we consider the ψ_is as known. Then, estimating $\tilde{\beta}$ and σ_u^2 using the restricted maximum likelihood we can derive the empirical best linear unbiased predictor (EBLUP)

$$\hat{\theta}_i^{FH} = \hat{\gamma}_i \hat{\theta}_i^{dir} + (1 - \hat{\gamma}_i)\mathbf{X}_i^T \hat{\beta}, \quad \hat{\gamma}_i = \frac{\hat{\sigma}_u^2}{\hat{\sigma}_u^2 + \psi_i}. \tag{3}$$

When the parameters in (2) are estimated we obtain the estimator (3) that has the following MSE

$$MSE(\hat{\theta}_i^{FH}) = \gamma_i \psi_i + (1-\gamma_i)^2 \mathbf{X}_i^T V(\hat{\beta}) \mathbf{X}_i + \psi_i^2 (\psi_i + \sigma_u^2)^{-3} V(\hat{\sigma_u^2}) \\ = g_{1i} + g_{2i} + g_{3i}, \tag{4}$$

where $g_{1i} = \gamma_i \psi_i$, g_{2i} is the contribution to the MSE from estimating β and g_{3i} is the contribution to the MSE from estimating σ_u^2. $V(\hat{\beta})$ and $V(\hat{\sigma_u^2})$ are the asymptotic variances of the estimators $\hat{\beta}$ of β and $\hat{\sigma}_u^2$ of σ_u^2. An estimator of (4) is as follows

$$mse(\hat{\theta}_i^{FH}) = \hat{g}_{1i} + \hat{g}_{2i} + 2\hat{g}_{3i}, \tag{5}$$

where $\hat{g}_{1i} = \hat{\gamma}_i \psi_i, \hat{g}_{2i} = (1-\hat{\gamma}_i)^2 \mathbf{X}_i^T \left[\sum_{i=1}^m \mathbf{X}_i \mathbf{X}_i^T / (\psi_i + \hat{\sigma}_u^2)\right]^{-1} \mathbf{X}_i, \hat{g}_{3i} = \psi_i^2 (\psi_i + \hat{\sigma}_u^2)^{-3}$ $2\left[\sum_{i=1}^m 1/(\hat{\sigma}_u^2 + \psi_i)^2\right]^{-1}$. More details concerning analytic MSE estimation for area level models can be found in Rao [13].

When the variable of interest is spatially correlated there can be a gain in terms of MSE using an EBLUP estimator that accounts for the spatial information. If this is the case, model (1) can be extended to allow for stationary spatially correlated area effects as follows. Let $\mathbf{v}$ be the result of a SAR process with unknown autoregression parameter ρ and proximity matrix $\mathbf{W}$ [4]

$$\mathbf{v} = (I_m - \rho \mathbf{W})^{-1} \mathbf{u}\ , \tag{6}$$

where $(\mathbf{I}_m - \rho \mathbf{W})^{-1}$ is supposed to be non-singular, $\mathbf{u} = [u_1, \ldots, u_m]^T \sim N(\mathbf{0}, \sigma_u^2 \mathbf{I}_m)$, $\mathbf{I}_m$ is an $m \times m$ identity matrix and $\mathbf{W}$ is defined in a row-standardized way. Putting together Eqs. (1) and (6) the model is

$$\hat{\theta}^{dir} = \mathbf{X}\beta + (I_m - \rho \mathbf{W})^{-1} \mathbf{u} + \mathbf{e}\ , \tag{7}$$

where $\hat{\theta}^{dir} = [\hat{\theta}_1^{dir}, \ldots, \hat{\theta}_m^{dir}]^T, \mathbf{e} = [e_1, \ldots, e_m]$ and $\mathbf{X} = [\mathbf{x}_1^T, \ldots, \mathbf{x}_m^T]$. The BLUP which follow from model (7) is

$$\tilde{\theta}_i^{SFH} = \mathbf{X}_i^T \tilde{\beta} + \mathbf{d}_i^T \mathbf{G} \mathbf{V}^{-1} [\mathbf{y} - \mathbf{X}\tilde{\beta}]\ , \tag{8}$$

where $\mathbf{V} = \mathbf{G} + diag(\psi_i)$ is the covariance matrix of (7), $\mathbf{G} = \sigma_u^2 [(\mathbf{I}_m - \rho \mathbf{W})^T (\mathbf{I}_m - \rho \mathbf{W})]^{-1}$ and $\mathbf{d}_i^T = [0, \ldots, 1, 0, \ldots, 0]$ is a selection vector with 1 in the ith position. Estimating the parameter $\tilde{\beta}$, ρ and σ_u^2 using REML we obtain the spatial EBLUP (SFH), $\hat{\theta}_i^{SFH} = \mathbf{X}_i^T \hat{\beta} + \mathbf{d}_i^T \hat{\mathbf{G}} \hat{V}^{-1} [\mathbf{y} - \mathbf{X}\hat{\beta}]$. Under normality and independence of random effects and errors, the MSE of the SFH can be decomposed as

$$MSE(\hat{\theta}_i^{SFH}) = g_{1i} + g_{2i} + g_{3i} = \mathbf{d}_i^T [\mathbf{G} - \mathbf{G}\mathbf{V}^1 \mathbf{G}] \mathbf{d}_i \\ + \mathbf{d}_i^T [\mathbf{I}_m - \mathbf{G}\mathbf{V}^1] \mathbf{X} (\mathbf{X}^\mathbf{T} \mathbf{V}^1 \mathbf{X})^1 \mathbf{X}^T [\mathbf{I}_m - \mathbf{V}^1 \mathbf{G}] \mathbf{d}_i + tr\{\mathbf{L}_i \mathbf{V} \mathbf{L}_i^T \mathscr{I}^{-1}\}\ , \tag{9}$$

where $\mathscr{I}$ is the information matrix and $\mathbf{L}_i = [\mathbf{d}_i^T(\mathbf{C}^{-1}\mathbf{V}^{-1} - \sigma_u^2\mathbf{V}^{-1}\mathbf{V}^{-1}\mathbf{C}^{-1}\mathbf{V}^{-1}),$ $\mathbf{d}_i^T(\mathbf{A}\mathbf{V}^{-1} - \sigma_u^2\mathbf{C}^{-1}\mathbf{V}^{-1}\mathbf{A}\mathbf{V}^{-1})]^T$ with $\mathbf{C} = (\mathbf{I}_m - \rho\mathbf{W})^T(\mathbf{I}_m - \rho\mathbf{W})$ and $\mathbf{A} = \sigma_u^2\mathbf{C}^{-1}$ $(\mathbf{W} + \mathbf{W}^T - 2\rho\mathbf{W}^T\mathbf{W})\mathbf{C}^{-1}$. To estimate the MSE we assume that under SAR models the covariance depends on a proximity matrix that specifies the proximity between the areas, so an approximately unbiased estimator of the MSE is [14] $mse(\hat{\theta}_i^{SFH}) = \hat{g}_{1i} + \hat{g}_{2i} + 2\hat{g}_{3i}$, where estimates of ρ and σ_u^2 obtained by REML have been plugged-in (9). Molina et al. [12] proposed some alternatives based on bootstrap to estimate the MSE in (9).

3 Application: Estimation of Local Welfare Indicators in Italy

3.1 The Data

As already discussed in Sect. 1, data coming from the EU-SILC can be used to compute indicators able to monitor the social protection performance in a given territory. Our aim is to estimate the ARPR before and after the social cash transfers in the 110 Italian provinces using the small area estimation techniques presented in Sect. 2. The choice of the province level was mainly data driven: indeed, only province level direct estimates were available for our analysis.[2] Nonetheless, though welfare policies in Italy are often managed at regional level (NUTS 2 level), some policies have an intra-regional component—i.e. provincial—so that is it also useful to provide within regions key statistics on these welfare policies. We also remark that the ARPR before and after the social transfers are two of the possible monetary poverty indicators that can be computed by using EU-SILC data. Nonetheless, the data and methods we used for the present application could be extended to include other indicators (i.e. inequality indicators).

We fit two separate small area models, one to estimate the ARPR before the social transfers and one for the ARPR including them. In the small area model for the ARPR before the social transfers the direct estimator $\hat{\theta}_i^{dir}$, $i = 1, \ldots, 110$, is computed using the total disposable household income before social transfers equivalised using the OECD modified scale [10]. As Baldini et al. [1] and Longford and Nicodemo [11] we decided to include in the social transfers also old-age and survivors' pensions. To compute the indicator we conventionally set the poverty line at the 60% of the total disposable household income after social transfers at Italian level. In Italy the

[2]We thank the ISTAT—*Ufficio regionale per la Toscana* for providing us the EU-SILC direct estimates of the ARPR before and after the social transfers at province level for the year 2013.

poverty line in 2013[3] was estimated to be 9134 euros. For a discussion on the use of national or regional poverty lines when computing local poverty indicators see Giusti et al. [8]. The direct estimator of the ARPR after the social transfers is computed using the same methodology, but adding to the total disposable household income all the cash transfers measured in the EU-SILC, that is unemployment, old-age, survivor', sickness and disability benefits, education-related and family/children related allowances, social exclusion not elsewhere classified and housing allowances.

As explained in Sect. 2, the small area methods used in the present application require covariate information available at the area-level, that is at provincial level. As potential sources of information we considered data coming from the Population and Housing census and from the Income Tax Office.

The 2011 Population and Housing census data considered as potential covariate information in this work include information such as the number of families, household size, tenure status, female-headed household quota. The mean per-capita income at provincial level (LAU 1) was instead derived from the administrative database of the Income Tax Office, the government agency devoted to oversight the tax compliance by citizens and enterprises, available at LAU 2 level. It is important to note that the definition of disposable income surveyed in EU-SILC and that of taxable income collected by Income Tax Office may significantly differ. However, the latter variable proved to be very powerful in the prediction of the ARPR at a province level.

3.2 Results

We now present the results of the estimation of the ARPR before and after the social transfers. Using AIC and BIC goodness of fit statistics we selected the best covariates in the two area level models. Since we obtained a rather high value of spatial correlation (values are reported in Table 2) and we rejected the hypothesis of non-stationarity,[4] we decided to use the SFH.

The selected covariates are summarized in Table 1. For the ARPR before the social transfers we selected the mean per-capita income at provincial level from the Income Tax Office database (*Per-capita income*) and the number of households living in the province (*N. households (th.)*) from the Population Census 2011. For the ARPR after the social transfers we selected instead the following covariates: the mean per-capita income at provincial level (*Per-capita income*) and some information from the Population Census 2011, namely the number of households living in the province (*N. households (th.)*), the mean household size at provincial level (*Household size*) and the share of households owning their house (*Household owning the house*). Table 1 also reports the summary of the sample size of EU-SILC at provincial level. Three quarters of the provinces have a sample size lower than 550

[3]EU-SILC 2013 income variables refer to 2012.

[4]To test the hypothesis of non stationarity we used the method proposed by Chandra et al. [3].

Table 1 Summary statistics of the covariates over the 110 provinces

	Min.	1st Qu.	Median	Mean	3rd Qu.	Max.
Sample size	10.00	176.00	274.50	405.70	542.50	2018.00
Per-capita income	13350.00	15970.00	18390.00	18020.00	19890.00	26170.00
N. households (th.)	24.63	100.10	152.70	223.70	243.70	1743.00
Household size	1.98	2.29	2.40	2.40	2.51	2.89
Household owning the house	56.65	71.12	73.42	73.31	75.63	83.37

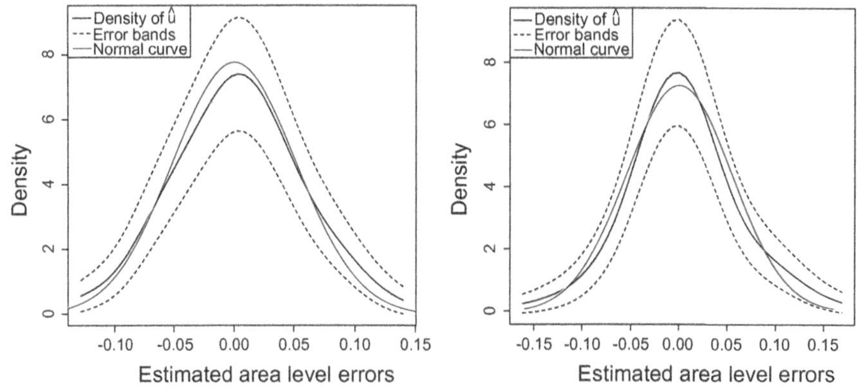

Fig. 1 Density estimates of $\hat{u}_i$ with a superimposed normal density for the ARPR before (left) and after (right) the social transfers

households, a quantity that does not allow for reliable direct estimation, particularly for non linear statistics such as the ARPR.

One of the assumptions of the model (7) is that $u_i \sim N(0, \sigma_u^2)$. We tested this assumption using $\hat{u}_i = \mathbf{d}_i \mathbf{G} \hat{\mathbf{V}}^{-1}(\mathbf{y} - \mathbf{X}\hat{\beta})$. Figure 1 represents the estimated density of the $\hat{u}_i$s for the ARPRs models. The density were estimated using the cross-validation approach suggested in Bowman et al. [2]. As we can see the normality assumption seems respected for both the two models. This is also confirmed by a Shapiro-Wilk normality test.

The regression parameters obtained fitting model (7) separately for the ARPR before and after the social transfers are presented in Table 2. As expected, the parameters referring to the *per-capita income* and to the *households owning the house* are negative, indicating an inverse linear relation with the direct estimates of ARPR. The other covariates show instead a positive value of the regression parameters.

Table 2 Estimates of parameters of small area level models for ARPR before and after the social transfers: βs, spatial autocorrelation (ρ) and model error standard deviation (σ_u)

	ARPR before		ARPR after	
	$\hat{\beta}$	p-value	$\hat{\beta}$	p-value
Intercept	1.1299	0.0000	0.7418	0.0041
Per-capita income (th.)	−0.0375	0.0000	−0.0238	0.0000
N. households (hu.th.)	0.0072	0.0164	0.0057	0.0516
Households size	–	–	0.1358	0.0231
Households owning the house	–	–	−0.0064	0.0014
	$\hat{\rho} = 0.47$	$\hat{\sigma}_u = 0.058$	$\hat{\rho} = 0.52$	$\hat{\sigma}_u = 0.054$

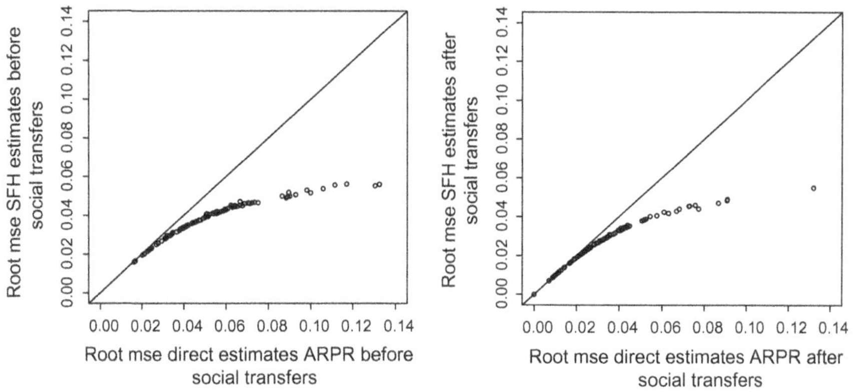

Fig. 2 Root *mse* of the direct estimates versus root *mse* of the model-based estimates for the ARPR before (left) and after (right) the social transfers

This means that an increase in the number of households living in a province results in a increase of the ARPR, and the same happens for an increase of the household size. It is important to remind that under the area-level approach we model direct estimates using area proportions and averages as covariates. Moreover, the main focus under the small area estimation approach is the predictive power of the model: it is the econometric approach that focus on the interpretation of the model parameters.

The main goals of the area-level model-based small area estimation techniques are to produce estimates similar to the direct ones and with a higher precision. In Fig. 2 the root *mse*s of the SFH estimates are compared with the root *mse*s of the direct ones. As we can see, the variability of the SFH estimates is always lower than the variability of the corresponding direct estimates. Moreover, the correlation of the SFH estimates with the corresponding direct ones is very high (about 0.95 for both the indicators). These analysis give strength and reliability to our estimates.

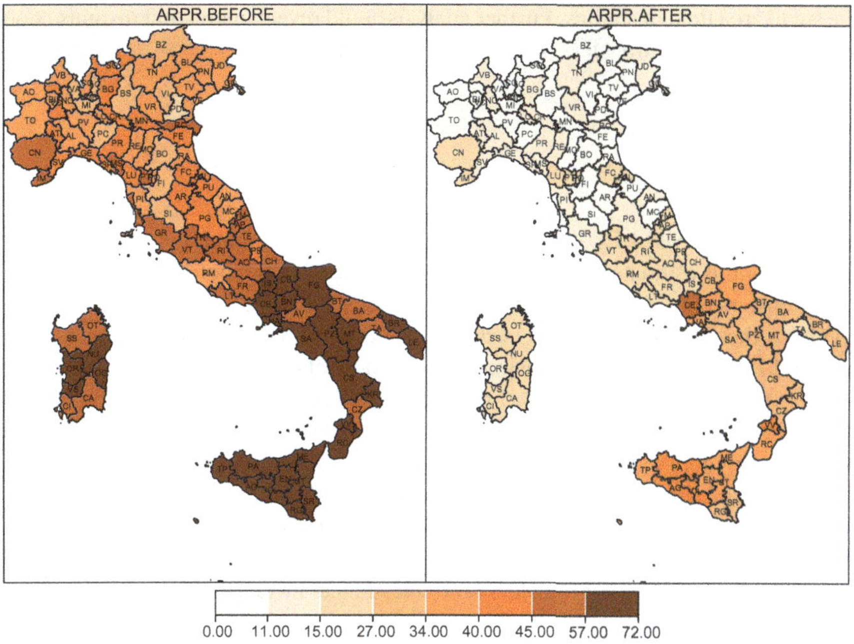

Fig. 3 SAE estimates of the ARPR before (left) and after (right) the social transfers

Figure 3 reports the map of the SAE estimates obtained for the two indicators. As we can see, both maps highlight that the poverty incidence is higher in the provinces located in the South of Italy and in the two main islands, Sicily and Sardinia. The gap is more evident when we look at the map of the ARPR after the social transfers (right map). As expected, we can notice that in all the provinces the ARPRs are reduced by the social transfers; however, the entity of the reduction is very changeable and does not seem to follow a geographical criterion. The reduction of the ARPR varies from −19% up to −48%. In five Sardinia provinces (Sassari, Nuoro, Oristano, Ogliastra and Medio Campidano) the reduction is between 33–48%, as in six provinces located in northern Italy (Biella, Vercelli, Pavia, Mantova and Ferrara). In the center of Italy only in four provinces (Massa, Livorno, Grosseto and Arezzo) the reduction is between 33–48%, while in the south (without Sardinia) this reduction is observed in five provinces (Terni, Ascoli, Teramo, Isernia and Trapani). The smallest reduction is observed in the province of Caserta (−19%), which remains one of the poorest provinces even after the social transfers are included as components of the household income. Other provinces where the reduction is limited (19–24%) are Verbania, Vasto, Milano, Lecco, Brescia, Bolzano, Vicenza, Padova, Prato, Firenze, Ravenna, Forlì, Roma, Pescara, Napoli, Barletta and Palermo.

The difference between the ARPR before and after social transfers has proved to be statistically significant for all the provinces by using a *t*-test for paired-samples. However, more methodological advances are needed to make tests on differences in the small area field of studies. Further analysis could reveal interesting insights about the impact of social transfers on the ARPR in Italy.

4 Conclusions

In this paper we used EU-SILC data, Income Tax Office data and the Population Census data to estimate the at risk of poverty rate (ARPR) before and after social cash transfers at provincial level by means of area level small area methods.

Given the presence of spatial correlation, we used spatial estimators, such as the Spatial FH, to improve the precision of the area-level direct estimates. Our findings suggest a good predictive power of the covariates selected in the small area models. The estimates show a spatial distribution that would have been lost by conducting the analysis at a more aggregated geographical level (e.g. Regional or NUTS 1 level). The provincial dimension chosen in this work is particularly relevant given that in Italy many welfare interventions have a local dimension. In this respect, the small area estimation techniques used in this work could fill the lack of good quality statistics on the impact of social transfers available at local level. However, uncertainty of estimates have to be considered when making comparisons.

Acknowledgements This work is part of the activities of the PO.WE.R. research project, supported by the University of Pisa (PRA 2016).

References

1. Baldini, M., Gallo, G., Reverberi, M., Trapani, A.: Social transfers and poverty in Europe: comparing social exclusion and targeting across welfare regimes. Technical Report No. 91, Department of Economics "Marco Biagi", University of Modena and Reggio Emilia—WP (2016)
2. Bowman, A., Hall, P., Prvan, T.: Bandwidth selection for the smoothing of distribution functions. Biometrika **85**, 799–808 (1998)
3. Chandra, H., Salvati, N., Chambers, R.: A spatially nonstationary Fay-Herriot model for small area estimation. J. Surv. Stat. Methodol. **3**(2), 109–135 (2015)
4. Cressie, N.: Statistics for Spatial Data. Wiley, New York (1993)
5. EC: Social protection performance monitor (SPPM)—methodological report by the indicators sub-group of the social protection committee. Technical Report. European Commission—Social Protection Committee (2012)
6. Fabrizi, E., Ferrante, M., Pacei, S.: A micro-econometric analysis of the antipoverty effect of social cash transfers in Italy. Rev. Income Wealth **60**(2), 323–348 (2014)
7. Fay, R., Herriot, R.: Estimation of income from small places: an application of James-Stein procedures to census data. J. Am. Stat. Assoc. **74**, 269–77 (1979)
8. Giusti, C., Masserini, L., Pratesi, M.: Local comparisons of small area estimates of poverty: an application within the Tuscany region in Italy. Soc. Indic. Res. (2016)
9. Giusti, C., Marchetti, S., Pratesi, M., Salvati, N.: Robust small area estimation and oversampling in the estimation of poverty indicators. Surv. Res. Methods **6**(3), 155–163 (2012)

10. Hagenaars, A., De Vos, K., Zaidi, M.: Poverty statistics in the late 1980s: research based on micro-data. Technical Report. Luxembourg: Official Pubblication of the European Communities (1994)
11. Longford, N., Nicodemo, C.: The contribution of social transfers to the reduction of poverty. Technical Report. IZA Discussion Paper no. 5223 (2010)
12. Molina, I., Salvati, N., Pratesi, M.: Bootstrap for estimating the mse of the spatial EBLUP. Comput. Stat. **24**, 441–458 (2009)
13. Rao, J.: Small Area Estimation. Wiley, New York (2003)
14. Zimmerman, D., Cressie, N.: Mean squared prediction error in the spatial linear model with estimated covariance parameters. Ann. Inst. Stat. Math. **44**, 27–43 (1992)

The manufacturer's authorised representative in the EU is Springer Nature Customer Service Centre GmbH, Europaplatz 3, 69115 Heidelberg, Germany. If you have any concerns regarding our products, please contact ProductSafety@springernature.com

Printed and bound by CPI Group (UK) Ltd, Croydon, CR0 4YY
30/09/2026
02225658-0012